Experimental Brain Research Supplementum 12

Ultradian Rhythms in Physiology and Behavior

Edited by H. Schulz and P. Lavie

With 90 Figures and 20 Tables

Springer-Verlag
Berlin Heidelberg New York Tokyo

Dr. Hartmut Schulz
Max-Planck-Institut für Psychiatrie
Kraepelinstrasse 10
8000 München 40, FRG

Peretz Lavie, Ph. D.
Associate Professor
Faculty of Medicine
Technion-Israel Institute of Technology
Haifa 32000, Israel

ISBN-13:978-3-642-70485-7 e-ISBN-13:978-3-642-70483-3
DOI: 10.1007/978-3-642-70483-3

Library of Congress Cataloging-in-Publication Data. Main entry under title: Ultradian rhythms in physiology and behavior. (Experimental brain research. Supplementum; 12) Bibliography: p. Includes index. 1. Ultradian rhythms. 2. Sleep. 3. Wakefulness. I. Schulz, H. (Hartmut), 1941– . II. Lavie, P. (Peretz), 1949– . III. Series. QP84.6.U45 1985 599'.0188 85-17241

Softcover reprint of the hardcover 1st edition 1985

2125/3140-543210

Symposium on Ultradian Rhythms in Physiology and Behavior

Participants

1 D.F. Kripke
2 P. Lavie
3 A. Borbély
4 M. Gerkema
5 T. Kobayashi
6 E. Knobil
7 K. Stephan
8 J. Aschoff
9 R. Lydic
10 G. Dirlich
11 S. Campbell
12 H.A. Baghdoyan
13 M.A. Corner
14 M.B. Sterman
15 R.W. McCarley
16 S. Daan
17 J. Zulley
18 R. Broughton
19 C. Wildgruber
20 I.Tobler
21 F. Kimura
22 H. Schulz
23 P. Naitoh
24 S. Honma
25 K. Honma
26 C. Pollak
27 D. Beersma
28 C. Gordon
29 W. Ehrenstein
30 S. Inoué

Contents

N. Kleitman
Introductory Note 1

Introduction 2

Section 1
Ultradian Rhythms in Physiology and Endocrinology . 7

M. P. Gerkema and S. Daan
Ultradian Rhythms in Behavior:
The Case of the Common Vole (Microtus Arvalis) . . . 11

E. Knobil and J. Hotchkiss
The Circhoral Gonadotropin Releasing Hormone (GnRH) Pulse Generator of the Hypothalamus and Its Physiological Significance 32

E. Van Cauter and E. Honinckx
Pulsatility of Pituitary Hormones 41

F. Kimura, C. Praputpittaya, N. Mitsugi, R. Hashimoto, and R. Suzuki
Relationship Between Ultradian Rhythms of the Sleep-Wakefulness Cycle and Growth Hormone and Corticosterone Secretion in Rats 61

K. Honma and S. Honma
Ultradian Rhythms in Locomotor Activity, Deep Body Temperature and Plasma Corticosterone Levels in Rats: Two Different Origins? 77

S. Honma and K. Honma
Interaction Between Circadian and Ultradian Rhythms of Spontaneous Locomotor Activity in Rats During the Early Developmental Period 95

C. R. Gordon and P. Lavie
Urinary Ultradian Rhythms in Dogs 110

R. Lydic, R. W. McCarley and J. A. Hobson
Timing Function of the Dorsal Raphe Nucleus and the Temporal Organization of the Ultradian Sleep Cycle . 125

Section 2
Sleep and Waking Rhythms 145

P. Lavie
Ultradian Rhythms: Gates of Sleep and Wakefulness . 148

H. Schulz
Ultradian Rhythms in the Nychthemeron of Narcoleptic Patients and Normal Subjects 165

M. B. Sterman
The Basic Rest-Activity Cycle Revisited: Some New Perspectives 186

D. F. Kripke, D. J. Mullaney and P. A. Fleck
Ultradian Rhythms During Sustained Performance . . 201

R. J. Broughton
Three Central Issues Concerning Ultradian Rhythms . 217

S. S. Campbell and J. Zulley
Ultradian Components of Human Sleep/Wake Patterns During Disentrainment 234

Section 3
Methods and Models 257

T. Kobayashi, Y. Tsuji, and S. Endo
Sleep Cycles as a Basic Unit of Sleep 260

P. Naitoh
In Search of REM Cycle in Short Sleep Record: Iterative Nonorthogonal r^2 Method 270

R. W. McCarley and S. Massaquoi
The REM Sleep Ultradian Rhythm: A Limit Cycle Mathematical Model 288

R. A. Wever
Modes of Interaction Between Ultradian and Circadian Rhythms: Toward a Mathematical Model of Sleep . . . 309

Epilogue 319

J. Aschoff and M. P. Gerkema
On Diversity and Uniformity of Ultradian Rhythms . . . 321

Subject Index 335

List of Contributors*

Aschoff, J. *321*[1]
Broughton, R.J. *217*
Campbell, S.S. *234*
Cauter, Van, E. *41*
Daan, S. *11*
Endo, S. *260*
Fleck, P.A. *201*
Gerkema, M.P. *11, 321*
Gordon, C.R. *110*
Hashimoto, R. *61*
Hobson, J.A. *125*
Honinckx, E. *41*
Honma, K. *77, 95*
Honma, S. *77, 95*
Hotchkiss, J. *32*
Kimura, F. *61*
Kleitman, N. *1*
Knobil, E. *32*
Kobayashi, T. *260*
Kripke, D.F. *201*
Lavie, P. *110, 148*
Lydic, R. *125*
Massaquoi, S. *288*
McCarley, R.W. *125, 288*
Mitsugi, N. *61*
Mullaney, D.J. *201*
Naitoh, P. *270*
Praputpittaya, C. *61*
Schulz, H. *165*
Sterman, M.B. *186*
Suzuki, R. *61*
Tsuji, Y. *260*
Wever, R.A. *309*
Zulley, J. *234*

* The address of the authors is given on the first page of each contribution

[1] Page, on which contribution commences

Introductory Note

N. Kleitman
Santa Monica, California, CA, USA

The interest in ultradian rhythmicity is demonstrated by the number of participants and the topics they chose to present.

The duration of the rhythms varies from 1-10 min to 4-5 h, with some being a multiple of others. Of particular significance are the 80 to 90-min cycles in adult humans, and corresponding values for infants and animals. These cycles are involved in self-preservation of the organism (hunger contractions of stomach) and continuity of the species (sexual excitation). It seems that these cycles are independent of the circadian cycles, even though they may be affected by the onset of sleep and, as pointed out by some of the participants, also influenced by the state of health of the subjects.

In addition to hunger and sexual excitation, several other physiologic variables - cardiac and respiratory rates, body motility, flow and osmolarity of urine - seem to show ultradian cyclicity.

It is true that the data in the various reports do not always run in a parallel fashion. For instance, one report shows the 90-min cycle in alertness, but not in kidney and stomach functions; in another, the ultradian cycle is evident in drinking and feeding, but not in performance.

The regulation of the ultradian cyclicity appears to be related to endocrine (pineal) activity as well as to neuronal mechanisms.

On the whole, in spite of the multioscillatory nature of the ultradian cyclicity, the outstanding feature is the 90-min cycle which manifests itself in both sleep and wakefulness. This basic rest-activity cycle obviates the necessity of ascribing a special function to the REM phase of the 90-min cycle during sleep.

Introduction

Historical Perspective

In 1842, in a paper entitled "Further Development of a General Law of Vital Periodicity", which was part of a series of seven Lancet papers on periodicities in health and disease, Thomas Laycock wrote: "As everything finite must have a period within which its existence is circumscribed, so every period so circumscribing the finite, being a measure of time, must be divisible into lesser periods. But it has always been found easier to reason from generals to particulars than to ascend from particulars to generals, especially in questions involving the phenomena of life" (p. 423). From a historical perspective, Laycock's insight indeed anticipated the progress of chronobiologic research. In spite of the abundant evidence pointing at the existence of short-term rhythms with periodicities much shorter than 24 hours, termed "ultradian" rhythms after Halberg (1964), it has generally been found much easier to investigate circadian rather than ultradian rhythms. In most cases, ultradian rhythms have been ignored, or dismissed as insignificant phenomena, even in cases where they could be easily "eyeballed" in the data. Laycock himself believed that the most basic periodicity in vital phenomena was 12 hours.

Short-term fluctuations in the levels of certain behaviors, which could not be accounted for by external stimulation or by internal stimuli, have been known to ethologists and other observers of animal behavior for many years. Szymanski (1916) who made some of the earliest animals' actographic recordings, reported that the rat had 16-19 rest-activity cycles in each 24-h period. He suggested that the basic difference between polyphasic and monophasic animals was their level of proficiency of the optic system. Animals with an inferior optic system, dependent mainly on the olfactory and tactile systems, were thought to be polyphasic.

In humans, evidence of spontaneous short-term periodic variations in behavioral and physiologic functioning can also be traced back to the beginning of the century. As early as 1922, Wada, working with C. Richter, recorded 1.5-h cycles in gastric motility of sleeping humans. This study was motivated by Richter's search for periodic stimuli that could explain the prominent short-term periodicities of about 2-4 h in rats' running behavior. Modern research confirmed these findings and extended them to waking humans as well (Hiatt and Kripke 1975; Lavie et al. 1978).

In spite of the abundant data, early observations on short-term periodicities have never been considered more than incidental phenomena and did not attract experimental attention. The turning point with respect to the experimental attitude toward ultradian phenomena was the discovery of rapid eye movement (REM) sleep in 1953 by Aserinsky and Kleitman. They and, four years later, Dement and Kleitman (1957) showed that sleep hides regular short-term cyclicity, which is known today as the REM-nonREM (NREM) cycle. This cycle reflects a rather dramatic change in the state of the central nervous system during sleep, involving many physiologic systems as well as psychologic phenomena. Although the sleep REM-NREM cycle has been the focus of intensive research in the last two decades, its functional significance, as well as the nature of its regulating mechanism, are still matters of controversy.

Some years before the cyclic nature of the sleep process was recognized two short articles on periodic processes during sleep were published by Ohlmeyer, Brilmayer and Hüllstrung (1944), and by Ohlmeyer and Brilmayer (1947). These authors who have a touch of what Robert K. Merton once called "anticipatory plagiarists" (Merton 1965, p. 23) described in the first article an ultradian (85-min) rhythm of penile erections in the sleep of human males, a phenomenon which has been confirmed later as a concomitant of REM sleep (Fisher et al. 1965). In the second article they report on a relationship between the occurrence of erections during sleep and spontaneous awakenings. Only many years later did it become an established fact that the tendency to wake up from sleep spontaneously is strongest during the REM episodes of sleep (Lavie et al. 1979; Campbell 1985).

Besides the discovery of the REM state and its complex physiologic manifestatons, Kleitman's (1961, 1963) theoretical speculation about the nature and origin of the REM-NREM cycle has been of singular importance vis-a-vis the investigation of ultradian rhythms in general. Kleitman postulated that the rhythmic occurrence of REM in sleep is only a nocturnal fragment of a broader biologic alertness rhythm which operates continuously in both sleep and wakefulness, and which he termed the basic rest-activity cycle (BRAC). The chapters of this book may give a solid base for further testing of this integrative hypothesis.

About the Book

The two editors of the present book, who have been involved in ultradian rhythms research for many years, have felt the need for such a book for quite some time. Although the number of researchers involved in the study of ultradian rhythms has grown in the last decade there has been no attempt to coordinate the different findings, or to critically evaluate them. There has been a definite need for "a starting point", a first meeting which would include the major researchers in the field. The opportunity came with the 7th congress of the European Sleep Research Society which took place in Munich in September 1984. Following the meeting, a workshop dedicated in its entirety to ultradian rhythms was organized at the Max-Planck-Institut für Verhaltensphysiologie at Seewiesen near Munich.

The list of participants included many of the major contributors to this emerging field of research. To our delight, nearly all of the invited participants were able to attend the meeting, and most were able to contribute to this book.

It is not surprising that in such a workshop issues such as proper terminology to describe various aspects of ultradian rhythms, definitions of terms, and the adequacy of different statistical procedures for the detection of ultradian rhythms, were enthusiastically debated and discussed. This occurred during the presentations, at coffee breaks and near the dinner table over pints of Bavarian beer. Several factors contributed, no doubt, to the centrality of the methodological issues. First, as is evident from the list of participants, the widest possible range of interests was represented, from pituitary control mechanisms to human performance and dog renal excretion. Second, unlike circadian rhythms and their faithful geophysical companions, ultradian rhythms have no discernible environmental counterparts. Furthermore, although some use the term "ultradian rhythms" only with respect to the 1.5-h sleep and waking rhythms, in fact the term denotes a wide range of frequencies. Nevertheless we felt it would be precipitant to add a glossary of terminology to this multi-authored book on ultradian rhythms. Since many terms of biologic rhythm research have the same meaning, independent of the frequency range studied, we refer the interested reader to glossaries in books on circadian rhythms (e.g., Wever 1979; Moore-Ede et al. 1982). Terminology of sleep research is put down in a special glossary (Czeisler et al. 1980) while statistical procedures applicable to the analysis of ultradian rhythms have been described recently by Kraemer et al. (1984).

We certainly hope that this book will indeed serve as a starting point in the study of ultradian rhythms and their interaction with circadian rhythms. Hopefully the exciting experimental data presented here will stimulate further research.

ACKNOWLEDGEMENTS. The production of the book was greatly supported by the invaluable help of the following persons. The editors are grateful to Mrs Marika Gedat for the secretarial work in preparing the workshop, Ms Monika Stucky, the typist, who was responsible for the different stages of the manuscripts, Dr. Scott Campbell, Dr. M. Wollman, and Ms June Pilcher for careful reading and linguistic shaping of the manuscripts, Mr. H. Barthelmes for his advice during the computerized editing of the text, and coworkers from Springer Publishing Company for their support. It was a great pleasure for us to have the fine advice of two eminent sleep and rhythm researchers, Prof. Nathaniel Kleitman and Prof. Jürgen Aschoff, while planning the workshop and the book.

Finally we wish to acknowledge the generous support of the Max Planck Society and the Schering Company in conducting the workshop.

REFERENCES

ASERINSKY E, KLEITMAN N (1953) Regularly occurring periods of eye motility and concomitant phenomena during sleep. Science 118:273

CAMPBELL SS (1985) Spontaneous termination of ad libitum sleep episodes with special reference to REM sleep. Electroencephalogr Clin Neurophysiol 60:237-242

CZEISLER CA, BORBELY AA, HUME KI, KOBAYASHI T, KRONAUER RE, SCHULZ H, WEITZMAN ED, ZIMMERMANN JC, ZULLEY J (1980) Glossary of standardized terminology for sleep-biological rhythm research. Sleep 2:287-288

DEMENT WC, KLEITMAN N (1957) Relation of eye movements during sleep to dream activity: objective method for study of dreaming. J Exp Psychol 53:339-346

FISHER C, GROSS J, ZUCH J (1965) Cycle of penile erection synchronous with dreaming (REM) sleep. Arch Gen Psychiat 12:29-45

HALBERG F (1964) Organism as circadian systems: temporal analysis of their physiologic and pathologic responses, including injury and death. In: Walter Reed Army Institute of Research Symposium, Medical aspects of stress in the military climate, April

HIATT JF, KRIPKE DF (1975) Ultradian rhythms in waking, gastric activity. Psychosom Med 37:320-325

KLEITMAN N (1961) The nature of dreaming. In: WOLSTENHOLME GEW, O'CONNOR M (eds) The nature of sleep. Churchill, London, p 349

KLEITMAN N (1963) Sleep and wakefulness. University of Chicago Press, Chicago, p 364

KRAEMER HC, HOLE WT, ANDERS TF (1984) The detection of behavioral state cycles and classification of temporal structures in behavioral states. Sleep 7:3-17

LAVIE P, KRIPKE DF, HIATT JF, HARRISON J (1978) Gastric rhythms during sleep. Behav Biol 23:526-530

LAVIE P, OKSENBERG A, ZOMER J (1979) "It's time you must wake now." Percept Mot Skills 49:447-450

LAYCOCK F (1842/43) Further development of a general law of vital periodicity. Lancet, October, 423-426

MERTON RK (1965) On the shoulders of giants. A Shandean postscript. Harcourt, Brace and World, New York

MOORE-EDE MC, SULZMAN FM, FULLER CA (1982) The clocks that time us. Physiology of the circadian timing system. Harvard University Press, Cambridge

OHLMEYER P, BRILMAYER H, HÜLLSTRUNG H (1944) Periodische Vorgänge im Schlaf. Pflügers Arch 248:559-560

OHLMEYER P, BRILMAYER H (1947) Periodische Vorgänge im Schlaf. II. Mitteilung. Pflügers Arch 249:50-55

SZYMANSKI JS (1916) Die Haupt-Tiertypen in Bezug auf die Verteilung der Ruhe- und Aktivitätsperioden im 24stündigen Zyklus. Biol Centralblatt 36:537-541

WADA T (1922) Experimental study of hunger and its relation to activity. Arch Psychol 8:1-65

WEVER RA (1979) The circadian system of man. Springer, Berlin Heidelberg New York

Section 1
Ultradian Rhythms in Physiology and Endocrinology

Although the first eight chapters of the book cover a wide variety of topics, they all focus on ultradian rhythms in physiology and endocrine functioning. They demonstrate that episodic secretions, bouts of locomotor activity, or episodic discharges of multiunit activity, are not random but orderly events which represent an essential mode of brain activity.

In the first chapter Gerkema and Daan discuss the diversity of ultradian rhythms, their possible function and underlying mechanisms. Initially they critically describe several methods to detect ultradian rhythms and demonstrate their limitations. Thereafter they go on to describe the feeding rhythm in microtine rodents. Their data show that the common vole has a robust and precise 2-3 h ultradian feeding rhythm which persists even in the absence of circadian rhythmicity. Further studies from Daan's laboratory pointed to the arcuate nucleus of the hypothalamus as a likely candidate for an ultradian pacemaker. As we shall see later, other experimental evidence also point to this structure as a possible pacemaker.

Episodic variations in plasma levels of pituitary hormones have been known for years. Their origin and physiological significance, however, have remained unknown. In Chapters 2 and 3 Knobil and Hotchkiss, and Van Cauter and Honinckx clearly demonstrate that episodic hormonal secretions with frequencies within the ultradian range are essential for proper endocrine functioning. Knobil and Hotchkiss provided perhaps the most convincing evidence for the essential role played by ultradian rhythms in endocrine secretory activity. They summarize a large number of studies demonstrating that the gonads are responsive only to pulses of gonadotropines with a frequency of about one per hour. Any deviation from this privileged frequency inhibits gonads' development and consequently causes an imminent failure of reproduction. The arcuate nucleus of the hypothalamus has been identified in further studies from Knobil's laboratory as one of the candidates for the underlying regulating mechanism.

The importance of episodic vs continuous mode of secretion is further illustrated in patients with hypercortisolism by Van Cauter and Honinckx (Chapter 3). In spite of a similar cortisol plasma level, secretory patterns of patients with Cushing syndrome and major endogenous depression are drastically different. The main focus of this chapter is on ultradian variations in the secretion of pituitary hormones other than gonadotropins.

In Chapters 4, 5 and 6 Kimura et al., K. Honma and S. Honma demonstrate that ultradian rhythms with frequencies from 8 c/h to 3 c/h predominate various aspects of rats' physiology and endocrine activity. In Chapter 4 Kimura et al. demonstrate the pervasiveness of the 3-h ultradian rhythms in growth hormone and cortisol secretion and explore their phase relationship with the sleep-wake cycle of rats. Their data suggest that although sharing the same basic frequency each cycle is controlled by a separate oscillator.

The Honmas, utilizing conditions which favor the expression of ultradian over circadian rhythms, such as continuous light, SCN lesions, and early ontogenetic developmental stages, investigated the interaction between circadian and ultradian rhythms. They suggest that circadian rhythms emerge from ultradian rhythms by both frequency and amplitude modulation.

A different physiologic system is explored in Chapter 7. Gordon and Lavie demonstrate high-amplitude circa 3-h rhythms in renal excretions in dogs. These were synchronized with the dogs' sleep-wake cycles and were independent of hormones regulating fluids balance. Pharmacologic and surgical denervation of the kidneys suggest the involvement of the sympathetic nervous system in the regulating mechanism.

In the last chapter (8) of this section Lydic et al. provide evidence that the dorsal raphe nucleus (DRN) participates in the regulation of the REM-NREM sleep cycle in the cat. By a technique called time-normalization of cycle duration, they demonstrate an impressive phase relationship between DRN firing rate and the behavioral states REM and NREM.

Ultradian Rhythms in Behavior: The Case of the Common Vole (Microtus Arvalis)

M. P. Gerkema[1] and S. Daan[2]
[1] Max-Planck-Institut für Verhaltensphysiologie, 8131 Andechs, FRG
[2] Zoological Laboratory, Groningen University, Haren, The Netherlands

ULTRADIAN RHYTHMS: DIVERSITY OF FUNCTION AND MECHANISM

Ultradian rhythms have been defined as "biological rhythms with a period shorter than that of a circadian rhythm" (HALBERG et al. 1965). While the definition of circa-rhythms (e.g., circadian, circannual rhythms) implies that they have evolved in adaptation to the daily, tidal, lunar and yearly environmental periodicities (ASCHOFF 1981), the definition of ultradian rhythms gives no key to their functional significance. It only delimits a class of phenomena in terms of a time scale.

Among the broad diversity of phenomena classified as ultradian rhythms, two well-known examples are the polyphasic activity patterns of small mammals (SZYMANSKI 1920; ASCHOFF 1962) and the alternation between rapid eye movement (REM) and no rapid eye movement (NREM) sleep stages in mammals (see later chapters in this volume). The periods of these ultradian cycles are positively correlated with body and brain mass within and among species. In contrast, circa-rhythms have frequencies unrelated to body mass (Fig. 1). Similar to other cycles (pulse, breathing, gut peristalsis, generation cycles) the period of ultradian rhythms can be approximated by the allometric equation $t = M^{+a}$, where t is period, M is body or brain mass and a is an exponent between 1/4 and 1/3 (DAAN and ASCHOFF 1982). During ontogeny the same relation between period length and body mass exists, as illustrated for the activity rhythms of voles and the human REM-NREM cycles (Fig. 1).

Allometric equations demonstrate the phenomenological definition of ultradian rhythms, since their range may extend beyond the ultradian time scale to periods longer than a day (entering the infradian domain). Thus, meal-digestion patterns in a shrew are ultradian rhythms,

Experimental Brain Research, Suppl. 12

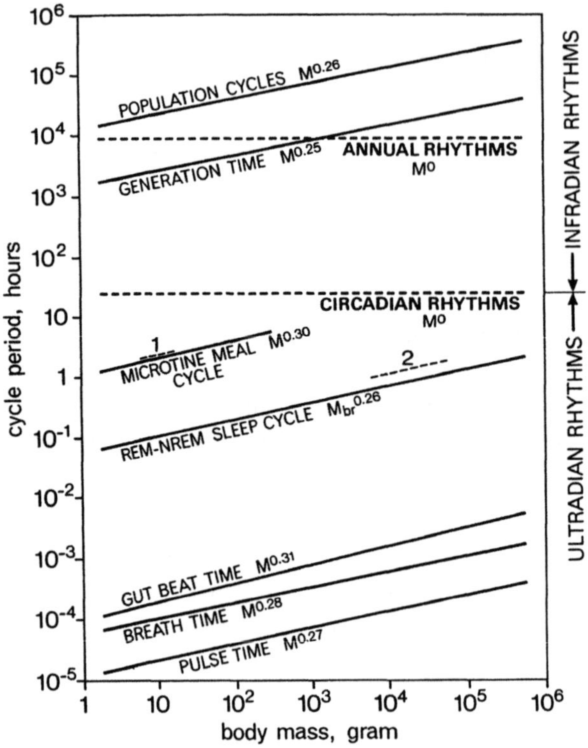

Fig. 1. Allometric scaling of cycle duration for a diversity of biological rhythms in mammals. Graphic representation of log-log linear regressions of cycle period on species body mass as compiled by DAAN and ASCHOFF (1982) with a few modifications. Population density cycles of birds and mammals according to PETERSON et al. (1984): $\tau = 1.18 . 10^4 M^{0.26}$ (M in grams, τ in h). Sleep cycle time regressed on mammal brain mass ($\tau = 0.18 * M_{br}^{0.26}$) by ALLISON et al. (1977), plotted against $M = 85 . M_{br}$. Stippled lines indicate variations with mass during ontogeny: 1. Microtus arvalis meal cycle (unpublished data). 2. Human REM-NREM sleep cycle with t = 1.80 h in adults (BEERSMA et al. 1984), t = 0.83 h in neonates (PARMELEE 1974). Note the corresponding slope of all except the periodicities adaptive to cosmic cycles (circadian, annual)

while those of a lion are infradian. The shrew's ultradian rhythm has functionally little correspondence with the human REM-NREM cycle although they may follow phenomenologically similar rules, e.g., in the association with the circadian cycles.

Along with the diversity in biological functions exhibiting ultradian rhythmicity, there is also a potentially wide variation in the processes generating such rhythmicity. In the simplest case, two mutually exclusive behaviors (A and B) alternate with each other; behavior B is triggered by the termination of behavior A, but its duration is independent of prior history. A system likely to generate ultradian rhythmicity in this fashion ("renewal process") is, for instance, that of a songbird gathering food for its nestlings, feeding its brood and foraging again in sustained alternation. A spectral analysis of a starling's foraging behavior obtained during parental care, indeed may show a significant periodicity of 6 min, and therefore it can be considered an ultradian rhythm. In a more complex model, two alternating behaviors are both under control of one endogenous oscillator. Presently, there is no documented case of such organization based on an ultradian pacemaker, although this model has now been generally accepted as a basis for circadian rhythmicity (ASCHOFF 1981). This model again will produce a significant peak in the spectral analysis.

Periodicity generated irrespective of mechanism can be detected using various statistical techniques (e.g., KRAEMER et al. 1984). SCHULZ et al. (1980) have maintained that mechanisms can be distinguished by statistical properties of the overt rhythm. They suggested that negative covariance of consecutive cycles would be characteristic of a periodically driven process, and not of renewal processes. However, for periodically driven processes like circadian activity rhythms of small mammals, it has been shown that the serial correlation coefficient may vary between 0.0 and -0.5 depending on the variances of the driving cycle and of the ancillary processes triggering the overt behavior (PITTENDRIGH and DAAN 1976a). Furthermore, a renewal process may well display negative covariance between consecutive cycles in at least three different ways.

1. The turning points in the renewal process are subject to error of measurement. Such errors affect the preceding and following cycle with opposite signs. This is certainly the case for the REM-NREM cycle where the commonly used criteria for distinguishing REM and NREM states are arbitrarily defined (RECHTSCHAFFEN and KALES 1968).

2. The renewal process generates external negative feedback. In the case of the starling parent described above, absences from the nest lasting longer than average may increase hunger of the brood and thereby stimulate the parent to return sooner than average the next time.

3. The renewal process may generate internal negative feedback, e.g., by interaction with a circadian system. This has been demonstrated in

a recent renewal model for human sleep (DAAN and BEERSMA 1984). When the period of an ultradian renewal process varies systematically with circadian phase we would expect both positive and negative serial correlations. Positive correlations occur when short cycles are clustered during activity time (α), and long cycles are clustered during rest time (ρ), as is typical of polyphasic rodents (ASCHOFF 1962, 1984). Negative correlations occur when the first and/or last cycle of a series is characteristically shorter than average, whereby short cycles are more frequent than random followed and/or preceded by long cycles. This has been demonstrated for human REM-NREM cycles (BEERSMA et al. 1984).

Thus, statistical evaluation of serial correlations does not identify the nature of the generating mechanism, neither for the REM-NREM cycle nor for other periodicities, and this criterion cannot be used for a further delimitation of the class of phenomena studied as ultradian rhythms. Experimental rather than descriptive analysis is a prerequisite for understanding the generating mechanism. Ultradian rhythms are probably as diverse in their physiological origin as in their functional meaning.

In circadian rhythms, generalizations concerning the stability, precision and homeostasis of frequency have been functionally related to the ecological needs of temporal organization with respect to predictable variations in the environment. Simultaneously, this had led to hypotheses about their physiological generation (PITTENDRIGH 1960; PITTENDRIGH and DAAN 1976b). Such generalizations are much less apparent in ultradian rhythms. The precision of ultradian rhythms in behavior (expressed as the reciprocal of the coefficient of variation in the cycle duration) is at best an order of magnitude less than the precision of circadian rhythms (Table 1). To understand the importance of ultradian temporal organization one should begin with the most clear-cut ultradian rhythms, although no claim for generality can be made. Here, we shall review the major facts available concerning the surface feeding rhythm of microtine rodents within the general context of vertebrate herbivores.

Table 1. Coefficients of variation

Circadian activity/rest cycles	$SD/\bar{\tau}$
Mesocricetus auratus	0.006[a]
Mus musculus	0.012[a]
Peromyscus leucopus	0.014[a]
Peromyscus maniculatus	0.026[a]
Man	0.053[b]
Ultradian cycles	
Microtus arvalis, feeding	0.149
Man, REM-NREM	0.246[c]
Parus major, incubation	0.314
Branta bernicla, feeding	0.373

[a]PITTENDRIGH and DAAN (1976a)

[b]WEVER (1979)

[c]BEERSMA et al. (1984)

FUNCTION OF ULTRADIAN ORGANIZATION

Unlike circadian rhythms, ultradian rhythms in behavior have no obvious counterpart in the environment. For a solitary individual having to alternate between two or more behaviors, the precise timing of these would therefore seem indifferent. The frequency of its feeding rhythm, e.g., of foraging alternating with digestive pauses and sleep, may be functionally adjusted to optimize energy processing (cf. SIBLY 1981), but no particular phase adjustment of the rhythm would be required. This is different for gregarious animals. Herbivores, living in herds need to make sure that their ultradian rhythms stay roughly in phase with each other. Rumination while the rest of the herd is grazing and moving along would isolate the individual ungulate from its conspecifics. Similarly, a foraging flock of geese would leave a sleeping out-of-phase individual behind, at the mercy of predators. Herds or flocks are temporally as well as spatially coordinated units. The synchrony of the rumination rhythms between members of a herd has been frequently documented in ruminants (e.g., cattle, HUGHES and REID 1951; roe deer, BUBENIK 1960; reindeer, ERIKSSON et al. 1981). Avian flocks of barnacle geese (Branta leucopsis) raising their young on Spitsbergen have similarly synchronized rhythms of grazing alternating with sleeping (PROP et al. 1978). Social rhesus monkeys are known to synchronize their ultradian rhythms when kept in pairs rather than individually (DELGADO-GARCIA et al. 1976). Thus, there is apparently an important external cycle - in the behavior of conspecifics - selectively favoring a specific phasing of the individual's ultradian rhythm.

Microtine rodents, although not living in herds, may benefit from ultradian synchrony with their family members and potentially with the population at large (DAAN and SLOPSEMA 1978; DAAN and ASCHOFF 1981). Voles alternate primarily between grazing and subterraneous sojourn in a 2 to 3-h rhythm with considerable individual precision. In nature their ultradian rhythms reveal a remarkable degree of synchrony in the population, at least in winter, as documented using both trapping and predator hunting yield (RAPTOR GROUP 1982; HOOGENBOOM et al. 1984). Such synchrony appears beneficial to the individual for three reasons.

1. Members of a family group may benefit from thermoregulatory energy gains by sleeping together. Meadow voles (Microtus pennsylvanicus) have been reported to use nests together in the field (WEBSTER and BROOKS 1981). From studies of small rodents (e.g., Apodemus agrarius) we know that simultaneous use of the nest may reduce metabolic rates by up to 32% (TERTIL 1972).

2. Within a family group, synchronous above-ground feeding would potentially allow the effective exchange of alarm calls in the event of predatory attack.

3. Population synchrony in surface feeding would, by varying predatory risk, selectively act against individuals exposing themselves out of phase with the population, unless high prey densities allow the predators to concentrate their activities with the prey (RAPTOR GROUP 1982).

While the functional merits of ultradian rhythmicity per se are presumably to be found in optimal coordination of physiological processes, the control of its phase, when important, is commonly related to social synchrony. On such functional grounds we speculate that the robust and precise timing with respect to the social environment is of adaptive significance and that endogenous oscillatory control could have evolved to enhance the temporal accuracy of renewal processes. The evidence presently available from an ongoing study on the ultradian rhythms in the behavior of the common vole is summarized below.

BEHAVIORAL ANALYSIS

Stability

One of the remarkable aspects in the case story of Microtus arvalis is illustrated in Fig. 2. When kept in continuous darkness (DD) in a registration cage with running-wheel and food rack, vole #318 initially showed a clear circadian rhythm in running-wheel activity. Feeding activity was diffuse during subjective night, whereas the ultradian feeding bouts were obvious during subjective day. After 20-25 days the circadian components appeared to disintegrate slowly and ultradian timing took control over the running-wheel activity. The same was shown earlier in DD by vole #319 (Fig. 2). These examples of spontaneous loss of circadian rhythmicity are not exceptions. Table 2 presents summary data on stability of rhythmicity in 146 intact voles; 32% of the animals lost their circadian activity pattern when held in continuous darkness. In some cases circadian rhythmicity disappeared even in a light-dark cycle. To date we have never observed spontaneous loss of ultradian rhythmicity in our laboratory studies. However, the expression of ultradian rhythmicity and its frequency is apparently modified by environmental conditions. HOOGENBOOM et al. (1984) documented seasonal changes in daily activity patterns in the common vole by (a) frequently trapping in the field, and (b) continuously recording caged voles exposed to natural meteorological conditions. The trap checks revealed a 2-h rhythm in daytime catches during most of the

Table 2. Rhythmic state in intact common voles

	Circadian	Ultradian
Rhythmic	111	146
Arrhythmic	35	0

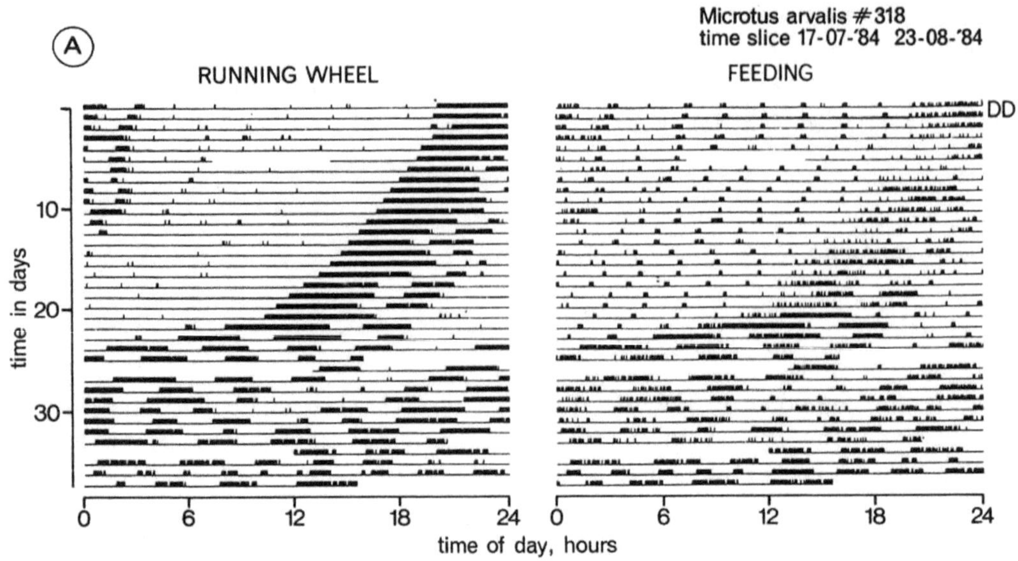

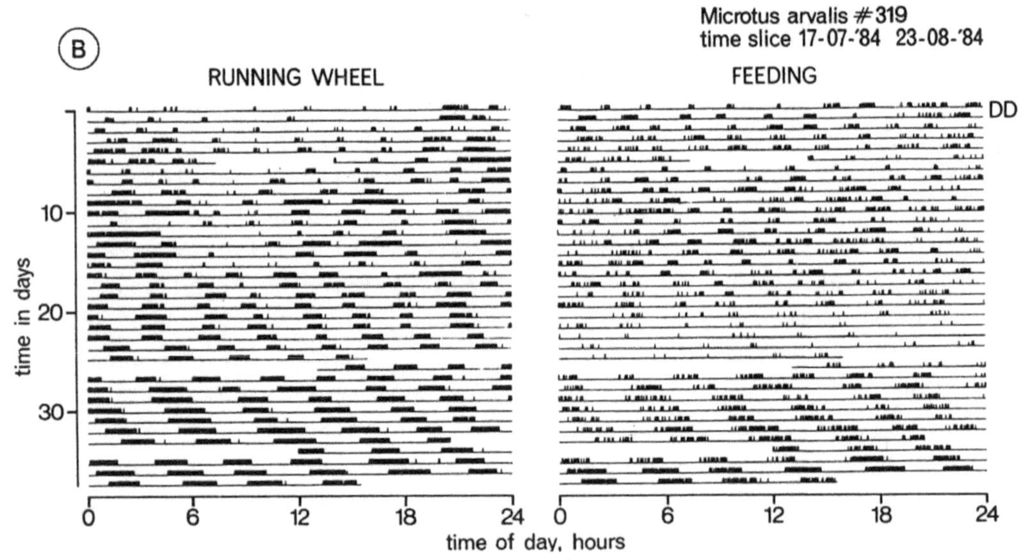

Fig. 2. Spontaneous loss of circadian rhythmicity. Simultaneous recording of wheel-running and feeding activity in two common voles in continuous darkness (DD)

year, but not during the summer. The feeding activity patterns of caged voles in summer were still organized in 2-h intervals, but the interindividual and day-to-day variation in phasing was larger than in other seasons. The lack of a short-term rhythm in trap catches during summer was adequately explained by an increase variability in phasing and a lack of synchrony of the population. Thus, we assume that the individual ultradian rhythmicity in the field persists the whole year around.

Food quality affects the frequency of the short-term feeding rhythm in voles. The ultradian period length (τ') of 2.0 h, derived from trapping data during the daytime, can also be obtained from caged voles, provided that the food supplied has a high raw fibre content (HOOGENBOOM et al. 1984). However, if voles are provided with standard rat chow this results in an increase of the feeding period to at least 2.3 h. The coefficient of variation in τ' in individuals is normally between 0.1 and 0.2. Between individuals ultradian periods vary over a large range (2.3-4.0 h).

Ambient temperature affects the length of the feeding bouts and meal size, rather than the frequency of the feeding rhythm (DAAN and SLOPSEMA 1978). The main effect of extremely low temperature (-15°C) is a nearly complete disappearance of running-wheel activity during the night, as shown in voles kept under a light-dark cycle (LD 12:12, J. VAN BELKUM personal communication). Under such conditions the ultradian feeding patterns persisted through the night. Ultradian feeding rhythms during the subjective night were observed also in DD experiments, where we blocked the running-wheel. We therefore surmise that the presence of a diffuse feeding pattern during the subjective night, concurring with intense running-wheel activity (Fig. 2, #318), is the result of interference of wheel-running behavior with the normal expression of ultradian rhythms.

In summary, the ultradian rhythm is a highly robust characteristic of the feeding system of the common vole. This short-term rhythm continues also in the absence of circadian rhythmicity. However, its expression may be modified both by environmental conditions and by circadian modulation.

Synchronization

The interaction between the ultradian and circadian systems in the common vole is of further interest with respect to the stability of ultradian phasing across the population. In the field the first morning peak in trap catches after sunrise kept a remarkably stable phase relation to dawn (RAPTOR GROUP 1982; HOOGENBOOM et al. 1984). This synchrony is obviously brought about by external entraining agents.

Social synchronization, while probably operating at the level of vole families, cannot explain phase stability in populations beyond the family size. Other ultradian zeitgebers in the environment are unknown. The fact that the ultradian population rhythm gradually fades out in the course of the day, suggests that resetting of the ultradian rhythms takes place in the early morning. The question may be raised whether such reset is a response of the ultradian system to dawn stimuli directly. Alternatively the response could be mediated via the

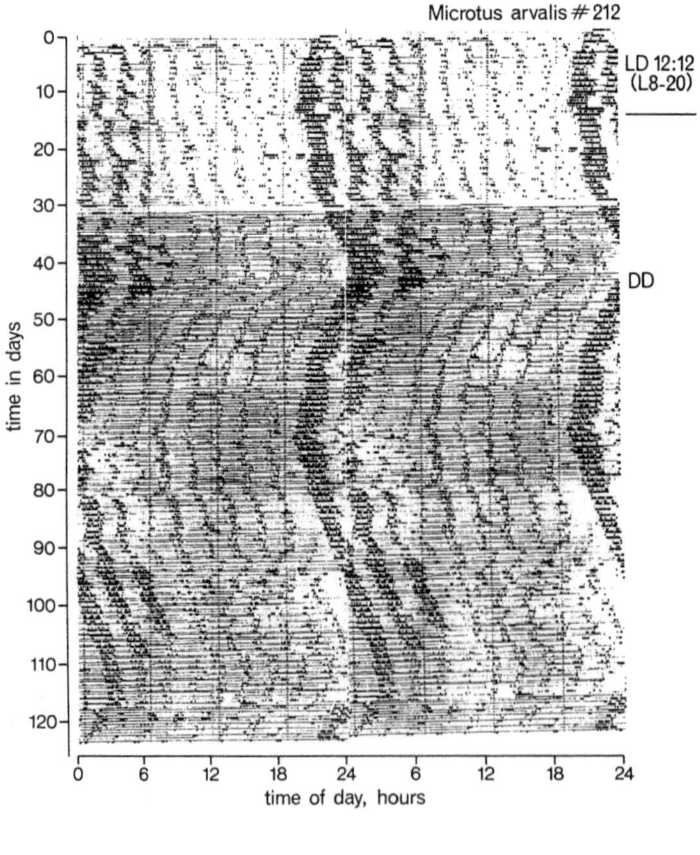

circadian system. Laboratory experiments suggest that indeed the circadian system is involved.

This is indicated in the first place by recorded activity patterns in which changes in circadian period occurred either spontaneously (Fig. 3) or in response to light pulses. In Fig. 3 circadian τ was longer than 24 h for the initial 25 days of the DD run. Spontaneous shortening of τ after 25 days was accompanied by simultaneous forward shifts in all the ultradian feeding bouts. A tight coupling of the ultradian and the circadian system is suggested. This does not provide evidence on the direction of coupling involved. Early morning synchronization in the field and in natural daylight recordings (HOOGENBOOM et al. 1984) proves that light resets at least one of the two systems which may mediate the reset to the other. The vole circadian system is responsive to light pulses, as are virtually all other circadian clocks. In preliminary short LD-cycle experiments we have been unable to entrain ultradian rhythms of voles with LD 1.0:1.0, LD 1.25:1.25, or LD 1.5:1.5. These results suggest that the resetting by

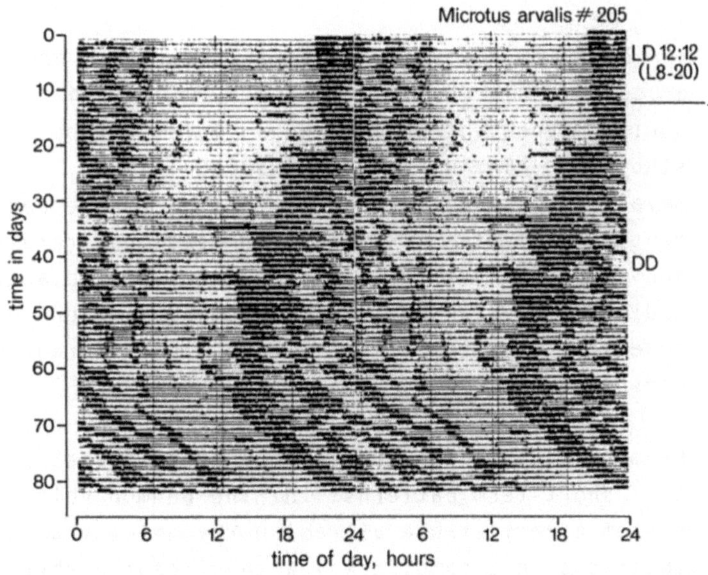

Fig. 4. Running and feeding record of a vole, initially in LD 12:12, then in DD. Records are plotted as in Fig. 3

◁ Fig. 3. Activity patterns of a vole in a light-dark cycle (L, 8-20; day 1-15) and in continuous darkness (DD; day 16-125). Running-wheel and feeding activity are plotted concurrently (running above and feeding below). Original records are plotted twice along the abscissa light is brought about by the circadian system and hence that the circadian system in turn controls the phase of the ultradian rhythm. This logic holds for early morning synchrony, and it does not exclude that at other phases of the circadian system there may be reverse coupling influences. There are a few cases, admittedly rare, where sudden "jumps" of the onset of running activity occur (Fig. 4). These jumps appear phase-locked to a "preceding" or a "succeeding" ultradian feeding bout, in a complex manner.

The absence of light entrainment of ultradian rhythms is indeed what we would expect on functional grounds. A vole comes above ground for feeding every 2 h and therefore itself generates the ultradian light cycle it perceives. Temporal adjustment to this cycle would seem useless. Likewise there is no functional reason for vole ultradian rhythms to synchronize with ultradian cycles of food availability, since the herbs and grass they feed on are continuously present. In ongoing experiments we have again not been able to establish entrainment by cycles of food availability. Nevertheless further experiments have to be done, using refined external ultradian cycles, as we cannot exclude the possibility that the ultradian periods offered were out of an ultradian range of entrainment.

In contrast we would expect social synchronization within family groups on functional grounds, as discussed earlier. The occurrence of social synchronization in a vole can be deduced from an enclosure study by LEHMANN and SOMMERSBERG (1980), in which passage counters revealed a synchronized rhythm of the experimental family. Experimental analysis of this phenomenon has rarely been undertaken. DAAN and SLOPSEMA (1978) compared the averaged activity patterns of three individual voles with the activity record of voles kept in a group together. In the latter case the ultradian rhythm was much more conspicuous.

From discussion of several phenomenological and functional aspects of the short-term patterns in the common vole, the emerging picture is one of a very stable ultradian rhythmicity that relates to feeding and interacts in a complex way with circadian rhythmicity.

GENERATION OF ULTRADIAN RHYTHMICITY

Two principally different mechanisms may cause ultradian behavioral rhythmicity. In one type, called a renewal or homeostatic process, the behavior studied is part of the generating causal loop. In the other type, the behavior studied is driven by an underlying oscillator which continues its motion even when the behavior is not expressed. Since many of the ultradian rhythms in behavior are connected with foraging and food intake, several authors have suggested that metabolic processes are involved in their generation (references in ASCHOFF 1962; STEBBINS 1975; DAAN and ASCHOFF 1981). Besides a classical physiological hunger-feedback system (see discussion in LE MAGNEN 1981), another homeostatic mechanism has been recently proposed by DAAN et al. (1984), based on a model for human sleep. In this model, a sleep-regulating variable S increases during activity and decreases during sleep. Two circadian thresholds determine the timing of sleep and wakefulness. Using this renewal model, one can simulate a vole-like pattern of rest and activity by reducing the distance between the two thresholds (DAAN et al. 1984). The nature of the underlying mechanisms can hardly be derived from statistical properties of the ultradian periods, such as their serial correlation. It is, however, possible to distinguish between these specific homeostatic mechanisms and a "clock-like" mechanism for ultradian rhythmicity by experimentally manipulating hunger and S-levels.

If it is assumed that a physiological manifestation of hunger determines the timing of foraging bouts, then food deprivation over some ultradian periods should lead to an irregular pattern of feeding attempts, probably with an increasing frequency. A hunger-independent clock model should result in an undisturbed ultradian pattern of food attempts (Fig. 5A). DAAN and SLOPSEMA (1978) have already reported that an intact vole did not change its ultradian feeding pattern during daytime food deprivation. In their experiments, a lucite screen was placed in front of the food rack, and feeding attempts (now unrewarded) were recorded as usual from pressing of the bar below the food rack. Using the same technique we deprived non-circadian voles, in conditions of constant darkness. Figure 6 shows the results for vole #046 following lesion of the nucleus paraventricularis hypothalami (PVNH; see also Fig. 8B). These results demonstrate that digestive feedback does not appear to be a prerequisite for ultradian rhythmicity.

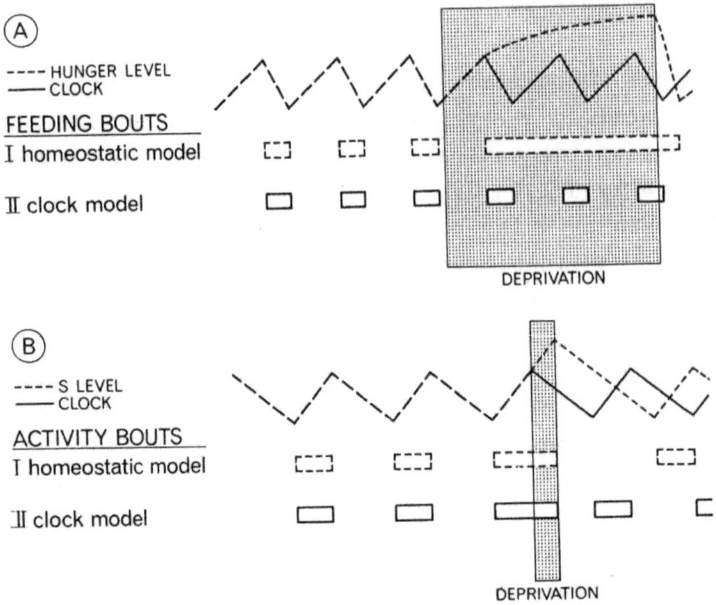

Fig. 5. Schematic representation of homeostatic timing of ultradian activity by hunger (panel A) and sleep propensity S (panel B), vs "clock" regulation.
A. Homeostatic model (dotted lines); hunger above certain threshold triggers a feeding bout. During food deprivation, the hunger level still increases (arbitrary rate chosen here), as does the intensity of feeding attempts. Clock model (dashed lines); no effect of deprivation on timing of feeding bouts occurs.
B. Homeostatic model (dotted lines); S builds up during activity, and sleep occurs above the upper threshold. During sleep, S declines until a lower threshold and activity starts. By experimentally lengthening the activity period (sleep deprivation), further increase in S causes lengthening of the succeeding rest episode. Clock model (dashed lines); lengthening of activity by sleep deprivation is compensated by shortening of the succeeding rest episode

During sleep deprivation, we expect an increase in S, on the basis of the sleep model of DAAN and BEERSMA (1984). In addition, the succeeding ultradian rest time (ρ') should lengthen proportionally. However, if the contrasting clock model is used, we expect the ultradian period to be stable and thus ρ' should shorten (Fig. 5B). To avoid the complications of a circadian modulation, we used voles that spontaneously had lost circadian activity, in continuous darkness in the sleep deprivation experiments. On the basis of the slight, but significant negative correlation between α' and ρ' for the uninfluenced data points (Fig. 7) we cannot distinguish between the two models, as argued already. The relatively low values of the ρ' succeeding the experimentally lengthened α' cannot be explained only by the homeostatic

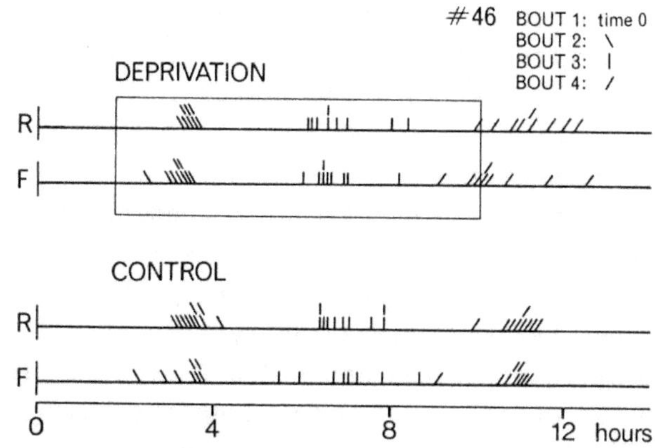

Fig. 6. Effect of food deprivation on onsets of ultradian wheel-running (R) and feeding bouts (F) in a non-circadian vole (#046, PVN lesion). Data for 9 days with food deprivation (during 8 h) and 9 control days. Onsets of bouts before deprivation standardized at time zero (bout 1)

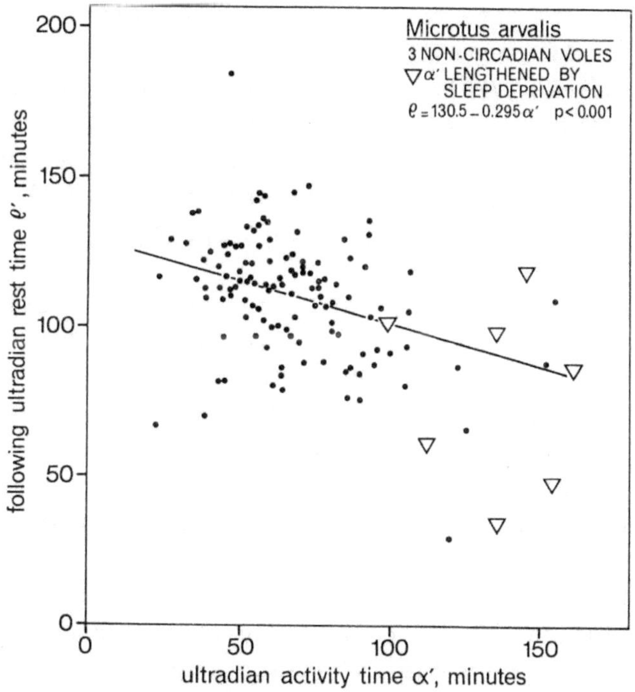

Fig. 7. Effects of sleep deprivation on the length of the succeeding ultradian rest time in three spontaneous non-circadian voles. Triangles, ultradian activity time lengthened by sleep deprivation; black dots, ultradian activity times before and after sleep deprivation

sleep model. Inevitably, these results lead to the conclusion that an ultradian oscillator is involved, without, however, excluding the possibility that this oscillator interacts with an S-renewal process.

Therefore, the postulated homeostatic mechanisms currently do not seem to apply to the ultradian feeding rhythmicity in the common vole. An endogenous central nervous oscillatory system has been proposed as an alternative mechanism, whereby the short-term rhythms could be generated by the circadian system itself, or by an autonomous ultradian oscillator (DAAN and ASCHOFF 1981). In mammals, the circadian pacemaker function of the suprachiasmatic nuclei (SCN) has been widely accepted (ASCHOFF et al. 1981). In rats, HONMA and HIROSHIGE (1978) showed the persistence of ultradian rhythmicity after disruption of the circadian system. If an ultradian pacemaker exists in the common vole, connections with the circadian system are likely because of the strong but complex relationships between ultradian and circadian rhythms already mentioned.

In the rat, the mediobasal hypothalamus (MBH), caudal from and intensively connected with the SCN, contains the ventromedial hypothalamic nuclei (VMH), which are involved in feeding regulation. The MBH also mediates several ultradian rhythms of hormonal secretion. The remarkably precise pulsatile release of gonadotropin releasing hormone (GnRH) in the nucleus arcuatus (Arc) is discussed in this volume by KNOBIL and HOTCHKISS (see Chap. 2).

A 3-h periodicity of growth hormone (GH) in rats is probably also mediated by the Arc (EIKELBOOM and TANNENBAUM 1983). RICHTER (1980) has related this GH rhythm to the ultradian feeding behavior. A relationship between ultradian patterns in behavior and hormone fluctuations in the rat has also been demonstrated by HIROSHIGE and HONMA (1979). They showed a phase relationship between ultradian running activity, corticosterone plasma titers and body temperature.

We have studied the involvement of hypothalamic structures in the generation of ultradian and circadian behavioral rhythms in the common vole in a series of lesion experiments. Criteria were derived for the loss of ultradian and circadian rhythmicity from Chi-square periodogram analysis (SOKOLOVE and BUSHELL 1980), in addition to visual inspection of the registrations. The interpretation of periodogram analysis for ultradian periods in the presence of circadian rhythms still poses some problems, which are probably alleviated when low-

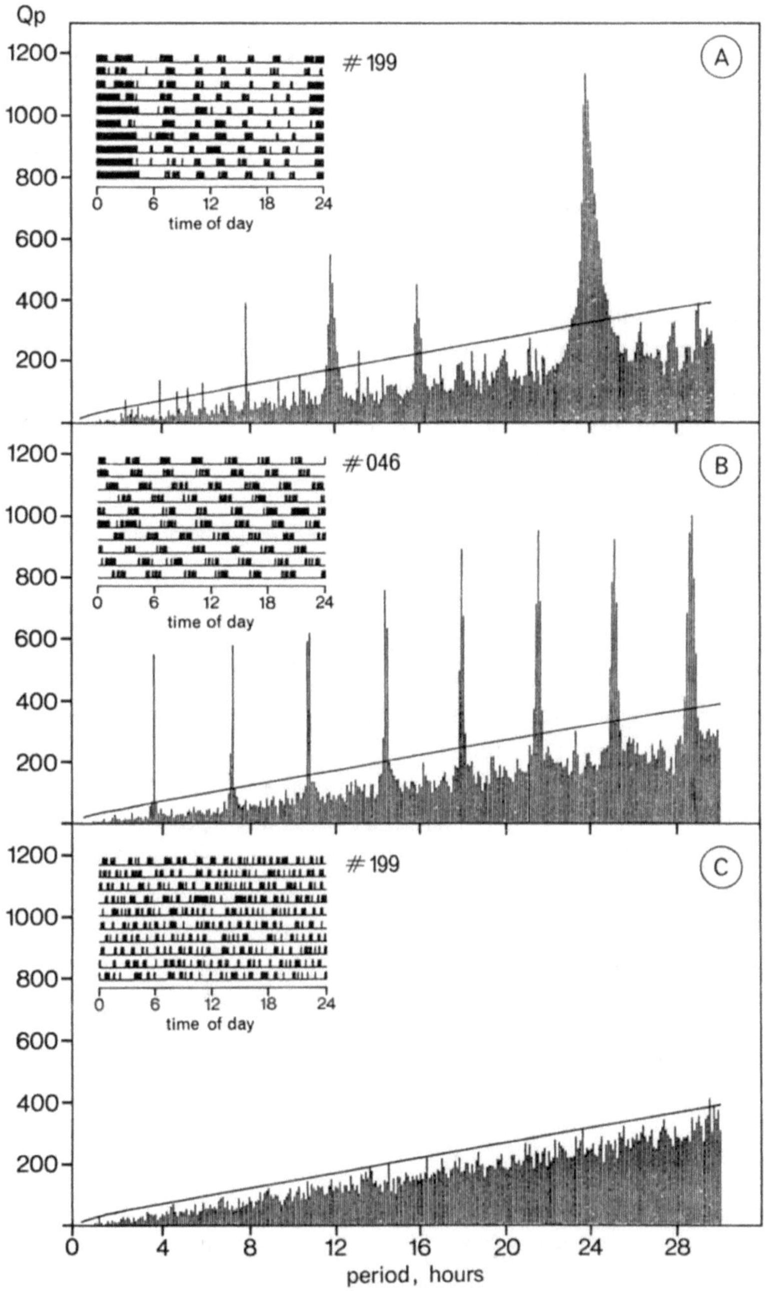

Fig. 8. Chi-square periodogram analysis of total activity (wheel-running and feeding); Qp values (index of rhythmicity, see SOKOLOVE and BUSHELL 1980) for periods from 0.2-30.0 h. Data for 10 days of 6' scores (DD registration). The sloping line designates p ≦ 0.005 level of significance. Inserts, 24-h actograms for the corresponding data set. A, vole #199, before lesioning; B, vole #046, PVN-lesioned; C, vole #199, RCA/Arc-lesioned

pass filter techniques are applied (DIRLICH and GERKEMA, in preparation; see Fig. 8A). Nonetheless, by using Chi-square periodogram analysis we can clearly show the loss of ultradian and circadian periodicity (Fig. 8B,C).

The preliminary results indicate that SCN is involved in the expression of circadian rhythmicity in Microtus arvalis as well as in many other species. We found also an absence of circadian rhythms in voles with lesions in the region dorsal to SCN, containing the periventricular and paraventricular diencephalic area (PeVN/PVNH; see also Fig. 8B) and in the rostral basal hypothalamus, encompassing the retrochiasmatic area (RCA) and Arc (see also Fig. 8A,C). These results are consistent with the pattern of efferent projections of SCN in the rat as described by STEPHAN et al. (1981). Circadian rhythmicity persisted after lesioning of the ventromedial and/or dorsomedial hypothalamic nuclei.

Destruction of the complete SCN does not result in loss of the ultradian rhythmicity. In the six cases of ultradian arrhythmicity in the common vole obtained to date, RCA was the common factor involved. We do not intend to suggest that RCA contains an ultradian oscillator. If any ultradian oscillator exists at all, Arc would probably be a more likely pacemaker candidate. We know that efferents from Arc are projected through RCA. These efferents may be identical with structures we destroyed in RCA which are obviously essential for the expression of ultradian rhythms of behavior. With the available evidence, this remains speculative. In the search for a central nervous oscillator responsible for ultradian rhythmicity, the feeding behavior of the common vole is a promising system.

REFERENCES

ALLISON T, GERBER SD, BREELOVE SM, DRYDER GL (1977) A behavioral and polygraphic study of sleep in the shrews Suncus murinus, Blarina brevicauda and Cryptotis parva. Behav Biol 20:354-366

ASCHOFF J (1962) Spontane lokomotorische Aktivität. Handb Zool 8:1-76

ASCHOFF J (ed) (1981) Handbook of behavioural neurobiology, vol 4. Biological rhythms. Plenum, New York

ASCHOFF J (1984) Circadian timing. Ann NY Acad Sci 423:442-468

BEERSMA DGM, DAAN S, VAN DEN HOOFDAKKER RM (1984) Distribution of REM latencies and other sleep phenomena in depression as explained by a single ultradian rhythms disturbance. Sleep 7:126-135

BUBENIK AN (1960) Le rhythme nycthéméral et le régime journalier des ongulés sauvages. Problèmes théoriques. Rhythme d'activité du chevreuil. Mammalia 29:1-58

DAAN S, ASCHOFF J (1981) Short term rhythms in activity. In: ASCHOFF J (ed) Handbook of behavioural neurobiology, vol 4. Biological rhythms. Plenum, New York, pp 491-498

DAAN S, ASCHOFF J (1982) Circadian contributions to survival. In: ASCHOFF J, DAAN S, GROOS GA (eds) Vertebrate circadian systems. Springer, Berlin Heidelberg New York, pp 305-321

DAAN S, BEERSMA DGM (1984) Circadian gating of human sleep and wakefulness. In: MOORE-EDE MC, CZEISLER CA (eds) Mathematical modelling of the circadian sleep-wake cycle. Raven, New York, pp 129-158

DAAN S, SLOPSEMA S (1978) Short term rhythms in foraging behaviour of the common vole, Microtus arvalis. J Comp Physiol 127:215-227

DAAN S, BEERSMA DGM, Borbely AA (1984) Timing of human sleep: recovery process gated by a circadian pacemaker. Am J Physiol 246:R161-R178

DELGADO-GARCIA JM, GRAU C, DeFEUDIS P, DELPOZO F, JEMINEZ JM, DELGADO JMR (1976) Ultradian rhythms in mobility and behavior of rhesus monkeys. Exp Brain Res 25:79-91

EIKELBOOM R, TANNENBAUM GS (1983) Effects of obesity inducing ventromedial hypothalamic lesions on pulsatile growth hormone and insuline secretion: evidence for the existence of a growth hormone releasing factor. Endocrinology 112:212-219

ERIKSSON LO, KALLQUIST ML, MOSSING T (1981) Seasonal development of circadian and short term activity in captive reindeer, Rangifer tarandus L.. Oecologia 48:64-70

HALBERG F, ENGELI M, HAMBURGER C, HILLMAN VD (1965) Spectral resolution of low-frequency, small-amplitude rhythms in excreted 17-ketosteroids: probably androgen-induced circaseptan desynchronization. Acta Endocrinol 103:1-54

HIROSHIGE T, HONMA K (1979) Internal and external synchronization of endogenous rhythms in the rat and involvement of brain biogenic amines. In: SUDA M, HAYAISHI O, NAKAGAWA H (eds) Biological rhythms and their central mechanism. Elsevier, Amsterdam, pp 233-245

HONMA KI, HIROSHIGE T (1978) Endogenous ultradian rhythms in rats exposed to prolonged continuous light. Am J Physiol 235:R250-R256

HOOGENBOOM I, DAAN S, DALLINGA H, SCHOENMAKERS M (1984) Seasonal change in the daily timing of behaviour in the common vole Microtus arvalis. Oecologia 6:18-31

HUGHES GP, REID D (1951) Studies on the behaviour of cattle and sheep in relation to the utilization of grass. J Agric Sci 41:360-366

KRAEMER HC, HOLE WT, ANDERS TF (1984) The detection of behavioral state cycles and classification of temporal structures in behavioral states. Sleep 7:3-17

LEHMANN U, SOMMERSBERG CW (1980) Activity patterns of the common vole Microtus arvalis. Autonomic recording of behaviour in an enclosure. Oecologia 47:61-75

LE MAGNEN J (1981) The metabolic basis of dual periodicity of feeding in rats. Behav Brain Sci 4:561-607

PARMELEE AH (1974) Ontogeny of sleep patterns and associated periodicities in infants. Mod Probl Paediatr 13:298-311

PETERSON RO, PAGE RE, DODGE KM (1984) Wolves, moose and the allometry of population cycles. Science 224:1350-1352

PITTENDRIGH CS (1960) Circadian rhythms and the circadian organization of living systems. Cold Spring Harbor Symp Quant Biol 25:159-184

PITTENDRIGH CS, DAAN S (1976a) A functional analysis of circadian pacemakers in nocturnal rodents. I. The stability and lability of spontaneous frequency. J Comp Physiol 106:223-252

PITTENDRIGH CS, DAAN S (1976b) A functional analysis of circadian pacemakers in nocturnal rodents. IV. Entrainment: pacemaker as clock. J Comp Physiol 106:291-331

PROP J, VAN EERDEN MR, DAAN S, DRENT RH, TINBERGEN JM, ST. JOSEPH AM (1978) Ecology of the barnacle goose (Branta leucopsis) during breeding season. Spitsbergen Symposium, Arctic Centre, University Groningen, pp 51-112

RAPTOR GROUP RUG/RIJP (1982) Timing of vole hunting in aerial predators. Mammal Rev 12:169-181

RECHTSCHAFFEN A, KALES A (1968) A manual of standardized terminology, techniques and scoring system for sleep stages of human subjects. UCLA Brain Information Service, Brain Research Institute, Los Angeles

RICHTER CP (1980) Growth hormone 3.6 h pulsatile secretion and feeding times have similar periods in rats. Am J Physiol 239:E1-2

SCHULZ H, DIRLICH G, BALTESKONIS S, ZULLEY J (1980) The REM-NREM cycle: renewal process or periodically driven process? Sleep 2:319-328

SIBLY RM (1981) Strategies of digestion and defecation. In: TOWNSEND CR, CALOW P (eds) Physiological ecology. Blackwell, Oxford, pp 109-139

SOKOLOVE PG, BUSHELL WN (1978) The chi-square periodogram: its utility for analysis of circadian rhythms. J Theor Biol 74:131-160

STEBBINS LL (1975) Short activity periods in relation to circadian rhythms in Clethrionomys gapperi. Oikos 26:32-38

STEPHAN FK, BERKLEY KJ, MOSS RL (1981) Efferent connections of the rat suprachiasmatic nucleus. Neuroscience 6:2625-2641

SZYMANSKI JS (1920) Aktivität und Ruhe bei Tieren und Menschen. Z Allg Physiol 18:105-162

TERTIL R (1972) The effect of behavioural thermoregulation on the daily metabolism of Apodenus agrarius. Acta Theriol 17:295-313

WEBSTER AB, BROOKS RJ (1981) Daily movements and short activity periods of freeranging meadow voles Microtus pennsylvanicus. Oikos 37: 80-87

WEVER RA (1979) The circadian system of man. Springer, Berlin Heidelberg New York

The Circhoral Gonadotropin Releasing Hormone (GnRH) Pulse Generator of the Hypothalamus and Its Physiological Significance

E. Knobil and J. Hotchkiss

Laboratory for Neuroendocrinology, The University of Texas Medical School at Houston, P.O. Box 20708, Houston, TX 77225, USA

In the course of validating a radioimmunoassay for luteinizing hormone (LH) in the rhesus monkey more than a decade ago, our attempt at measuring this hormone in the plasma of ovariectomized females met with very erratic results. We initially assumed that imperfections in the assay were responsible for our unpredictable measurements, but, after much additional work, we were eventually persuaded that our assay was not at fault and that the surprising variations in hormone levels were real and, somehow, a function of our animals (ATKINSON et al. 1970). We then proceeded to take blood samples every 10-20 min and found to our astonishment that the large fluctuations in plasma LH concentrations were not random, as we had initially assumed, but were strikingly rhythmic. The time courses of these elevations in plasma LH concentrations suggested that they resulted from abrupt pulsatile discharges of the hormone which then declined exponentially with the half-time of this polypeptide in the monkey circulation (DIERSCHKE et al. 1970). Because the frequency of the LH pulses was approximately once every 60 min, we described this intermittent secretory mode as being "circhoral" (DIERSCHKE et al. 1970). Shortly thereafter, a rhythmic pulsatile pattern of gonadotropin secretion was described in a number of other mammals, including the human (NANKIN and TROEN 1971; KATONGOLE et al. 1971; GAY and SHETH 1972; BUTLER et al. 1972; YEN et al. 1972).

We originally proposed that the intermittent discharges of gonadotropin were the consequence of packets of gonadotropin releasing hormone (GnRH) released into the pituitary portal circulation (DIERSCHKE et al. 1970). This notion was strengthened by the subsequent finding that the i.v. administration of antisera to synthetic GnRH to ovariectomized rhesus monkeys causes an abrupt reduction in plasma gonadotropin concentrations (McCORMACK et al. 1977), that a variety of α-adrenergic blocking agents promptly interrupts the pulsatile dis-

Experimental Brain Research, Suppl. 12

charges of gonadotropin in the monkey (BHATTACHARYA et al. 1972), and that the concentration of immunoreactive GnRH in pituitary stalk blood of monkeys appears to fluctuate with a frequency consonant with the circhoral mode of gonadotropin secretion (CARMEL et al. 1976). Furthermore, discrete lesions in the arcuate nuclei of female rhesus monkeys (PLANT et al. 1978a) result in abrupt reductions in plasma gonadotropin levels with time courses similar to those which follow the administration of α-adrenergic blocking agents. The assumption that these drugs acted at the neural level rather than directly on the pituitary was eventually verified by additional direct experimental evidence (PLANT et al. 1978b; KAUFMAN et al. 1985).

Electrical recording of multiunit activity in the medial basal hypothalamus of ovariectomized rhesus monkeys has suggested the existence of electrophysiological correlates of the circhoral, pulsatile pattern of GnRH release and, thereby, of a neuronal oscillator with an approximate period of 1 h, as illustrated in Fig. 1, panel A (DUFY et al. 1979; WILSON et al. 1984).

While the rhythmic, pulsatile pattern of gonadotropin secretion is most striking and frequently illustrated, no physiologic significance could be ascribed to this phenomenon until recently when it was recognized that the functioning of the hypophysiotropic control system which directs gonadotropin secretion is obligatorily intermittent. Continuous GnRH infusion into female monkeys with hypothalamic lesions that destroy the arcuate nuclei and abolish endogenous GnRH production is incapable of reinitiating sustained gonadotropin secretion, whereas intermittent infusion of the decapeptide (1 pulse per hour) reestablishes normal pituitary function (NAKAI et al. 1978). Furthermore, continuous infusion of GnRH inhibits gonadotropin secretion established by the pulsatile administration of the decapeptide to such animals (BELCHETZ et al. 1978). This inhibition of gonadotropin secretion by continuous GnRH infusion is reversed when the pulsatile replacement regimen is reinstituted. That this phenomenon is the consequence of the pattern of hypophysiotropic stimulation rather than of the total mass of decapeptide delivered to the gonadotrophs is suggested by the results of experiments in which the infusion rate of GnRH was varied from 0.001 to 1.0 μg/min. None of these infusion rates was effective in restoring gonadotropin secretion in monkeys with hypothalamic lesions when given in the continuous mode (BELCHETZ et al. 1978).

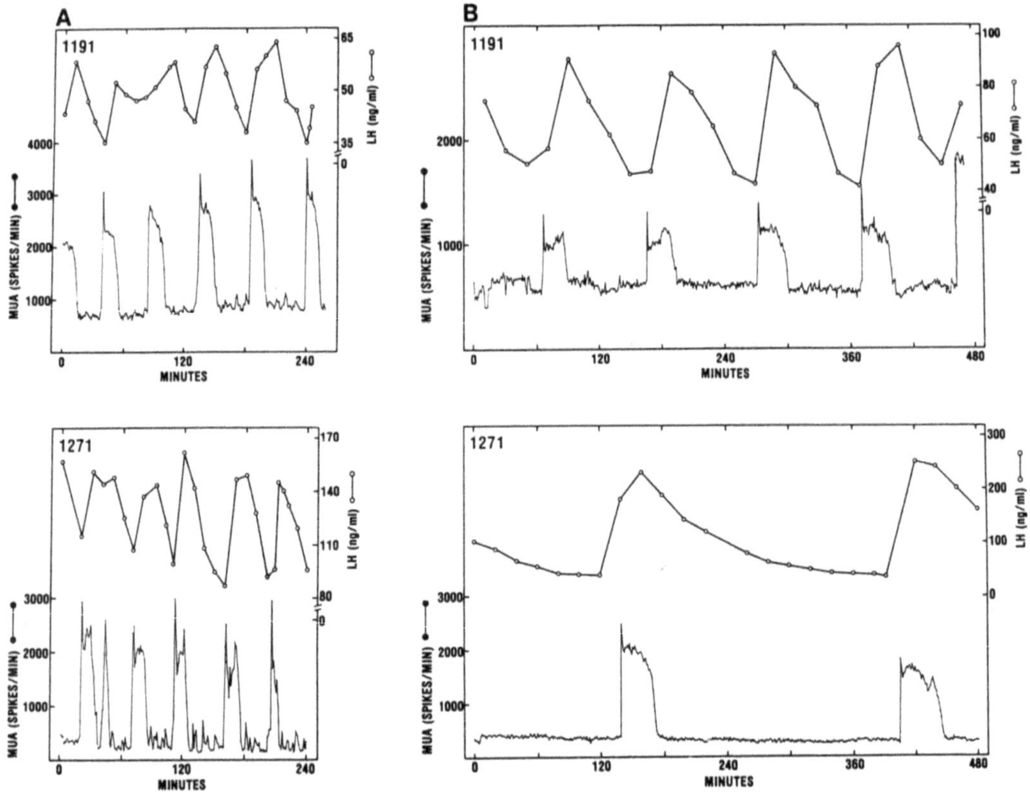

Fig. 1. Electrophysiologic correlates of pulsatile LH secretion in the ovariectomized rhesus monkey. A. In two awake restrained animals, increases in plasma LH and hypothalamic multiunit activity (MUA) occur spontaneously at approximately hourly intervals. Note the synchrony between the initiation of both events. B. Recordings from the same two animals anesthetized by a continuous infusion of thiopental. Note the reduction in frequency and the prolongation of the increments in hypothalamic electrical activity and of the coincident LH pulses in the peripheral circulation. (Reprinted with permission from WILSON et al 1984)

In view of the recognition that temporal patterns of hypophysiotropic stimulation have profound effects on pituitary response, we initiated a study designed to determine the effects of various frequencies of GnRH administration on the secretion of LH and follicle-stimulating hormone (FSH). The arcuate-lesioned ovariectomized monkey model was again employed. LH and FSH secretion was reestablished in such animals by administration of GnRH at a rate of 1 μg/min for 6 min once every hour (standard frequency) using suitably programmed infusion pumps. Once this had been achieved, the frequency of the GnRH pulses was either increased or decreased, each experimental frequency being maintained for 14 days. Each experimental period was followed by a reinstitution of the standard hourly frequency to verify the continued viability of the experimental preparations.

A frequency of five pulses of GnRH per hour extinguishes gonadotropin secretion in the same manner as that observed when an hourly frequency was followed by a continuous infusion of the decapeptide. A frequency of three pulses per hour has the same outcome, but the rate of decline in plasma LH and FSH concentrations is somewhat diminished. This inhibitory phenomenon is observed even when the frequency of hypophysiotropic stimulation is increased from the physiologic frequency of one pulse of GnRH per hour to two pulses per hour. These reductions in plasma gonadotropin levels are the consequence of reduced pituitary responses to each GnRH stimulus (WILDT et al. 1981).

Reducing the frequency of GnRH pulses from one per hour to one pulse every 3 h yielded rather striking results. In most instances this regimen did not have much of an effect on mean plasma LH concentrations, but FSH levels gradually increased in every one of the 13 experiments performed (WILDT et al. 1981). This shift in LH to FSH ratios is attributable to the lower metabolic clearance rate of FSH when compared with LH (YEN et al. 1970; PECKHAM unpublished) and to the fact that the magnitude of the discharge in response to the GnRH pulse is considerably larger at the lower frequency than at the standard frequency of one pulse per hour. It appears that the increase in the size of the LH bolus is matched by the increase in the time available for its clearance. This is not the case for FSH, permitting the glycoprotein with the longer half-life to accumulate. The foregoing study permits the important conclusion that not only can a change in the frequency of GnRH stimulation dramatically alter the concentration of LH and FSH in the circulation, but it can also have major effects on the ratio of LH to FSH.

Varying the amplitude of the GnRH pulses (leaving the frequency constant) has little regulatory effect in the monkey (WILDT et al. 1981).

These initial studies were performed in ovariectomized animals where pulsatile gonadotropin secretion had first been noted. Of particular physiologic interest, however, is the more recent finding (KNOBIL 1980; KNOBIL et al. 1980) that the circhoral discharge of GnRH is the only neuroendocrine signal required to subserve the entire 28-day menstrual cycle in the animal with ovaries in situ. This was demonstrated in animals with hypothalamic lesions which abolished endogenous GnRH production as evidenced by undetectable circulating gonadotropin and unresponsiveness to the positive feedback action of estradiol. The administration of the standard pulsatile GnRH replacement regimen, using our "hypothalamic prosthesis", caused the reestablishment of recurring ovulatory menstrual cycles of normal duration. When the GnRH was discontinued, all reproductive hormones again became undetectable, even after bilateral ovariectomy, and the positive feedback action of estradiol was again undemonstrable. In collaboration with M. Ferin and his colleagues, identical results have been obtained in stalk-sectioned monkeys on the same GnRH replacement regimen (RICHARDSON et al. 1983).

As in the ovariectomized monkey, small changes in the frequency of exogenous GnRH administration in the intact animal with a "hypothalamic prosthesis" have a profound influence on gonadotropin secretion, and upon resulting ovarian function. For instance, a decrease in the frequency of GnRH pulses from the physiologic rate of one per hour to one every 90 min reduces the incidence of ovulatory menstrual cycles. A regimen of one pulse every 2 h induces marginal functional development without ovulation, while GnRH administered once every 3 h is incapable of inducing any measurable follicular development (POHL et al. 1983).

While the entire menstrual cycle can devolve in the presence of an unvarying circhoral hypophysiotropic stimulus, acting in a permissive manner, there can be no doubt that, in a physiologic context, the neural component of the control system is profoundly influenced by higher neural centers as well as by changes in hormonal milieu. Progesterone, for example, markedly reduces the frequency of the pulse generator during the luteal phase of the ovarian cycle in several species including man (YEN et al. 1972; SANTEN and BARDIN 1973), cow (RAHE et al. 1979), sheep (HAUGER et al. 1977; BAIRD 1978), and monkey (NORMAN et al. 1984).

The sensory input associated with suckling appears to inhibit this neuronal mechanism completely (BAIRD et al. 1979; PLANT et al. 1980). When these inhibitory influences wane, FSH secretion is resumed first, followed by that of LH (ROLLAND et al. 1975; PLANT et al. 1980). This observation is consonant with the view that hypothalamic activity is reinitiated with a very low frequency, which causes FSH to accumulate while LH does not, and gradually accelerates to approximately one pulse per hour, leading to a convergence of LH and FSH concentrations in the plasma.

While pulsatile hypothalamic activity (as judged by consequent pituitary and gonadal hormone secretions) is evident for several months in the newborn rhesus monkey (PLANT 1982) and human (WALDHAUSER et al. 1981) such activity is not maintained. The reinitiation of gonadal function and sexual development at puberty can best be explained as the result of the reestablishment of pulsatile GnRH secretion by the hypothalamus, as precocious puberty and ovulatory menstrual cycles can be induced in immature female monkeys by the circhoral administration of GnRH using the "hypothalamic prosthesis" (WILDT et al. 1980). The longitudinal pattern of the gonadotropic hormones in the course of puberty in the human, where FSH appears first in the circulation followed by LH with a gradual convergence of the two hormones (BURR et al. 1970; SIZONENKO et al. 1970), can also be explained in terms of a gradual acceleration of the hypothalamic oscillator from total inactivity to nocturnal activation (BOYAR et al. 1972) to the final circhoral adult frequency.

Sleep onset is associated with striking increases in the frequency and amplitude of LH pulses in the pubertal human (BOYAR et al. 1972) and in some patients with specific reproductive dysfunctions (WILDT et al. 1983). In contrast, barbiturate anesthesia in the ovariectomized monkey decreases the frequency, and prolongs the duration, of both LH pulses (WEHRENBERG and FERIN 1981) and the coincident hypothalamic multiunit activity, as shown in Fig. 1, panel B (WILSON et al. 1984). A reduction in frequency, or complete cessation, of these hormonal and electrical events can also be brought about by the acute administration of α-adrenergic blocking agents (phentolamine, phenoxybenzamine, prazocin) or the antidopaminergic agent metoclopramide (KAUFMAN et al. 1985).

Challenging tasks for the future include description of the locus and neuronal circuitry of the hypothalamic pulse generator and the manner of its operation as well as its relation to other ultradian rhythms.

ACKNOWLEDGEMENTS. The studies from the author's laboratory have been generously supported by grants from the National Institute of Child Health and Human Development, the Ford Foundation, and the A.W. Mellon Foundation.

REFERENCES

ATKINSON LE, BHATTACHARYA AN, MONROE SE, DIERSCHKE DJ, KNOBIL E (1970) The effects of gonadectomy on plasma LH concentration in the rhesus monkey. Endocrinology 87:847-849

BAIRD DT (1978) Pulsatile secretion of LH and ovarian estradiol during the follicular phase of the sheep estrous cycle. Biol Reprod 18: 359-364

BAIRD DT, McNEILLY AS, SAWERS RS, SHARPE RM (1979) Failure of estrogen-induced discharge of luteinizing hormone in lactating women. J Clin Endocrinol Metab 49:500-506

BELCHETZ PE, PLANT TM, NAKAI Y, KEOGH EJ, KNOBIL E (1978) Hypophysial responses to continuous and intermittent delivery of hypothalamic gonadotropin releasing hormone. Science 202:631-633

BHATTACHARYA AN, DIERSCHKE DJ, YAMAJI T, KNOBIL E (1972) The pharmacologic blockade of the circhoral mode of LH secretion in the ovariectomized rhesus monkey. Endocrinology 90:778-786

BOYAR RM, FINKELSTEIN J, ROFFWARG H, KAPEN S, WEITZMAN E, HELLMAN L (1972) Synchronization of augmented luteinizing hormone secretion with sleep during puberty. N Engl J Med 287:582-586

BURR IM, SIZONENKO PC, KAPLAN SL, GRUMBACH MM (1970) Hormonal changes in puberty. I. Correlation of serum luteinizing hormone and follicle stimulating hormone with stage of puberty, testicular size and bone age in normal boys. Pediatr Res 4:25-35

BUTLER WR, MALVEN PV, WILLETT LB, BOLT DJ (1972) Patterns of pituitary release and cranial output of LH and prolactin in ovariectomized ewes. Endocrinology 91:793-801

CARMEL PW, ARAKI S, FERIN M (1976) Pituitary stalk portal blood collection in rhesus monkeys: evidence for pulsatile release of gonadotropin-releasing hormone (GnRH). Endocrinology 99:243-248

DIERSCHKE DJ, BHATTACHARYA AN, ATKINSON LE, KNOBIL E (1970) Circhoral oscillations of plasma LH levels in the ovariectomized rhesus monkey. Endocrinology 87:850-853

DUFY B, DUFY-BARBE L, VINCENT JD, KNOBIL E (1979) Etude électrophysiologique des neurones hypothalamiques et régulation gonadotrope chez le singe rhésus. J Physiol (Paris) 75:105-108

GAY VL, SHETH NA (1972) Evidence for a periodic release of LH in castrated male and female rats. Endocrinology 90:158-162

HAUGER RL, KARSCH FJ, FOSTER DL (1977) A new concept for control of the estrous cycle of the ewe based on the temporal relationships between luteinizing hormone, estradiol and progesterone in peripheral serum and evidence that progesterone inhibits tonic LH secretion. Endocrinology 101:807-817

KATONGOLE CB, NAFTOLIN F, SHORT RV (1971) Relationship between blood levels of luteinizing hormone and testosterone in bulls, and the effects of sexual stimulation. J Endocrinol 50:457-466

KAUFMAN JM, KESNER JS, WILSON RC, KNOBIL E (1985) Electrophysiologic manifestation of LHRH "pulse generator" activity in the rhesus monkey: influence of alpha-adrenergic and dopaminergic blocking agents. Endocrinology (in press)

KNOBIL E (1980) The neuroendocrine control of the menstrual cycle. Rec Prog Horm Res 36:53-88

KNOBIL E, PLANT TM, WILDT L, BELCHETZ PE, MARSHALL G (1980) Neuroendocrine control of the rhesus monkey menstrual cycle: permissive role of the hypothalamic gonadotropin releasing hormone (GnRH). Science 207:1371-1373

McCORMACK JT, PLANT TM, HESS DL, KNOBIL E (1977) The effect of luteinizing hormone releasing hormone (LHRH) antiserum administration on gonadotropin secretion in the rhesus monkey. Endocrinology 100: 663-667

NAKAI Y, PLANT TM, HESS DL, KEOGH EJ, KNOBIL E (1978) On the sites of the negative and positive feedback actions of estradiol in the control of gonadotropin secretion in the rhesus monkey. Endocrinology 102: 1008-1014

NANKIN HR, TROEN P (1971) Repetitive luteinizing hormone elevations in serum of normal men. J Clin Endocrinol Metab 33:558-560

NORMAN RL, LINDSTROM SA, BANGSBERG D, ELLINWOOD WE, GLEISSMAN P, SPIES HG (1984) Pulsatile secretion of luteinizing hormone during the menstrual cycle of rhesus macaques. Endocrinology 115:261-266

PLANT TM (1982) Pulsatile luteinizing hormone secretion in the neonatal male rhesus monkey (Macaca mulatta). J Endocrinol 93:71-74

PLANT TM, KREY LC, MOOSSY J, McCORMACK JT, HESS DL, KNOBIL E (1978a) The arcuate nucleus and the control of gonadotropin and prolactin secretion in the female rhesus monkey (Macaca mulatta). Endocrinology 102:52-62

PLANT TM, NAKAI Y, BELCHETZ P, KEOGH E, KNOBIL E (1978b) The sites of action of estradiol and phentolamine in the inhibition of the pulsatile, circhoral discharges of LH in the rhesus monkey (Macaca mulatta). Endocrinology 102:1015-1018

PLANT TM, SCHALLENBERGER E, HESS DL, McCORMACK JT, DUFY-BARBE L, KNOBIL E (1980) The influence of suckling on gonadotropin secretion in the female rhesus monkey (Macaca mulatta). Biol Reprod 23:760-766

POHL CR, RICHARDSON DW, HUTCHISON JS, GERMAK JA, KNOBIL E (1983) Hypophysiotropic signal frequency and the functioning of the pituitary-ovarian system in the rhesus monkey. Endocrinology 112: 2076-2080

RAHE CH, OWENS RE, NEWTON HJ, FLEEGER JL, HARMS PG (1979) Dependence of the pattern of circulating luteinizing hormone (LH) upon the stage of the estrous cycle. Fed Proc 38:981

RICHARDSON DW, WILDT L, HUTCHISON JS, POHL CR, ANTUNES JL, FERIN M, KNOBIL E (1983) Induction of ovulatory menstrual cycles by an unvarying pulsatile GnRH infusion in pituitary stalk-sectioned rhesus monkeys. Proc Endocrine Soc Abst 196

ROLLAND R, LEQUIN RM, SCHELLEKENS LA, DE JONG FH (1975) The role of prolactin in the restoration of ovarian function during the early post partum period in the human female. I. A study during physiological lactation. Clin Endocrinol 4:15-25

SANTEN RJ, BARDIN CW (1973) Episodic luteinizing hormone secretion in man. Pulse analysis, clinical interpretation, physiologic mechanisms. J Clin Invest 52:2617-2628

SIZONENKO PC, BURR IM, KAPLAN SL, GRUMBACH MM (1970) Hormonal changes in puberty. II. Correlation of serum luteinizing hormone and follicle stimulating hormone with stages of puberty and bone age in normal girls. Pediatr Res 4:36-45

WALDHAUSER F, WEISSENBACHER G, FRISCH H, POLLAK A (1981) Pulsatile secretion of gonadotropins in early infancy. Eur J Pediatr 137:71-74

WEHRENBERG WB, FERIN M (1981) Alteration of pulsatile LH secretion in monkeys by pentobarbital anesthesia. Proc Soc Exp Biol Med 168:286-289

WILDT L, MARSHALL G, KNOBIL E (1980) Experimental induction of puberty in the infantile female rhesus monkey. Science 207:1373-1375

WILDT L, HAUSLER A, MARSHALL G, HUTCHISON JS, PLANT TM, BELCHETZ PE, KNOBIL E (1981) Frequency and amplitude of gonadotropin-releasing hormone stimulation and gonadotropin secretion in the rhesus monkey. Endocrinology 109:376-385

WILDT L, SCHWILDEN H, WESNER G, ROLL C, BRENSING KA, LUCKHAUS J, BAHR M, LEYENDECKER G (1983) The pulsatile pattern of gonadotropin secretion and follicular development during the menstrual cycle and in women with hypothalamic and hyperandrogenemic amenorrhea. In: LEYENDECKER G, STOCK H, WILDT L (eds) Brain and pituitary peptides. II. Pulsatile administration of GnRH in hypothalamic failure: basic and clinical aspects. Karger, Basel, pp 28-57

WILSON RC, KESNER JS, KAUFMAN JM, UEMURA T, AKEMA T, KNOBIL E (1984) Central electrophysiologic correlates of pulsatile luteinizing hormone secretion in the rhesus monkey. Neuroendocrinology 39:256-260

YEN SSC, LLERENA LA, PEARSON OH, LITTELL AS (1970) Disappearance rates of endogenous follicle-stimulating hormone serum following surgical hypophysectomy in man. J Clin Endocrinol Metab 30:325-329

YEN SSC, TSAI CC, NAFTOLIN F, VANDENBERG G, AJABOR L (1972) Pulsatile patterns of gonadotropin release in subjects with and without ovarian function. J Clin Endocrinol Metab 34:671-675

Pulsatility of Pituitary Hormones

E. Van Cauter and E. Honinckx
Institute of Interdisciplinary Research, Erasme Campus, Free University of Brussels, 808 Route de Lennik, 1070 Brussels, Belgium

INTRODUCTION

During the last decade, the development of high-frequency blood sampling techniques and of sensitive and specific radioimmunoassays has made possible the demonstration of spontaneous temporal fluctuations in plasma levels of pituitary and other hormones. Under normal conditions, two types of variations have been identified in the complex secretory patterns of most hormones:

1. A periodic component synchronized with or determined by the sleep-wake cycle, the light-dark cycle, and/or exogenous factors associated with the daily changes in the environment. This periodic component will be referred to hereafter as a "circadian rhythm", irrespective of its wave shape and endogenous or exogenous origin, because of its regular occurrence at time intervals of approximately 1 day.

2. Episodic fluctuations of variable duration and magnitude, referred to in the literature as secretory "episodes", "spikes", "pulses", or "bursts". The repetition of such "episodes" has been referred to as an "ultradian rhythm" because they occur at frequencies much greater than one cycle per 24 hours (i.e., one cycle per 1-4 hours). However, it should be noted that, despite the use of the word "rhythm", this phenomenon does not appear to be strictly periodic. In what follows, the terms "episodic" and "pulsatile" will be used as synonyms.

Episodic variations of plasma levels have been observed for all pituitary hormones and for a large number of other hormones. Despite this ubiquity, their origin and physiological significance remain largely unknown. The pulsatility of gonadotropin secretion stands out as an

Experimental Brain Research, Suppl. 12

exception since intensive research has revealed its involvement in sexual maturation, its essential role in the adequate functioning of the pituitary-gonadal axis, and its origin in hypothalamic luteinizing hormone-releasing hormone (LHRH) releasing "pacemakers". These advances are presented in a separate chapter of this volume (Chap. 2). The present chapter focuses on ultradian variations in levels of pituitary hormones other than gonadotropins, with special emphasis on the pituitary-adrenal axis. Episodic variations of plasma cortisol will also be analyzed as an easily measurable index of corticotropin (ACTH) pulsatility. After introducing the quantitative characteristics of ultradian hormonal variations, we discuss hypotheses which have been suggested for their origin and present some of the factors responsible for their modulation. The last section of this chapter will describe preliminary data on adrenocortical pulsatility which suggest that episodic, rather than continuous, hormonal release may be of functional importance for peripheral endocrine systems other than those of the hypothalamo-pituitary-gonadal axis.

QUANTITATIVE CHARACTERISTICS

Figure 1 illustrates, as an example, the 24-h profiles of plasma corticotropin (ACTH), prolactin (PRL), luteinizing hormone (LH), thyroid stimulating hormone (TSH), and growth hormone (GH), obtained simultaneously in a healthy 28-year-old male subject. Blood was sampled every 15 min. Except for GH, the plasma levels are expressed in percent of their respective 24-h mean to allow comparison of the range of fluctuations in these various pituitary hormones.

Circadian and ultradian variations result in changes in concentration roughly ranging from 0%-200% of the 24-h mean for ACTH and PRL (upper panels). For TSH and LH (middle panels), the ranges of these combined variations are more limited (approximately 75%-150% of the mean for TSH and 50%-200% of the mean for LH). The 24-h profile of plasma GH (bottom panel) differs strikingly from the patterns observed for the other pituitary hormones. Indeed, while episodic spikes occur throughout the 24-h span for ACTH, PRL, TSH, and LH, the profile of plasma GH levels in young adults consists of only a few spikes interrupting consistently low levels. The major secretory event is generally a large peak starting shortly after sleep onset. Expressing the changes

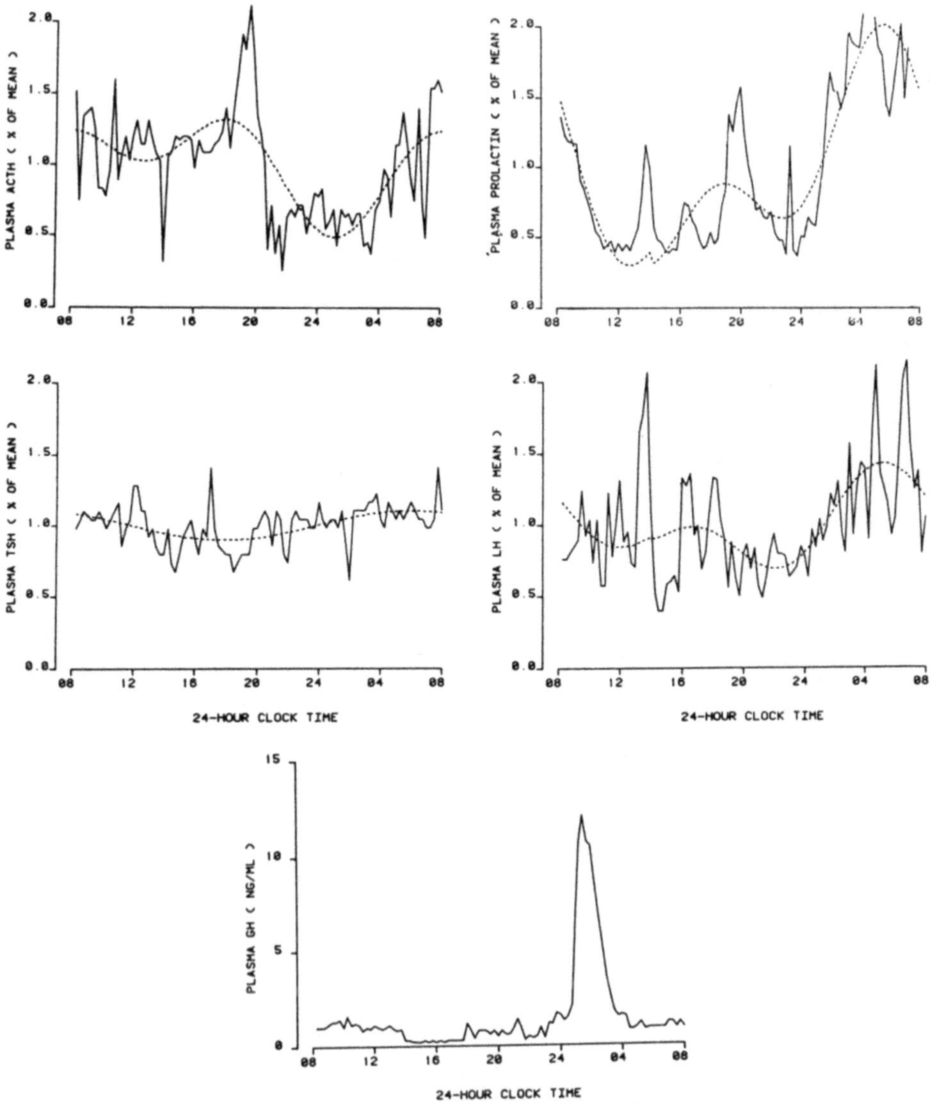

Fig. 1. 24-h profiles of plasma ACTH, PRL, TSH, LH, and GH of a healthy male 28-year-old subject. Blood was sampled every 15 min. Dashed line is a best-fit curve quantifying the circadian rhythm

in plasma GH concentration in percent of their 24-h mean is usually meaningless because basal levels are below or around the limit of detectability of most assays.

Many early studies of episodic hormonal variations were simply based on a visual examination of the data. Later, a number of quantitative procedures for the characterization of secretory spikes was proposed. In our method (VAN CAUTER 1981), a spike is considered significant if both the increase and the decrease exceed, in relative terms, twice the intraassay coefficient of concentration in the relevant range of concentration. The procedure takes into account differences in the accuracy of the hormone assays in various ranges of concentration. This point may be of particular importance since the plasma levels of some hormones (e.g., ACTH and GH) are close to the limit of sensitivity during certain intervals of the 24-h span. Every significant spike is characterized in terms of total duration, duration of the ascending and declining portions, absolute increment (difference between level at the peak and level at the preceding trough), relative increment or "magnitude" (absolute increment divided by level at the preceding trough) and apparent half-life (calculated from the slope, in semilog coordinates, of the least-square regression line for the declining portion of the spike). Using this procedure, we analyzed a set of 24-h profiles of ACTH, cortisol, PRL, TSH, and GH obtained with a 15-min blood sampling interval in normal male adults, studied under basal conditions. For each hormone, all secretory spikes obtained across individual studies were pooled to obtain the statistical values listed in Table 1.

The assay procedures for ACTH, cortisol, PRL, TSH, and GH were similar to those described in VIRASORO et al. (1974), DESIR et al. (1981), BADAWI et al. (1974), GOLSTEIN and VANHAELST (1973), and VIRASORO et al. (1971), respectively. Our data bank does not allow as yet a similar evaluation of the spike characteristics for the gonadotropins. From the means and standard deviations (SD) reported in Table 1, the non-Gaussian nature of the distributions of spike characteristics is obvious. Therefore, the medians, rather than the means, provide the better estimation of the average duration, increment, and apparent half-life of the spikes. The results given in Table 1 indicate that the characteristics of ultradian variations vary considerably from one pituitary hormone to another. This variability could reflect differences at the level of the mechanisms generating the pulsatility, but it could also be due to differences in factors modulating the expres-

Table 1. Quantitative characteristics of secretory spikes

Hormone	Number of 24-h profiles analyzed	Total no. of significant spikes	Average no. of spikes per 24-h span	Total duration (min)		Increment		Apparent half-life (min)	
ACTH	20	306	15	median:	60	median:	121%	median:	29
				mean:	64	mean:	242%	mean:	37
				SD:	41	SD:	372%	SD:	41
CORTISOL	19	288	15	median:	60	median:	70%	median:	47
				mean:	78	mean:	146%	mean:	54
				SD:	50	SD:	205%	SD:	33
PRL	25	298	12	median:	75	median:	68%	median:	49
				mean:	91	mean:	106%	mean:	66
				SD:	63	SD:	125%	SD:	54
TSH	25	455	18	median:	45	median:	44%	median:	48
				mean:	62	mean:	69%	mean:	70
				SD:	39	SD:	189%	SD:	86
GH	25	95	4	median:	75	median:	6.4 ng/ml	median:	28
				mean:	96	mean:	9.7 ng/ml	mean:	32
				SD:	48	SD:	9.6 ng/ml	SD:	17

sion of the pulsatility such as metabolic degradation and feedback by peripheral hormones.

From a methodological point of view, it should be emphasized that the quantitative characterizations of hormonal spiking activity depend upon the interval of blood sampling and upon the precision of the assay. Thus, comparisons between episodic variations in hormonal profiles should be restricted to investigations using the same sampling interval and assays of similar precision.

ORIGIN OF EPISODIC HORMONAL VARIATIONS

When episodic hormonal variations were first discovered, the possibility that they may represent random fluctuations of no physiological significance ("biological noise"), possibly associated with parallel changes in plasma dilution, was considered. Figure 2 illustrates typical results that have led to the rejection of this hypothesis. In the left side of the figure, the 24-h variations of plasma cortisol and total protein levels, expressed in percent of their respective 24-h means, are plotted using the same scale for the ordinate. It clearly appears that, because of their much higher magnitude, episodic cortisol variations are unlikely to be due to changes in plasma dilution and a detailed analysis of these data actually shows that the fluctuations in these two blood constituents are uncorrelated. On the right panel, the relationship between the magnitude (relative increment) and the total duration of significant spikes of plasma cortisol, PRL, and total proteins is illustrated. For cortisol and PRL, the magnitude and the duration of the spikes are positively correlated. Thus, higher spikes last longer, as one might expect if a peak in plasma concentration indeed corresponds to a burst of secretion. For total proteins, the durations of the spikes are independent of their magnitudes. Moreover, when these rapid fluctuations of total plasma protein levels are submitted to "white noise tests" (VAN CAUTER 1979), it appears that they may be considered as purely random variations.

Another possible explanation for the presence of episodic variations in plasma hormone levels is that such changes occur as responses to identifiable stimuli. Short-term increases in pituitary hormone concentrations may indeed be elicited by various physiological challenges

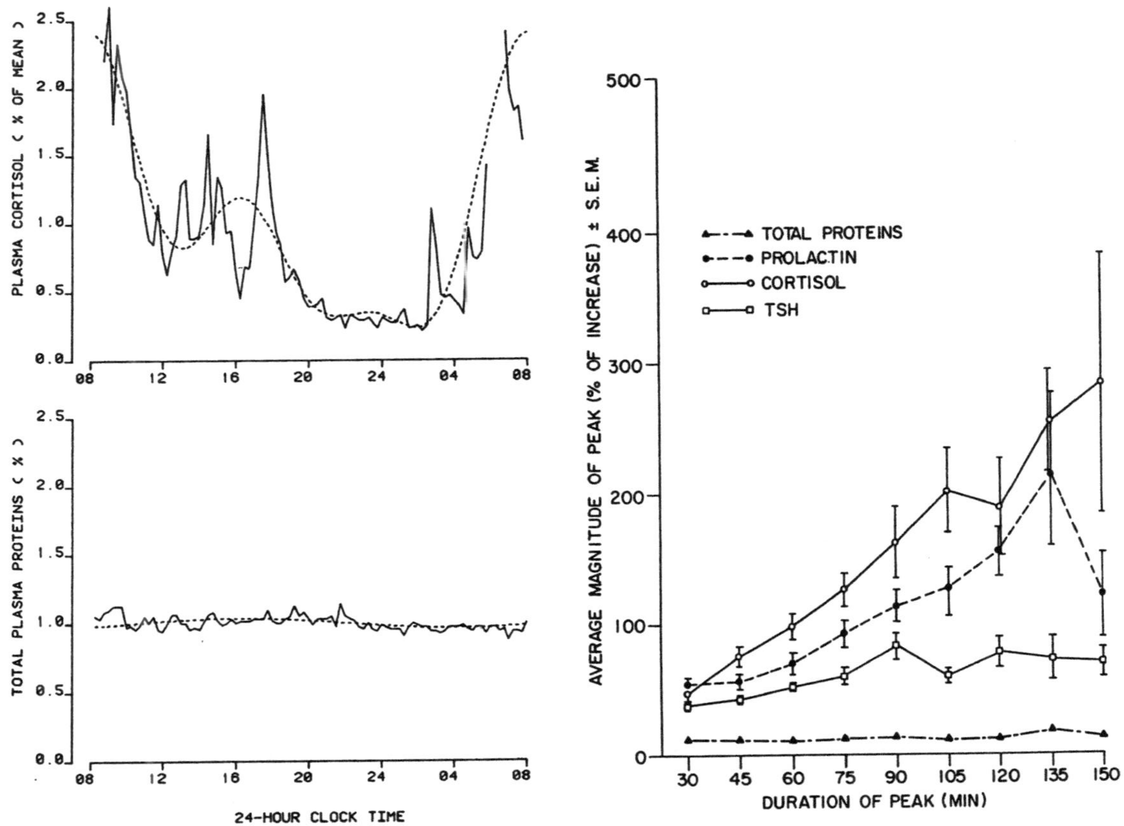

Fig. 2. Left: 24-h profiles of plasma cortisol (upper panel) and total proteins (lower panel) obtained simultaneously at 15-min intervals. Dashed line is a best-fit curve quantifying the circadian variation. Right: Relationship between average magnitude and average duration of episodic spikes of plasma cortisol, prolactin, TSH, and total proteins

including exercise, stress, and food ingestion. Figure 3 shows two examples of hormonal spikes evoked either by stress (upper panel) or by food intake (lower panel). Within 15 min after their occurrence, catheter insertion and food intake both result in a consistent burst of cortisol secretion. Studies in which the activities of the subjects, including meals, sleep, naps, snacks, and potentially stressful events, were carefully monitored, have identified "evoked" secretory peaks and demonstrated the existence of spontaneous episodic variations occurring in the absence of any recognizable stimulus.

Once the nonrandom and spontaneous nature of the ultradian hormonal fluctuations were established and variations in plasma dilution excluded as their possible origin, episodic spikes in plasma levels of pituitary hormones were assumed to result from secretion in discrete bursts in response to pulsatile stimulation (and/or inhibition) by hypothalamic factors. Extensive in vivo and in vitro data support this concept in the case of the pituitary-gonadal axis (see Chap. 2 of this volume). Though no direct evidence so far exists for the pulsatile release of corticotropin-releasing factor (CRF), thyrotropin-releasing hormone (TRH), or other putative or demonstrated hypothalamic factors, the hypothesis that these factors are released in a pulsatile fashion is now generally accepted. Investigations where plasma ACTH and cortisol levels are measured simultaneously clearly show that the pulsatile variations of a stimulating hormone (i.e., ACTH) may be transmitted to the hormone secreted in response by its target tissue (i.e., cortisol); Fig. 4 illustrates such results.

The overall temporal correlation between the major secretory spikes of both hormones is evident, even though there does not appear to be a direct correspondence in magnitude or duration. Moreover, two consecutive ACTH spikes may result in a single cortisol spike and clear increases in ACTH level may occur without a concomitant cortisol rise.

Differences in metabolic degradation rates of these two hormones, possible distortions introduced by protein binding, and modulation by feedback mechanisms probably account for these discrepancies. Such "masking" effects may be more pronounced as, for example, in the case of LH and testosterone, where identification between LH pulse and resulting testosterone increase can only be achieved for major spikes (ROWE et al. 1975).

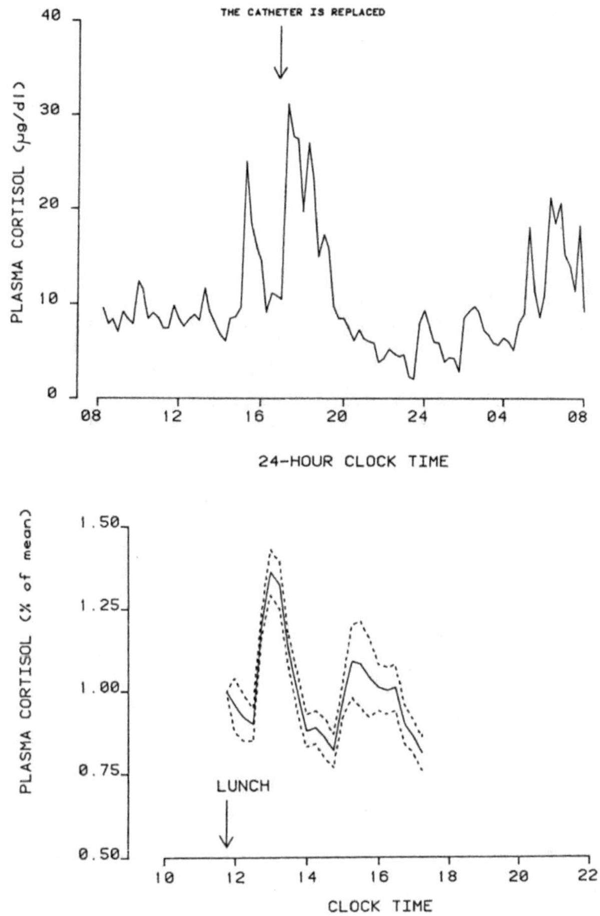

Fig. 3. Top: Effect of catheter replacement on plasma cortisol levels. Bottom: Post-lunch cortisol rise (mean ± SD of eight individuals)

Assuming a central nervous system (CNS) origin for pituitary pulsatility, further speculations along this line may be summarized by the following propositions:

1. The "single ultradian pacemaker" hypothesis: all releasing and/or inhibiting factors are secreted in a pulsatile fashion under the control of a common hypothalamic mechanism.

2. The "multiple ultradian pacemakers" hypothesis: the pulsatility of each releasing or inhibiting factor is controlled by a specific hypothalamic mechanism, presumably inherent to the neural structures responsible for its discharge.

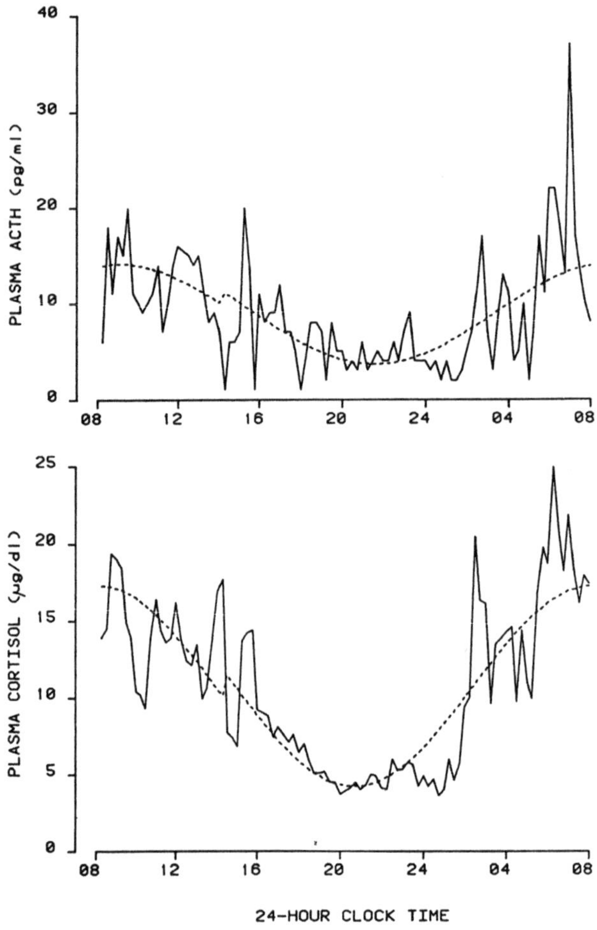

Fig. 4. 24-h profiles of ACTH and cortisol obtained simultaneously at 15-min intervals. Dashed line is a best-fit curve quantifying the circadian variation

3. The "central ultradian clock" hypothesis: the pulsatility of hypothalamic releasing or inhibiting factors is produced by an ultradian clock, possibly located outside of the hypothalamus, and also responsible for the generation of other ultradian rhythms, such as the REM-NREM cycle in sleep.

4. The "extra-hypothalamic modulation" hypothesis: the hypothalamic ultradian pacemaker(s) is (are) influenced by ultradian rhythms of extrahypothalamic origin, such as the REM-NREM cycle.

Some of these propositions may apply to certain, but not all, hypothalamic factors.

In a recent and as yet unpublished study, we have found evidence in favor of partial synchronization of pulsatility along two different axes, the pituitary-adrenal axis and the pituitary-gonadal axis. The 24-h profiles of plasma cortisol and testosterone were obtained at 15-min intervals in 11 healthy males. Due to technical difficulties, blood sampling had to be interrupted after 20 h in one of the subjects. The number and characteristics of significant episodic spikes were estimated in each individual profile. A total of 173 cortisol spikes and 206 testosterone spikes were detected. Spikes in the two hormones were considered to be concomitant if the corresponding maxima in plasma levels occurred at exactly the same sampling time or, when the durations of the two spikes overlapped by at least 15 min, at the same sampling time plus or minus 15 min. Thus, the maxima included in such "doublets" of concomitant cortisol and testosterone spikes were never separated by more than 30 min. For each hormone, the probability of observing a spike in a given 30-min interval under the hypothesis of random distribution over the 24-h span was obtained by dividing the total number of spikes detected by the total number of 30-min intervals included in the ten 24-h profiles and the one 20-h profile collected. The resulting probabilities were 0.3327 for cortisol and 0.3962 for testosterone. If spikes of cortisol and testosterone were to occur at random without interrelationship, the probability of finding a doublet in any given 30-min interval would be 0.3327 x 0.3962. The probability of finding a cortisol spike without testosterone spike would be 0.3327 x (1 - 0.3962), and, similarly, the probability of finding a testosterone spike without cortisol spike would be 0.3962 x (1 - 0.3327). To obtain a valid estimation of the significance of concomitance, the probability of finding no cortisol spike and no testosterone spike in a given 30-min interval has also to be calculated and is (1 - 0.3327) x (1 - 0.3962). The number of observed events (doublets, isolated spikes, and intervals without spikes) may then be compared with their expected frequency under the hypothesis of purely random occurrence and lack of interhormonal correlation using the X^2 test as described in SIEGEL (1956). The results are summarized in Table 2.

The observed frequency of occurrence of concomitant spikes of cortisol and testosterone is significantly higher than the expected frequency

Table 2. Concomitance analysis of episodic cortisol and testosterone spikes

Type of event	Probability of occurrence on the basis of chance	Expected frequency of occurrence	Observed frequency of occurrence	X^2 Value
Doublet	0.1318	67	92	9.33
Isolated cortisol spike	0.2009	103	87	2.49
Isolated testosterone spike	0.2645	135	121	1.45
30-min interval without spike	0.4029	206	211	0.12
Total	1.0001	511	511	13.39

(p $\leq$ 0.005), indicating the existence of synchronized episodic spikes of adrenal and gonadal secretions and suggesting that a common CNS mechanism may control, at least partially, the pulsatility of both hormones.

A role of the REM-NREM cycle in the generation or modulation of episodic pituitary secretion was suggested by the consistent temporal association of the major burst of GH release with the first slow wave (SW) stage in sleep, by an early report of elevàted LH levels during REM sleep as compared with NREM sleep in pubertal boys, and by the claim that episodic nadirs and peaks of plasma PRL were related to REM and NREM stages (WEITZMAN 1976; PARKER et al. 1974). However, this latter result was not confirmed in subsequent studies and we demonstrated its artifactuality in a detailed spike-by-spike analysis of 24 night-time profiles of PRL collected at 15-min intervals (VAN CAUTER et al. 1982). More recently, WEITZMAN and coworkers (1983) reported the existence of a sleep-dependent component in the 24-h profile of cortisol, consisting of a drop in plasma level shortly after sleep onset. Taken together, the available evidence indicates that increases or decreases of pituitary secretion may be "evoked" by sleep onset or the occurrence of SW stages. The hypothesis of a consistent entrainment of nocturnal pulsatile release by the REM-NREM cycle has little support at the present time.

MODULATION OF PULSATILE VARIATIONS

Modulation of hormonal pulsatility may occur at two levels:

1. Direct influence on the characteristics of the pulses of secretion (e.g., feedback effects, modulation by circadian rhythmicity)

2. Influence on the way pulses of secretion are reflected in measurable changes in plasma concentration (e.g., protein binding, metabolic degradation)

Experiments with pharmacological doses of steroid analogues such as clomiphene citrate (SANTEN and RUBY 1979) and dexamethasone (REFETOFF et al. 1985) have indicated that steroid hormones exert their negative feedback effects by modulating the frequency and/or the height of pulses of pituitary secretion. Little is known at the present time about the mechanisms involved in this modulation. It is also likely that circadian rhythmicity is one of the factors modulating episodic hormonal release, probably through mechanisms operating at the hypothalamic level. Preliminary studies described in the next section of this chapter indicate that the height of cortisol secretory spikes is regulated by a circadian rhythm.

For hormones which bind to specific proteins in plasma, the parameters of the binding process will also influence the episodic variations of total (i.e., bound and free) plasma concentrations. In a previous study (VAN CAUTER et al. 1975) we have shown that the relative amplitude of the circadian rhythm of total plasma cortisol is dampened compared with that of free cortisol. Protein binding is likely to have a similar dampening effect on the episodic variations of cortisol secretion. However, the major factor influencing, in a direct way, the episodic variations of plasma hormonal levels is the metabolic degradation rate of the hormone. Computer simulations showing how pulses of secretion by the gland are reflected in the peripheral plasma levels help us understand the importance of this phenomenon. Such simulations are easy to perform since reasonable models for the disappearance rate of many hormones are currently available. For example, it has been shown that the disappearance rates of cortisol, GH, LH, and FSH follow double exponential curves from which a short and a long half-life can be estimated (BRADLEY and WATERHOUSE 1966; REFETOFF and SÖNKSEN 1970; YEN et al. 1968; YEN et al. 1970). The kinetics of secretion, distribution, and degradation may thus be represented by a two-compartment model of the type illustrated in Fig. 5.

S is the rate of secretion of the hormone, k_0 and k_1 are the exchange rates between the two pools, and k_2 is the rate of disappearance from the intravascular pool. The parameters k_0, k_1, and k_2 may be directly obtained from the short and long half-lives and the volume of distribution of the hormone. The time evolution of the plasma concentration x is obtained by integrating a simple system of two linear differential equations. In the computer simulation illustrated in Fig. 6, parameters characteristic of cortisol secretion, distribution, and metabolism were used (short half-life, 5 min; long half-life, 90 min; BRADLEY and WATERHOUSE 1966) and secretion (S) was assumed to occur as a succession of rectangular pulses separated by intervals without any release.

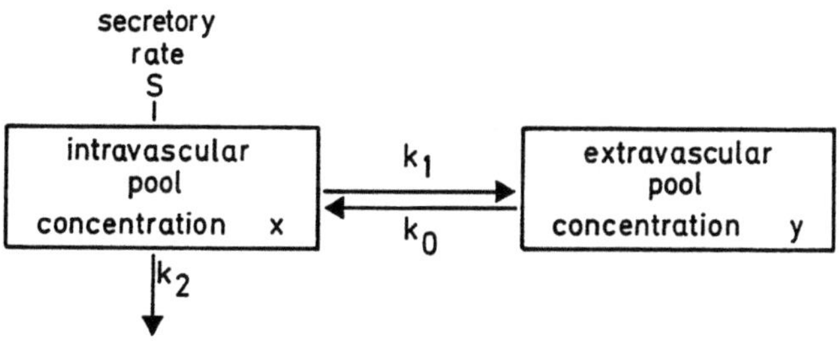

Fig. 5. Two-compartment model for secretion, distribution, and metabolism

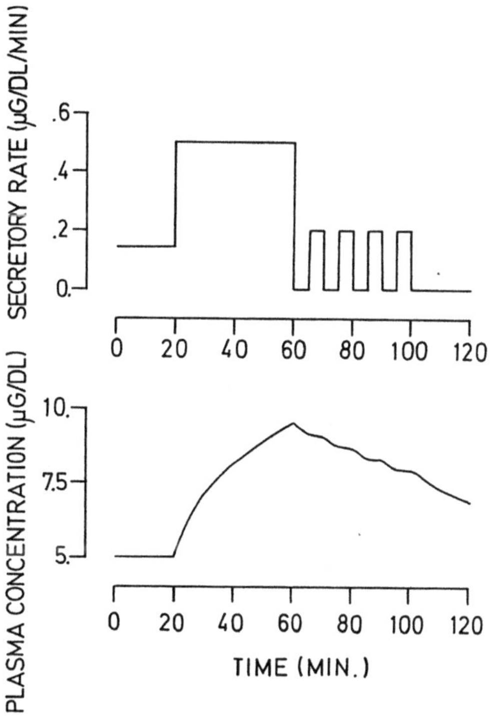

Fig. 6. Top: Computer simulation of a succession of five rectangular pulses of cortisol secretion. Bottom: Calculated variation of plasma cortisol concentration resulting from the profile of secretory pulses illustrated in the upper panel

A secretory pulse of 0.5 µg/dl min lasting 40 min followed by four pulses of 0.2 µg/dl min, each lasting 5 min (upper panel), results at the level of plasma concentrations (lower panel) in a single "spike" with a progressive increase above the basal level and a slow decline, with very small rises and falls which would remain undetectable, even with the most sensitive cortisol assay currently available. The apparent half-life which may be deduced from a semilogarithmic plot of the declining portion of the simulated spike is of several hours, thus considerably longer than the actual half-life time used in the model.

Longer than expected apparent half-lives were deduced from the backside of LH (SANTEN and BARDIN 1973) and PRL (VAN CAUTER et al. 1981) spikes. In both cases, these observations were interpreted as resulting from the superimposition of continuous and episodic secretion. Alternatively, these observations could also be accounted for by a sequence of secretory pulses too rapid to result in detectable fluctuations of plasma levels.

The method used to produce the computer simulations illustrated in Fig. 6 may also be applied to the calculation of the profile of secretory rate from the profile of observed plasma concentrations by assuming that the secretory rate remains constant between sampling times and by deducing its level, step by step, for each sampling interval. The theoretical profile of plasma concentrations implied by the calculated profile of secretory rates is then computed and compared with the actual data. The short and long half-life times of cortisol are adjusted within the range 3-6 min and 60-85 min, respectively, until the theoretical plasma profile and the actual data differ by less than 10%.

IMPORTANCE OF EPISODIC VS CONTINUOUS RELEASE

As shown above, the combination of secretion, distribution, and metabolic degradation may result in a profile of plasma levels suggesting the existence of continuous secretion when the actual release is strictly an on-and-off process. Current concepts on pulsatile variations "superimposed" on a tonic release may thus have to be reevaluated. We describe hereafter a preliminary study in which, using the method described in the preceding paragraph, we attempted (a) to determine whether the secretion of cortisol in normal subjects is entirely episodic (i.e., there are time intervals without any secretion) or includes a continuous component (i.e., there are no time intervals without secretion) and (b) to evaluate the relative importance of continuous vs episodic cortisol release in patients with Cushing's disease and in patients with hypercortisolism associated with (major) endogenous depression. Figure 7 illustrates typical results from this preliminary study. Data for the healthy subject are shown on the left side panel. It appears that the normal secretion of cortisol is entirely episodic, without any detectable continuous component. The height of the secretory pulses is modulated by a circadian rhythm. About one third of the 24-h span is spent without secreting any detectable amount of cortisol. In Cushing's disease (middle panel), cortisol release is uninterrupted. In contrast, in hypercortisolism associated with depression (right side panel), cortisol secretion remains episodic with large bursts separated by intervals without detectable secretory activity. It is noteworthy that the mean plasma cortisol levels in the patient with Cushing's disease (13.8 ± 3.7 µg/dl) and in the depressed subject (12.8 ± 5.6 µg/dl) are very similar. Nevertheless, the Cushing's patient presented the full clinical picture of hypercortisolism (truncal obesity, hypertension, mild diabetes, easy bruising, etc.) while the depressed subject had none of these physical symptoms.

In the patient with affective illness, the alternation of periods of depression and hypercortisolism with periods of euthymia and normal cortisol lévels may explain the absence of clinical signs of Cushing's syndrome. However, the data presented here also suggest that, in the ultradian range, the intermittent nature of cortisol hypersecretion in depression may not allow the development of the syndrome which would only appear when peripheral organs are exposed to continuously high levels of steroids.

If further investigations were to confirm this hypothesis, they would demonstrate the functional importance of episodic, rather than continuous, hormonal release for peripheral endocrine systems other than those of the hypothalamo-pituitary-gonadal axis.

ACKNOWLEDGEMENTS. The authors thank Professor F.W. Turek for reviewing the manuscript and Mrs. Y.W. Richmond for secretarial help. Data illustrated on the lower panel of Fig. 3 were collected by Drs D. Robbins and J. Jaspan. Data shown on the lower middle panel of Fig. 7 were obtained by Dr. P. Linkowski. This work was supported by a grant "Actions Concertees" of the Ministère Belge de la Politique Scientifique and by APMO Foundation.

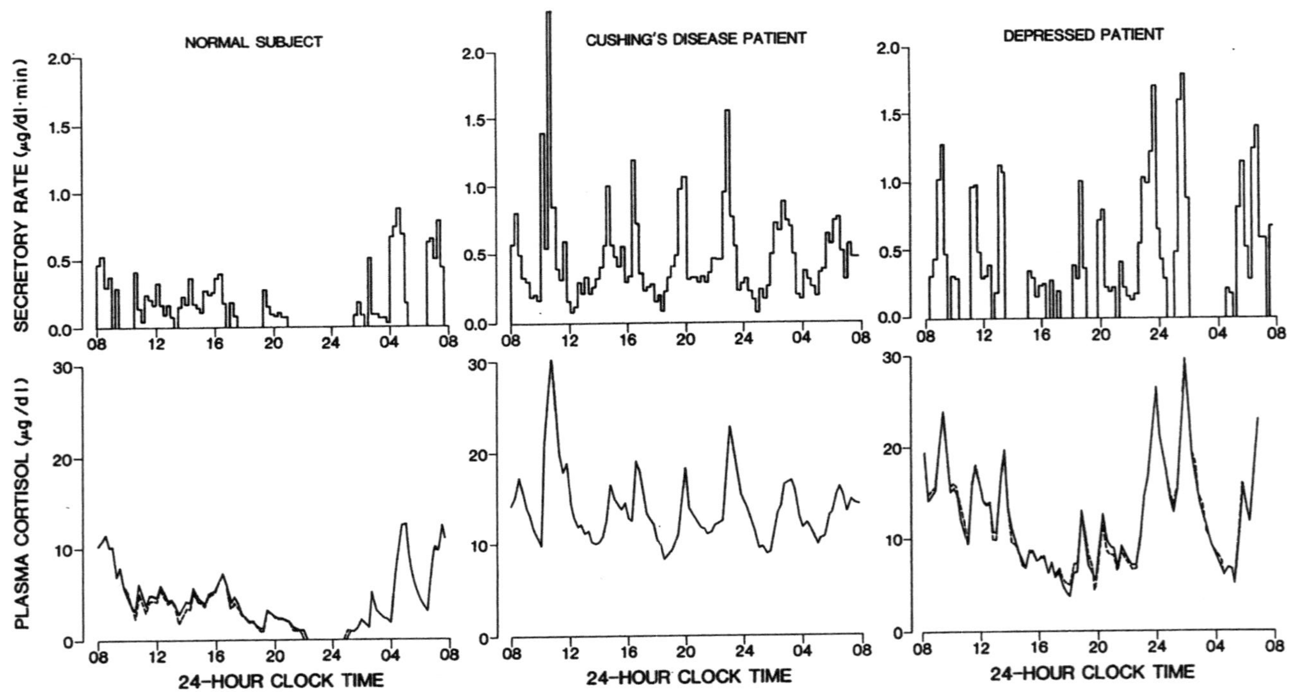

Fig. 7. Top: Theoretical profiles of cortisol secretory rates computed from measured plasma levels. Bottom: Profiles of plasma cortisol concentration implied by computed pattern of secretory rates illustrated in upper panels. Dashed lines indicate the measured 24-h profile of cortisol. In the patient with Cushing's disease, theoretical and experimental data were superimposed

REFERENCES

BADAWI M, BILA S, L'HERMITE M, PEREZ-LOPEZ FR, ROBYN C (1974) Comparative evaluation of radioimmunoassay methods for human prolactin using anti-ovine and anti-human prolactin sera. In: Radioimmunoassay and related procedures in medicine, vol 1. International Atomic Energy Agency, Wien, pp 411-422

BRADLEY EM, WATERHOUSE C (1966) Effect of estrogen administration on extravascular cortisol. J Clin Endocrinol Metab 26:705-714

DESIR D, VAN CAUTER E, FANG V, MARTINO E, JADOT C, SPIRE JP, NOEL P, REFETOFF S, COPINSCHI G, GOLSTEIN J (1981) Effects of "jet lag" on hormonal pattern. I. Procedures, variations in total plasma proteins and disruption of adrenocorticotropin-cortisol periodicity. J Clin Endocrinol Metab 52:628-641

GOLSTEIN J, VANHAELST L (1973) Influence of thyrotropin-free serum on the radioimmunoassay of human thyrotropin. Clin Chim Acta 49:141-146

PARKER DC, ROSSMAN LG, VANDERLAAN EF (1974) Relation of sleep-entrained human prolactin release to REM-non REM cycles. J Clin Endocrinol Metab 38:646-651

REFETOFF S, SÖNKSEN PH (1970) Disappearance rate of endogenous and exogenous human growth hormone in man. J Clin Endocrinol Metab 30:386-392

REFETOFF S, VAN CAUTER E, FANG VS, LADERMAN C, GRAYBEAL MI, LANDAU RL (1985) The effect of dexamethasone on the 24-h profiles of ACTH and cortisol in Cushing's syndrome. J Clin Endocrinol Metab (in press)

ROWE PH, RACEY PA, LINCOLN GA, ELLWOOD M, LEHANE J, SHENTON JC (1975) The temporal relationship between the secretion of luteinizing hormone and testosterone in man. J Endocrinol 64:17

SANTEN RJ, BARDIN CW (1973) Episodic luteinizing hormone secretion in man. J Clin Invest 52:2617-2628

SANTEN RJ, RUBY EB (1979) Enhanced frequency and magnitude of episodic luteinizing hormone-releasing hormone discharge as a hypothalamic mechanism for increased luteinizing hormone secretion. J Clin Endocrinol Metab 48:315-319

SIEGEL S (1956) Nonparametric statistics for the behavioral sciences. McGraw Hill, New York

VAN CAUTER E (1979) Method for the characterization of 24-hour temporal variations of blood components. Am J Physiol 237:E255-E264

VAN CAUTER E (1981) Quantitative methods for the analysis of circadian and episodic hormone fluctuations. In: VAN CAUTER E, COPINSCHI G (eds) Circadian and episodic variations. Nyhoff, The Hague, pp 1-25

VAN CAUTER E, GOLSTEIN J, VANHAELST L, LECLERCQ R (1975) Effects of oral contraceptive therapy on the circadian patterns of cortisol and thyrotropin (TSH). Europ J Clin Invest 5:115-121

VAN CAUTER E, L'HERMITE M, COPINSCHI G, REFETOFF S, DESIR D, ROBYN C (1981) Quantitative analysis of spontaneous variations of plasma prolactin in normal man. Am J Physiol 241:E355-E363

VAN CAUTER E, DESIR D, REFETOFF S, SPIRE JP, NOEL P, L'HERMITE M, ROBYN C, COPINSCHI G (1982) The relationship between episodic variations of PRL in plasma and REM-non-REM cyclicity is an artifact. J Clin Endocrinol Metab 54:70-75

VIRASORO E, COPINSCHI G, BRUNO ED, LECLERCQ R (1971) Radioimmunoassay of human growth hormone using a charcoal-dextran separation procedure. Clin Chim Acta 31:294-297

VIRASORO E, COPINSCHI G, BRUNO OD (1974) Degradation of labelled hormone in radioimmunoassay of ACTH. In: Radioimmunoassay and related procedures in medicine, vol 1. International Atomic Energy Agency, Wien, p 323

WEITZMAN ED (1976) Circadian rhythms and episodic hormone secretion in man. Ann Rev Med 27:225

WEITZMAN ED, ZIMMERMAN JC, CZEISLER CA, RONDA J (1983) Cortisol secretion is inhibited during sleep in normal man. J Clin Endocrinol Metab 56:352-358

YEN SSC, LLERENA O, LITTLE B, PEARSON OH (1968) Disappearance rates of endogenous luteinizing hormone and chorionic gonadotropin in man. J Clin Endocrinol Metab 28:1763-1767

YEN SSC, LLERENA LA, PEARSON OH, LITTELL AS (1970) Disappearance rates of endogenous follicle-stimulating hormone in serum following surgical hypophysectomy in man. J Clin Endocrinol Metab 30:325-329

Relationship Between Ultradian Rhythms of the Sleep-Wakefulness Cycle and Growth Hormone and Corticosterone Secretion in Rats

F. Kimura, C. Praputpittaya, N. Mitsugi, R. Hashimoto, and R. Suzuki
Department of Physiology, Yokohama City University, School of Medicine, 2-33 Urafune-cho, Minami-ku, Yokohama 232, Japan

INTRODUCTION

In the field of neuroendocrinology, rhythmic phenomena of anterior pituitary hormone secretions are important subjects. It appears that elucidation of the mechanism for the rhythmicity contributes greatly to understanding the control mechanism of anterior pituitary hormone secretion.

Among many kinds of secretory rhythms, our recent investigations have focussed mainly on the 3-h ultradian rhythm of growth hormone (GH) secretion. The GH secretory rhythm which was first demonstrated by TANNENBAUM and MARTIN (1976), fascinated us because of its strict accuracy of period and because of the earlier finding that in humans GH secretion occurs in relation with sleep (TAKAHASHI et al. 1968). The correlation of GH secretion with sleep in rats was questioned by the group who discovered the rhythm (WILLOUGHBY et al. 1976), but we were able to support the correlation initially in the immature rat (KIMURA and KAWAKAMI 1981; KAWAKAMI et al. 1983, 1984), and subsequently in the adult rat (KIMURA and TSAI 1984). From our experiments, we have obtained a concept that different physiological functions have common ultradian rhythms with similar periods. They may be governed by pacemakers that are independent of each other, but are entrained by a central pacemaker. In this chapter, we will describe the outline of these experimental results.

Experimental Brain Research, Suppl. 12

MATERIALS AND METHODS

All experiments were done using immature and mature rats of Wistar strain which were maintained in a room controlled both for light (on from 0500-1900) and temperature (24°-26°C). For the cortical EEG recording, stainless steel electrodes were implanted over the cerebral cortex at least 1 week before the blood-sampling experiment. Electrodes for recording multiple unit activity (MUA) in the arcuate nucleus (Arc) and/or the ventromedial nucleus (VMH) of the hypothalamus and/or the suprachiasmatic nucleus (SCN), as well as those for cortical EEG, were implanted according to the method described previously (KAWAKAMI et al. 1982) 1-2 days before the sampling experiment. Intracardiac cannulae for either the blood sampling or the anesthesia with 0.4% thiopental sodium were implanted the day before the experiment.

Cortical EEGs were recorded for all animals, except for those who served as controls, simultaneously with the blood sampling which took place at 10-min intervals. GH and corticosterone (CS) were measured in whole blood. In the first series of experiments in which the relationship between GH and CS secretions and sleep-wakefulness cycles was examined, EEGs were scored into wakefulness and sleep, and the amount of sleep for every 10 min was plotted against the GH and CS values as assessed using radioimmunoassay and protein-binding assay, respectively. In the second series of experiments, in which effects of long-term anesthesia with thiopental on GH and CS secretions were examined, and in the third, in which MUA was examined under the same anesthesia in relation with GH secretion, the rate of thiopental infusion was readjusted frequently in order to maintain a stable EEG sleep pattern throughout the experimental period.

All the time series of GH, CS values, the amount of sleep, and the spike numbers from the MUA recordings were analyzed, using power spectrum analysis, for periodical characteristics. The spectrum was computed by means of a Fourier transform of the autocovariance function of the series (harmonic analysis). The precise periodicity was identified by using the least-squares spectrum method (HALBERG et al. 1964). Cross-correlation analysis was performed between two time series, where necessary.

RESULTS AND DISCUSSION

Relationship Between GH and CS Secretion and the Sleep-Wakefulness Cycle

The initial study which was designed to examine possible correlations between the sleep-wakefulness cycle and GH secretion was done in the immature, 29 to 31-day-old rat. From the serial blood sampling from 1200-1800, it was found that the control rats had the most dominant GH secretory rhythm with a mean (± SE) periodicity of 3.12 ± 0.09 h (n = 14) (KAWAKAMI et al. 1983). When periodicities were studied for each of the other 6-h periods over a day, i.e., 0000-0600, 0600-1200, and 1800-2400, an overall mean (± SE) periodicity of 3.15 ± 0.06 h (n = 31) was obtained (KAWAKAMI et al. 1984).

In almost all (96%) of the 27 animals in which EEG was recorded simultaneously with blood samplings for the 6-h period from 1200-1800, there were significant positive correlations between the GH value and the amount of sleep which occurred during the preceding 10-20 min. The time series of the amount of sleep and the GH values had mean (± SE) periodicities of 3.09 ± 0.17 and 3.05 ± 0.13 h, respectively (n = 27) (KAWAKAMI et al. 1983). Figure 1 illustrates typical individual sleep-wakefulness cycle profiles plotted against GH profiles in two animals. The existence of a 3-h ultradian rhythm of sleep propensity in these data was confirmed in the analysis of 24-h EEG data obtained from 27 to 31-day-old rats (KAWAKAMI et al. 1984). These findings in the immature rat indicate that the rat GH secretory rhythm with a 3-h periodicity, relating to the 3-h ultradian rhythm of the sleep cycle, develops by 30 days of age. Even so, there was a possibility that this correlation of GH secretion with sleep would disappear or alter due to certain developmental changes, since WILLOUGHBY et al. (1976) failed to find it in the adult rat. Subsequently, we performed similar experiments using adult rats.

As in the immature rats, the time series of GH concentrations in the control rats revealed that the mean (± SE) periodicity was approximately 3 h; i.e., 2.93 ± 0.10 h (n = 5) for the period from 0000-1200 and 2.85 ± 0.06 h (n = 8) for the period from 1200-2400 (KIMURA and TSAI 1984). For the animals in which EEGs were recorded simultaneously with blood samplings from 1200-2400, the cross-correlation analysis performed between the time series of the amount of sleep and GH values

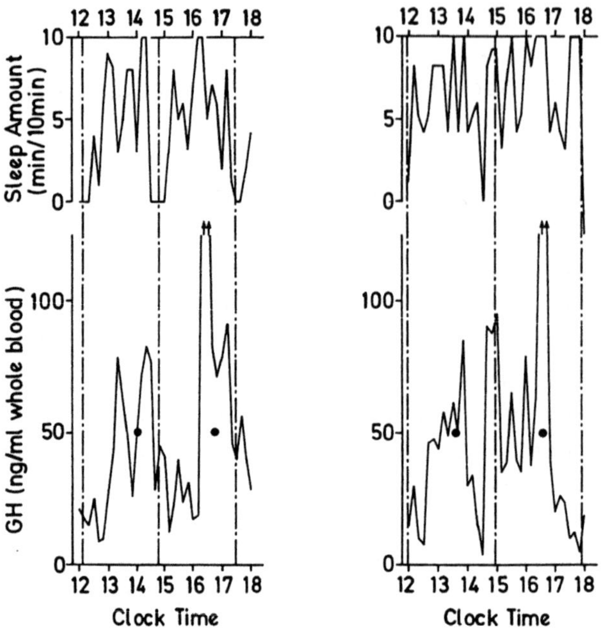

Fig. 1. Sleep-wakefulness cycle profiles plotted against GH secretory profiles in two representative immature rats. Arrows indicate values exceeding 128 ng/ml whole blood, the upper limit of the assay system. In order to show the phase relationship between the time series of GH concentrations and the amount of sleep clearly, vertical dashed lines defining each sleep cycle are drawn across both time series, on the basis of periodicity and acrophase estimated for the time series of the amount of sleep. Acrophases for the GH secretory rhythms are indicated by solid circles on the time series of GH. (From KAWAKAMI et al. 1983)

revealed that the GH values were significantly correlated with the amount of sleep during the preceding 10-20 min in 42% and the preceding 30-60 min in 29% of animals. The number of animals having a positive correlation at short time lags appeared relatively less than in immature rats, but in most animals there was a definite relationship in the way that each GH peak occurred during the early phase of the sleep cycle, from 40-70 min following the onset of the sleep cycle which recurred in approximately 3-h periods. In Fig. 2, the left side shows an example of individual time series of GH concentration against which the time series of the amount of sleep is plotted. According to these data, GH secretion in the adult rat also seems to be correlated with sleep, although somewhat differently than in the immature rat.

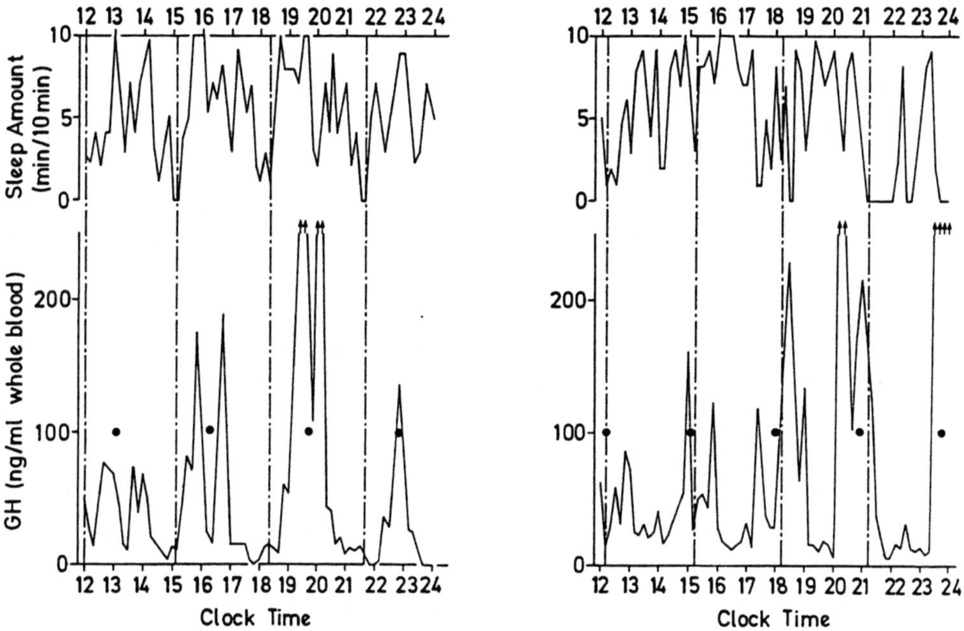

Fig. 2. Sleep-wakefulness cycle profiles plotted against GH secretory profiles in the adult rat. The left panel shows an example from a rat in which the phase relationship between sleep cycle and GH rhythms is consistent. The right panel shows an example from a rat in which the consistent phase relationship is absent. Arrows indicate values exceeding 256 ng/ml whole blood, the upper limit of the assay system. Other details as in Fig. 1. (From KIMURA and TSAI 1984)

In addition, the close relationship between sleep and GH secretion was confirmed by testing the effect of sleep deprivation on GH secretion. For the immature rat it has been shown (KAWAKAMI et al. 1983) that deprivation of sleep for 2 h, from 1300-1500, was effective in preventing GH secretory bursts which otherwise should occur during this time of day. In the mature rat (KIMURA and TSAI 1984), the GH secretory bursts which occurred during sleep deprivation from 1300-1600 were of small amplitude, indicating that a decrease in the amount of sleep was associated with an attenuation of GH secretion.

Although the majority of rats showed an almost uniform relationship between GH secretion and sleep, there was a minority (approximately 30%) that showed quite peculiar features. An example is shown in the right panel of Fig. 2. In these rats, periodicities in the time series of GH concentrations and the amount of sleep differed considerably from each other, and thus no consistent relationship of the GH peak time to the onset of the sleep cycle was obtained. This temporal dissociation of the two physiological functions suggests the possible existence of independent pacemakers for sleep and GH secretion. This is in good agreement with the occurrence of GH secretory bursts, even of small amplitude, under sleep deprivation, as mentioned above. It seems quite probable that sleep only intensifies the inherent activity of the GH-releasing mechanism, and is not able to induce the activation itself.

The subsequent experiments studying the temporal correlation of GH and CS secretion with the sleep-wakefulness cycle (MITSUGI and KIMURA 1985) offer further evidence for the hypothesis stated above. Time series of CS concentration, as well as those of GH and the amount of sleep, had uniform ultradian rhythms with periodicities of 1.5 h and its multiples. Although the most prominent CS rhythm was the 1.5-h period, CS secretion occurring with a 3-h period had such a phase relationship with the 3-h sleep rhythm that the secretion began in the later phase of a sleep cycle and reached its peak around the time of wakefulness. In accordance with this, significantly negative cross-correlation coefficients were obtained between the amount of sleep and CS values in most animals. Considering the amplitude of the CS secretory bursts, it was evident that the bursts, being components of either 3-h or 1.5-h rhythms, were prominently high from the late afternoon to the evening. In view of the well-known CS circadian rhythm, these bursts were considered to constitute the peak components of the circadian rhythm. These characteristics are apparent in Fig. 3, in which individual profiles for GH, CS, and sleep-wakefulness cycles in two animals are shown. In contrast, a rather unusual relationship was seen in one of the seven rats examined. In this animal, the periods of the ultradian rhythms of sleep, GH, and CS differed so much that no consistent relationship between sleep, GH, and CS secretory rhythms was found.

Effect of Long-Term Anesthesia with Thiopental on GH and CS Secretion

From the preceding studies, one may conclude that each mechanism for GH and CS secretion and for the sleep-wakefulness cycle has its own ultradian pacemaker. These seem to be coupled mutually with a fixed phase relationship. If that is the case, it would be of interest to examine whether the secretions of GH and CS would still manifest their

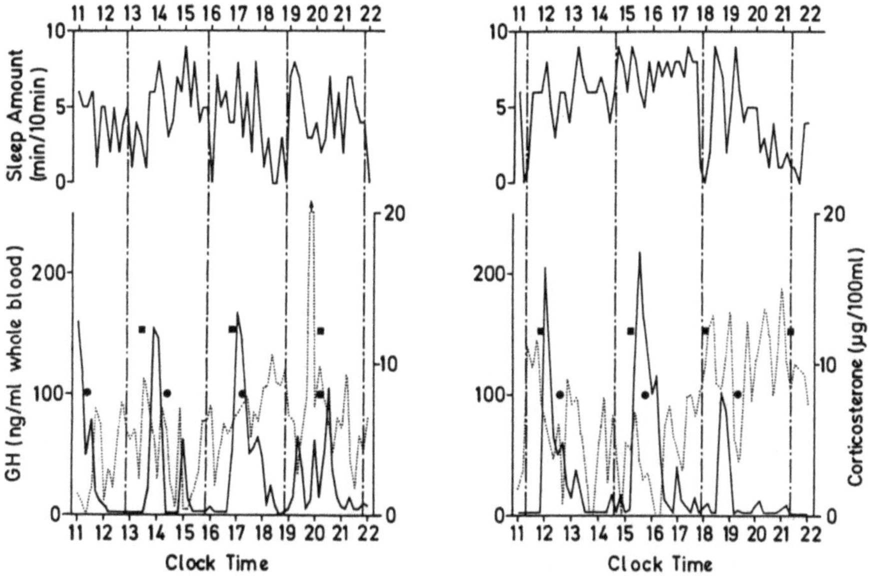

Fig. 3. Sleep-wakefulness cycle profiles plotted against GH (solid lines) and CS (dotted lines) secretory profiles in the adult rat under long-term thiopental anesthesia. Acrophases for the GH and CS are indicated by solid circles and solid squares, respectively. Other details as in Fig. 1. (From MITSUGI and KIMURA 1985)

ultradian rhythms under conditions where the rhythmicity of sleep-wakefulness is absent.

Of 41 animals subjected to long-term anesthesia, 32 (78%) showed apparent fluctuations in the GH levels (PRAPUTPITTAYA and KIMURA 1985). Among them, 27 (84%) had periodicities within the range obtained for the unanesthetized but similarly sampled animals, from 2.40 to 3.85 h. The mean value was 3.01 $\pm$ 0.06 h. The results clearly demonstrate that the GH-releasing mechanism continued to operate with a 3-h periodicity under this condition without rhythmic sleep occurrence, indicating that the GH-releasing mechanism possesses its own pacemaker.

In subsequent experiments, with lighter anesthesia, CS as well as GH were measured in six rats (MITSUGI et al. unpublished observations).

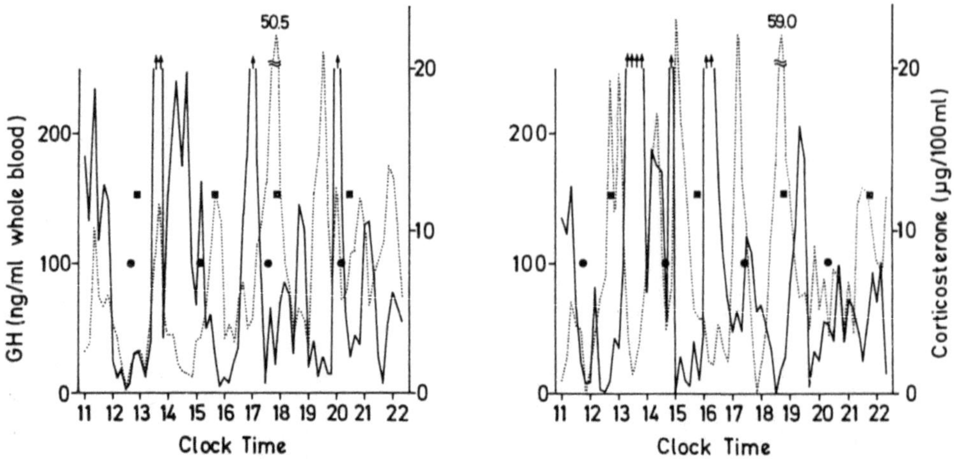

Fig. 4. GH (solid lines) and CS (dotted lines) secretory profiles in two adult rats under long-term thiopental anesthesia. Arrows indicate GH values exceeding 256 ng/ml whole blood, the upper limit of the assay system. Acrophases for the GH and CS are indicated by solid circles and solid squares, respectively, on each time series. (MITSUGI, PRAPUTPITTAYA and KIMURA unpublished observation)

As is apparent in Fig. 4, in which two individual secretory profiles of GH and CS under stable EEG sleep state due to thiopental anesthesia are shown, the secretion of CS as well as that of GH still occurred with ultradian rhythms. Period analysis showed almost the same characteristic as found in the unanesthetized animals. The time series of CS and GH revealed three major ultradian rhythms with periods of approximately 1.5 h and its multiples. As in the unanesthetized, freely moving animals, the most prominent was the 1.5-h period for CS, and the 3-h period for GH. Probable circadian components appeared in the CS profiles, as well. However, it could be argued that the definite phase relationships which were observed between GH and CS rhythms in most of the unanesthetized animals seemed to be lost. Only one of six rats examined to date kept the similar phase relation as the GH and CS rhythms had when the sleep-wakefulness rhythms were present. This case is illustrated in Fig. 4, right panel. A possible interpretation is that under long-term thiopental anesthesia pacemakers for individual rhythms would maintain their function while a putative central pacemaker which controls their phase relationship might lose its normal function.

MUA in the Hypothalamus in Relation to GH Secretion under Long-Term Anesthesia with Thiopental

The series of endocrinological experiments described above indicate that the neural elements involved in the rhythmic secretion of GH and CS operate even under long-term anesthesia with thiopental sodium. This means that the GH and CS releasing mechanisms as well as the pacemakers for them operate normally even under the continuous sleep state, though the phase relationships among them are obscure. In the present study, we recorded neuronal electrical activities under anesthesia with thiopental in order to further elucidate the mechanism of hormonal secretion.

In Fig. 5, the individual time series of the number of spike discharges per min in the Arc under anesthesia with thiopental are illustrated in five of the 12 rats examined. The spike number 3840 represents the upper limit in the figure. It is evident simply by inspecting these time series that the MUA in the Arc fluctuates in a pulsatile fashion. Spike discharges as low as approximately 500-1000 per min at the basal level rise abruptly to levels higher than 3840 per min intermittently and form an episodic discharge burst. Period analysis of the time series of the number of spike discharges revealed different ultradian rhythms in each MUA record. Since they appear to cluster in groups, Table 1 was compiled by tentatively classifying them into six such groups, i.e., 1-10, 11-20, 21-40, 41-60, 61-120, and 121-300 min. From the calculation of the mean periodicity for each rhythm classified, it was suggested that each MUA recorded from the Arc had the potential to fluctuate with ultradian rhythms having mean periodicities of 7.1, 16.4, 31.5, 52.5, 86.7, and 189.8 min. It is noticeable that the periodicity of these rhythms seems to be set based on "15 min" or "30 min" as the unit length. The rhythms, with 1.5-h and 3-h periods, conform to the rule, as a matter of course.

The persistence of circadian rhythmicity usually seen in the hypothalamic electrical activity was also confirmed. As seen in Fig. 6, the baseline rate of the MUA in the Arc showed an elevation in the afternoon. At the same time the amplitude of the discharge bursts which are superimposed on the baseline level is also increased. Further, as seen in Fig. 6, the MUAs recorded in the vicinity of the SCN showed a clear-cut afternoon elevation. Based on previous findings in the unanesthetized, freely moving rats (KAWAKAMI et al. 1973; KAWAKAMI and KIMURA 1975, 1978; INOUYE and KAWAMURA 1979) those elevations are

Table 1. Ultradian rhythms estimated in individual MUA time series. Classification by periods (min)

MUA	1-10	11-20	21-40	41-60	61-120	121-300
# 125	4	16	34	47	68	215
# 160	4,10	-	38	59	-	213
# 168	6	17	29	56	83	153
# 184	5	17	28	-	70,104	175
# 191	10	17	32	44	110	176
# 194	6	16	32	52	75	137
# 202	5	16	35	60	84	166
# 216	-	-	21	48	71	269
# 225	-	17	36	-	-	200
# 232	-	19	25	56	76	142
# 237	9	15	36	50	96	242
Mean	7.1	16.4	31.5	52.4	86.7	189.8
±SE	±0.8	±0.4	±1.6	±1.9	±4.5	±12.7

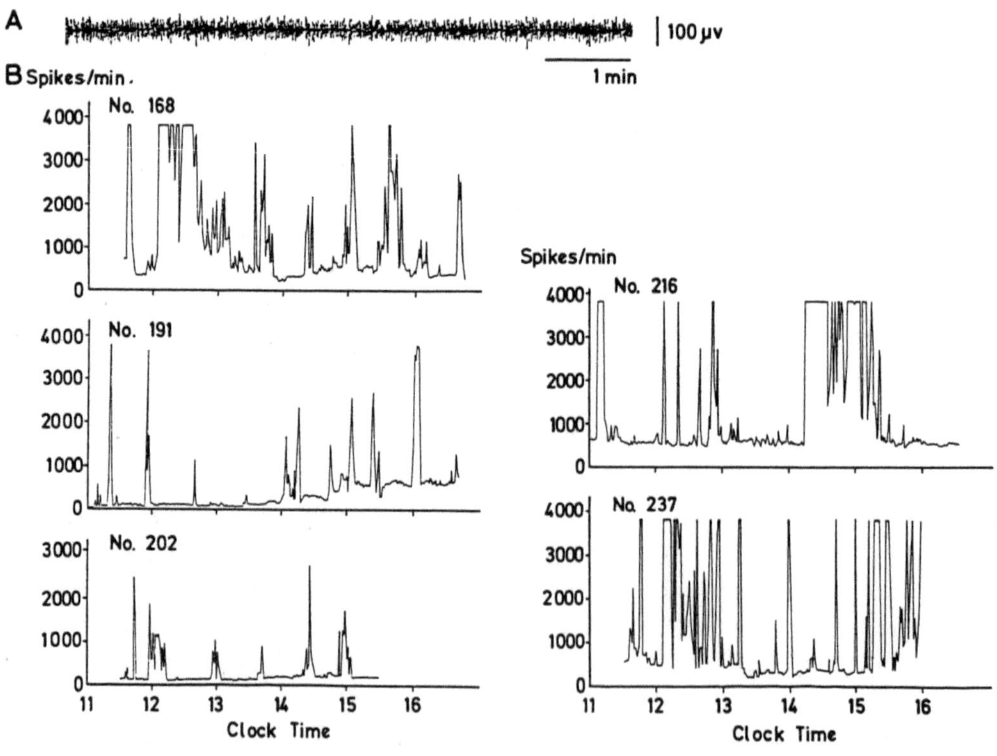

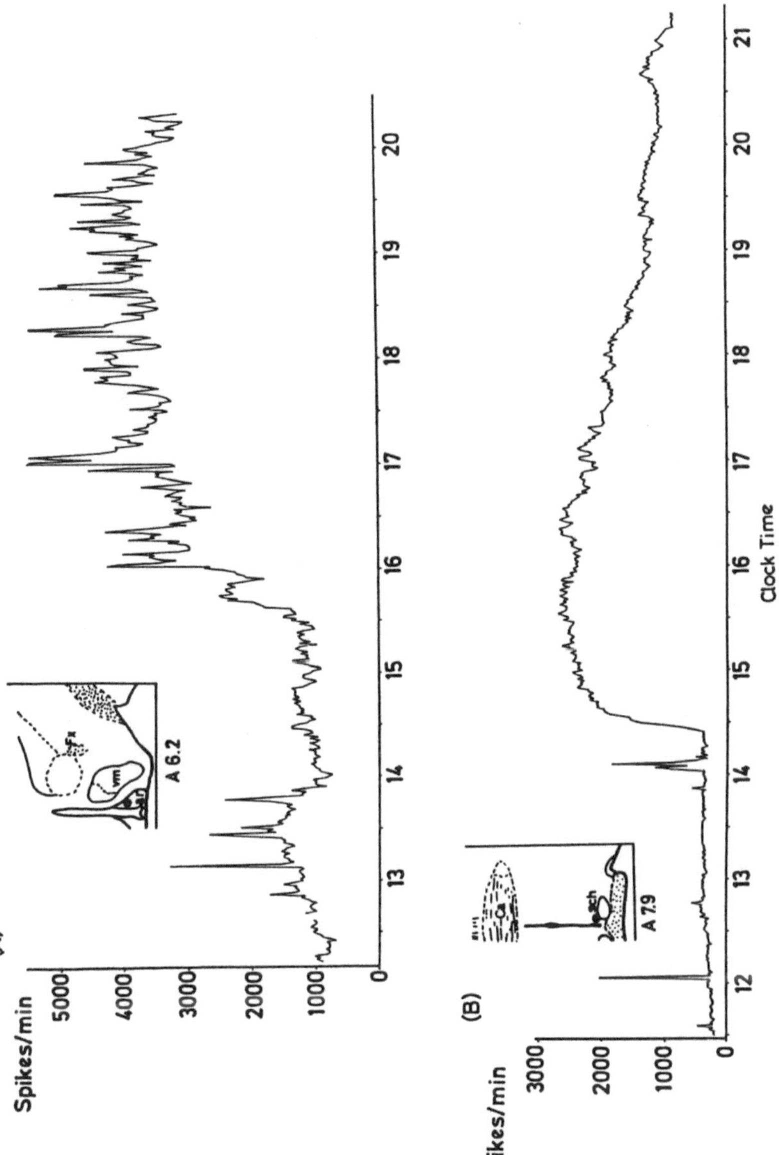

Fig. 6. MUA profiles in the Arc (A) and in the vicinity of the SCN (B) in the adult rat under long-term thiopental anesthesia. The sites recorded are indicated by solid circles on the frontal sections of the rat brain. Other details as in Fig. 5

◁ Fig. 5. EEG record from the frontal cortex and MUA profiles in the Arc in the adult rats under long-term thiopental anesthesia. Thiopental sodium was infused carefully during MUA recording in order to maintain a stable sleep pattern as shown in (A). Five individual examples of time series of the number of spike discharges per min are illustrated (B) considered to express circadian rhythmicity. Our results indicate that these rhythmicities in neural activity are not abolished by anesthesia with thiopental.

On the basis of the above finding that long-term thiopental anesthesia does not largely interfere with the rhythmical neuronal activity and also on the recent circumstantial evidence that a GH-releasing factor (GRF) mechanism plays a predominant role in the spontaneous GH secretory burst (WEHRENBERG et al. 1982; WILLOUGHBY et al. 1983) and that GRF neurons are probably situated in both the Arc and VMH (for review, see MARTIN 1976), we tried to correlate GH secretory bursts with electrical phenomena of the Arc-VMH neurons. Although the majority of the Arc and VMH MUAs did not show correlations with GH fluctuations in blood, about 17% of the MUAs had fluctuations that paralleled the increase or decrease of GH concentrations. Figure 7 illustrates simultaneously recorded MUAs in the Arc and VMH, which show apparent correlations with GH values.

These results suggest that some populations of the Arc and VMH neurons show changes in electrical activity which correlate with GH secretion. It has not been ascertained whether the neurons that show parallel changes with GH levels are the GRF neurons themselves or the ones that control them. However, our aim, confirmation of whether the hypothalamic neurons involved in the secretion of GRF directly or indirectly function with rhythmicities under the continuous sleep state, has been attained.

CONCLUSION

The results of the present series of experiments strongly suggest that different physiological functions have common ultradian rhythms which are controlled by different pacemakers. It is assumed that each pacemaker is fixed in specific phase relations with others, possibly by a central pacemaker.

The concept of a multiplicity pacemaker is based on the following observations. (a) A decrease in the amount of sleep by the sleep-deprivation procedure induces only an attenuation of GH secretion that occurs at the anticipated time in the 3-h period rhythm. (b) There are a minority of animals in which a temporal dissociation of three ultra-

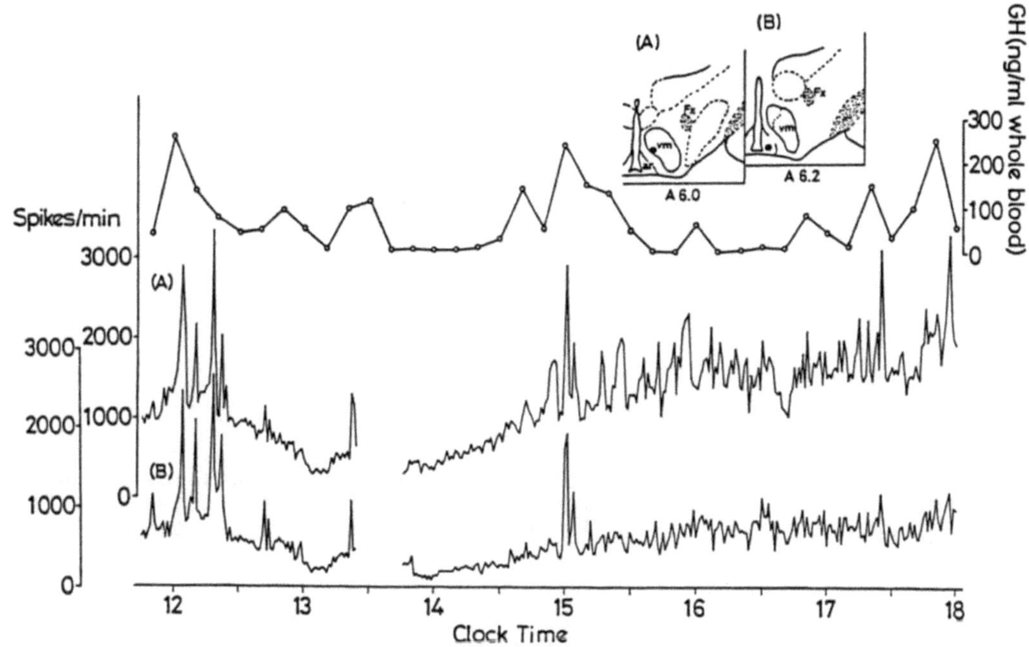

Fig. 7. MUA profiles in the VMH (A) and Arc (B) plotted against GH secretory profiles. All the recordings and blood samplings were done simultaneously in a rat under long-term thiopental anesthesia. Other details as in Figs. 5 and 6

dian rhythms, i.e., sleep, GH, and CS, is apparent. (c) Under the conditions of a continuous sleep state lacking in the rhythmicity due to long-term thiopental anesthesia, the ultradian rhythms of GH and CS secretions are still maintained with almost the same periodicities as in the freely moving animals.

We can raise two other possibilities from the results of the present experiments. The first one is that sleep and wakefulness themselves, which are under the control of a pacemaker, govern the rhythmic secretion of GH and CS. The second one is that one central pacemaker governs all three functions. The facts that the activity of the GH-releasing mechanism will be intensified during sleep and that the adrenocorticotropic hormone (ACTH) releasing mechanism will be intensified during wakefulness strongly support the first hypothesis. In addition, CS secretion has been reported to be related to locomotor activity (HONMA and HIROSHIGE 1978). On the other hand, the present observation of GH and CS ultradian rhythms under thiopental anesthesia conflicts with the first hypothesis. In addition, it is inconceivable

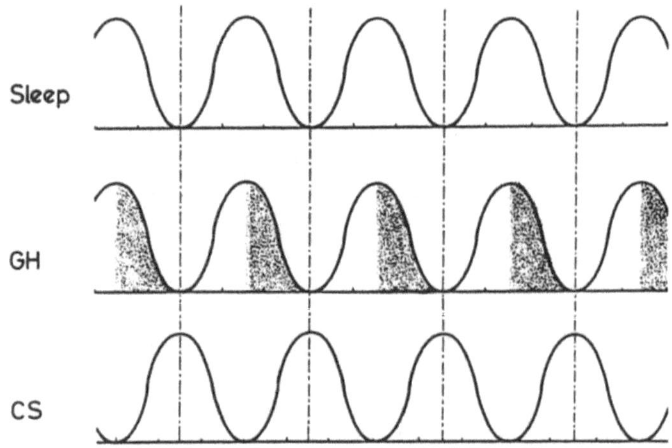

Fig. 8. Schematic illustration showing phase relationships among 3-h rhythms in sleep, GH and CS. Vertical dashed lines are drawn to define each sleep cycle, as in Fig. 1. For further details see text

that the pacemaker for the sleep-wakefulness cycle is functional under long-term anesthesia and governs the rhythmic secretion of GH and CS without the mediation by sleep-wakefulness, the direct effector.

The second observation mentioned above, namely, the temporal dissociation of sleep, GH, and CS secretion in a minority of animals, may contradict the second hypothesis. The potential dissociation in the phase relationships among the three functions is incompatible with the assumption of one central pacemaker governing all three functions.

The phase relationships among the 3-h period ultradian rhythms, which we regard as one of the unit rhythms in three physiological functions, are assumed to be fixed as schematically illustrated in Fig. 8. The 3-h secretory rhythm of GH has a 0° phase angle difference with the sleep rhythm. Since the sleep state has the potential to intensify the GH secretion, the relationship between sleep and the GH rhythm undergoes certain modifications. This intensification becomes restricted to the early stage of a sleep cycle as the animal matures. Thus, in the adult rat, the later part of a sleep cycle loses an apparent correlation with GH secretion. This is indicated by the shaded areas in Fig. 8. The 3-h CS rhythm seems to manifest a 180° phase angle difference to the rhythms of sleep and GH, at least in the adult animal. The phase relationships of the GH and CS secretory profiles to the 3-h sleep cycle are similar to those in humans during night sleep (TAKAHASHI et al. 1968).

While we have reached a tentative conclusion, as described above, we need further evidence for the multiplicity pacemaker hypothesis and further knowledge of the mode for fixing the phase relationships among the pacemakers.

ACKNOWLEDGEMENT. The authors thank Ms Y. Nakamura for her help in preparing the manuscript. The present series of studies was supported by a Grant-in-Aid for Scientific Research, from the Ministry of Education, Japan.

REFERENCES

HALBERG F, ENGELI M, HAMBURGER C, HILLMAN D (1964) Spectral resolution of low-frequency small-amplitude rhythms in excreted 17-ketosteroids: probable androgen-induced circaseptan desynchronization. Acta Endocrinol (Suppl) Copenh 103:5-54

HONMA K, HIROSHIGE T (1978) Internal synchronization among several circadian rhythms under constant light. Am J Physiol 235:R243-R249

INOUYE SIT, KAWAMURA H (1979) Persistence of circadian rhythmicity in a mammalian hypothalamic "island" containing the suprachiasmatic nucleus. Proc Natl Acad Sci USA 76:5962-5966

KAWAKAMI M, KIMURA F (1975) Possible roles of CNS estrogen-neuron systems in the control of gonadotropin release. In: STUMPF WE, GRANT LD (eds) Anatomical neuroendocrinology. Karger, Basel, p 216

KAWAKAMI M, KIMURA F (1978) The limbic forebrain structures and reproduction. In: BRAMBILLA F, BRIDGES PK, ENDCRÖCZI E, HEUSER G (eds) Perspectives in endocrine psychobiology. Akademiai Kiado, Budapest, p 101

KAWAKAMI M, TERASAWA E, KIMURA F, HIGUCHI T, KONDA N (1973) Changes in multiunit electrical activity (MUA) in rat brain during the estrous cycle and after administration of sex steroids. Prog Brain Res 39:125

KAWAKAMI M, UEMURA T, HAYASHI R (1982) Electrophysiological correlates of pulsatile gonadotropin release in rats. Neuroendocrinology 35:63-67

KAWAKAMI M, KIMURA F, TSAI CW (1983) Correlation of growth hormone secretion to sleep in the immature rat. J Physiol (Lond) 339:325-337

KAWAKAMI M, KIMURA F, TSAI CW (1984) Relationship between the three-hour-period sleep-wakefulness cycle and growth hormone secretion in the immature rat. J Physiol (Lond) 348:271-283

KIMURA F, KAWAKAMI M (1981) Serum hormone levels during sleep and wakefulness in the immature female rat. Neuroendocrinology 33:276-283

KIMURA F, TSAI CW (1984) Ultradian rhythm of growth hormone secretion and sleep in the adult male rat. J Physiol (Lond) 353:305-315

MARTIN JB (1976) Brain regulation of growth hormone secretion. In: MARTINI L, GANONG WF (eds) Frontiers in neuroendocrinology, vol 4. Raven, New York, p 129

MITSUGI N, KIMURA F (1985) Simultaneous determination of blood levels of corticosterone and growth hormone in the male rat: relation to sleep-wakefulness cycle. Neuroendocrinology (in press)

PRAPUTPITTAYA C, KIMURA F (1985) Growth hormone secretion during long-term thiopental anesthesia in the adult male rat. Neuroendocrinology 40:63-67

TAKAHASHI Y, KIPNIS DM, DAUGHADAY WH (1968) Growth hormone secretion during sleep. J Clin Invest 47:2079-2091

TANNENBAUM GH, MARTIN JB (1976) Evidence for an endogenous ultradian rhythm governing growth hormone secretion in the rat. Endocrinology 98:562-570

WEHRENBERG WB, BRAZEAU P, LUBEN R, BOHLEN P, GUILLEMIN R (1982) Inhibition of the pulsatile secretion of growth hormone by monoclonal antibodies to the hypothalamic growth hormone releasing factor (GRF). Endocrinology 111:2147-2148

WILLOUGHBY JO, MARTIN JB, RENAOUD LP, BRAZEAU P 1976) Pulsatile growth hormone release in the rat: failure to demonstrate a correlation with sleep phases. Endocrinology 98:991-996

WILLOUGHBY JO, KOBLAR S, JERVOIS PM, MENADUE MT, OLIVER JR (1983) Evidence that the regulation of growth hormone secretion is mediated predominantly by a growth hormone releasing factor. Neuroendocrinology 36:358-363

Ultradian Rhythms in Locomotor Activity, Deep Body Temperature and Plasma Corticosterone Levels in Rats: Two Different Origins?

K. Honma and S. Honma
Department of Physiology, Hokkaido University, School of Medicine, Sapporo 060, Japan

INTRODUCTION

The temporal patterns of many physiological and behavioral functions in the albino rat (Rattus norvegicus albicus) are characterized by both circadian and ultradian rhythmicities. The circadian rhythms of nocturnal rodents are regulated by self-sustained oscillators, presumably located within the suprachiasmatic nucleus (SCN) in the brain (RUSAK and ZUCKER 1979; INOUYE and KAWAMURA 1979). On the other hand, the origins and functions of ultradian rhythms are not well understood. At present, an ultradian rhythm is arbitrarily distinguished from random fluctuations by a period analysis when there is a significant period shorter than circadian. However, the periods of ultradian rhythms are less stable than those of circadian rhythms, ranging from several minutes to a few hours.

Although ultradian rhythms can be detected in intact adult rats, they are much more pronounced during the early developmental stage, before the emergence of the circadian rhythms (HONMA and HIROSHIGE 1977). Ultradian rhythms are also more pronounced under prolonged exposure of an animal to constant illumination (HONMA and HIROSHIGE 1978c), and after destruction, or isolation, of the SCN (STEPHAN and ZUCKER 1972; HONMA et al. 1984). Several explanations for the origin of ultradian rhythms and for their relationship to circadian rhythms have been proposed. These explanations were based on two main ideas: (a) the existence of a separate ultradian oscillator whose expression is modified by the circadian rhythm (KLEITMAN 1961), and (b) the fact that ultradian rhythms are the result of a desynchronization among a large population of circadian oscillators (PITTENDRIGH 1960).

The main focus of this chapter is on ultradian rhythms in rats which appear under prolonged exposure to constant light (LL) or after the

Experimental Brain Research, Suppl. 12

destruction, or isolation, of SCN. These include rhythms in locomotor activity, deep body temperature, and plasma cortisol levels. We examined the properties of these rhythms, especially their periods and the phase relationships between different variables.

INTERNAL COUPLING OF TWO CIRCADIAN OSCILLATORS IN ALBINO RATS

Before going into the details of ultradian rhythms, the functional structure of the circadian pacemaking system in rats will be briefly discussed. Recently, we (HONMA et al. to be published) established the phase response curve (PRC) of the circadian locomotor rhythm of male rats by applying single light pulses (300 lx, 30 min duration) to animals whose rhythms were free-running in constant darkness (DD). Two types of PRCs were obtained: (a) immediate PRCs based on the phase shifts measured immediately after the light pulse, and (b) steady-state PRCs based on the phase shifts measured after a new steady state was established. Each could be calculated in two different ways, according to the onset or the offset of the circadian activity phase.

The PRCs had different shapes. The immediate PRC, based on the onset of activity, lacked the phase advance (A) area, while the immediate PRC, based on the offset of activity, showed both the phase advance and phase delay (D) areas. The shapes of the two steady-state PRCs were similar except for the range of the light-sensitive area. This was wider in the PRC based on the onset of activity than in the PRC based on the offset of activity.

The differences in the shape of the PRCs as referred to onset and offset of activity, as well as between the immediate and steady-state PRCs, are adequately explained by the assumption that the circadian locomotor rhythm in nocturnal rodents is regulated by a two-oscillator pacemaking system (PITTENDRIGH and DAAN 1976). According to this model, onset and offset of activity represent phases of two separate circadian oscillators with different sensitivity to light. The two oscillators are mutually coupled and behave as a single circadian oscillator in the steady state. The period of the system (τ), the shape of its PRC, and its activity time are related to the coupling state between the two oscillators. Figure 1 illustrates the steady-state PRCs based on the onset of activity for rhythms with short and long τs (HONMA et al. to be published). The light-sensitive range was wider in

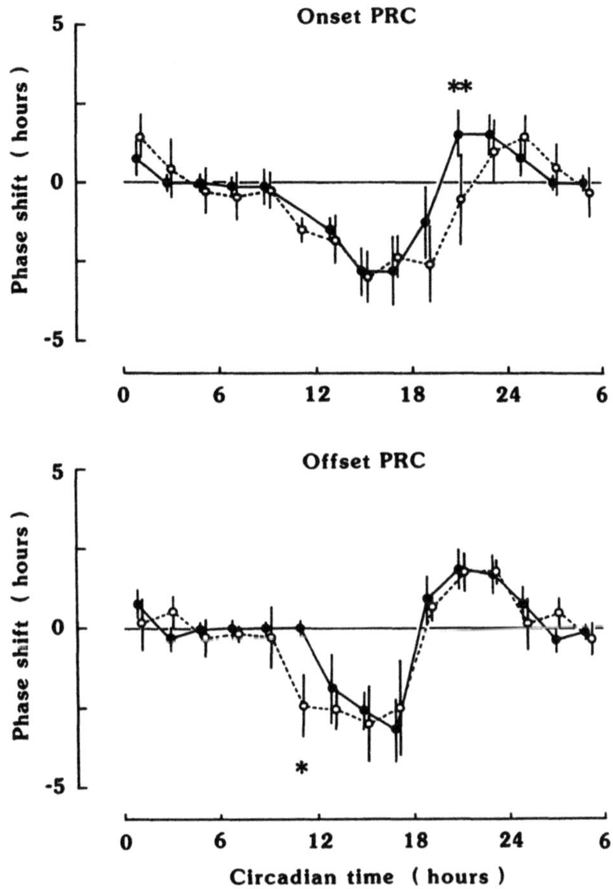

Fig. 1. Dependence of onset and offset PRCs on τ. PRCs are based on steady-state phase shifts. Closed circles with solid lines indicate PRCs of the short τ group (mean τ = 24.33 h) and open circles with broken lines indicate those of the long τ group (mean τ = 24.53 h). A vertical bar on each point represents SD. Asterisks indicate statistically significant difference between two points at the same circadian time (*, p ≦ 0.05; **, p ≦ 0.01, using two-way analysis of variance). (HONMA et al. to be published)

rhythms with long τs than in rhythms with short τs, and in rhythms with long τs the cross-over point of the PRCs occurred later during the subjective night. The maximum phase shifts were not significantly different for rhythms with long and short τs, neither in the advance area nor in the delay area. Surprisingly, the position of the advance and delay areas in the PRC differed between the two groups by 2 h, while the difference in τ was only 0.2 h. This change in the phase relationship was presumed to reflect the phase-angle difference between the two oscillators (PITTENDRIGH and DAAN 1976). Therefore, it was surmised that under a condition which increases τ the phase-angle difference of the two oscillators increases and the circadian rhythm tends to split into two components.

ULTRADIAN RHYTHMS OBSERVED IN CONSTANT LIGHT

It has long been recognized that the circadian rhythm is dampens out in constant light (BÜNNING 1973; ASCHOFF 1979). This disintegration of the circadian locomotor rhythm in rats has been reported to be independent of light intensity (ALBERS et al. 1981; EASTMAN and RECHTSCHAFFEN 1983), duration of light exposure, and the period of the free-running circadian rhythm (ALBERS et al. 1981).

Figure 2 illustrates a double-plotted record of locomotor activity of a rat exposed to 200 lx in LL. Locomotor activity was measured using an Animex apparatus (HONMA and HIROSHIGE 1978b) and expressed as counts per 15 min. During the first 2 weeks in LL, the activity rhythm ran free with a period longer than 24 h. After approximately 20 days exposure to LL, the phase relation of the activity onset and activity offset of the circadian rhythm changed, and splitting of the circadian rhythm occurred. After a short transitional period, the two activity components again merged into a single activity band. Splitting and merging of locomotor activity appeared to occur alternately thereafter.

Finally, the circadian rhythmicity was completely replaced by disintegrated and multiple short-lasting activity bursts. The results of the period analysis of the first 3 days and the last 3 days in LL condition (12 weeks after exposure to LL) are demonstrated on the right side of the locomotor chart. The period analysis was based on a least-squares method (HONMA and HIROSHIGE 1978c). In the first 3-day inter-

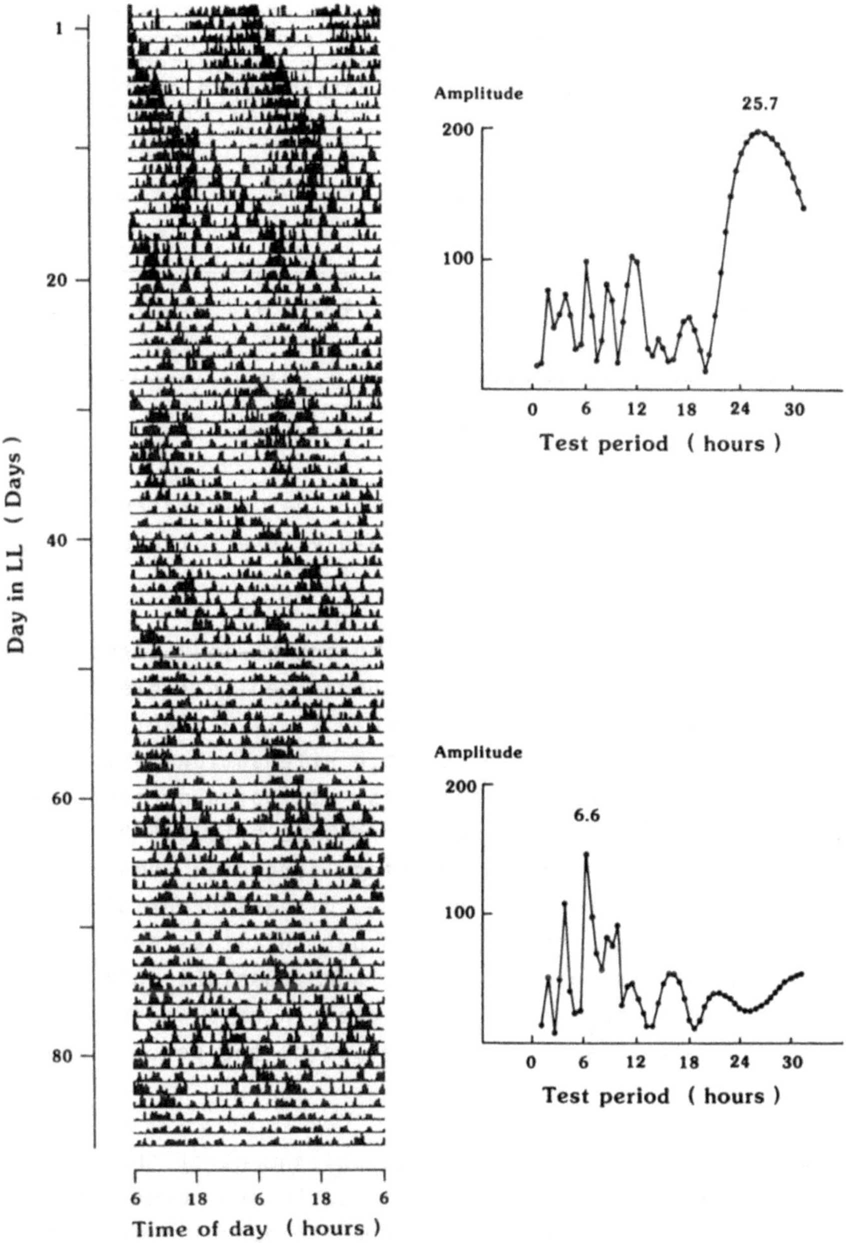

Fig. 2. Spontaneous locomotor activity of a rat exposed to 200 lx constant light (LL) and the result of a period analysis. The locomotor activity is expressed by histogram and the record is double-plotted. The rat was entrained by the 24-h LD cycle (light 0600-1800) before exposure to LL. The upper spectrum indicates the result for the first 3-day record in LL and the lower spectrum for the last 3 days. Numbers on the top of the predominant peak in each spectrum indicate the period of significant periodicity

Table 1. Circadian and ultradian periods of the locomotor activity in rats under the prolonged LL. The periods were obtained by a least-square method and their significance was tested by analysis of variance. When several significant periods were obtained, the most prominent period was demonstrated; absence of significant peaks is indicated by a dotted line

Rat No.	1	2	3	4	5	6	7	8	9
Ultradian (h)	6.6	4.5	4.2	4.3	4.2	4.8	6.6	4.3	6.1
Circadian (h)	26.2	---	25.8	---	---	25.2	---	---	---

val a significant period within the circadian range was detected (25.7 h), whereas in the last 3-day interval a significant ultradian period of 6.6 h appeared.

The effects of the prolonged LL on the period of spontaneous locomotor rhythms in individual rats are summarized in Table 1. Although in some rats circadian periodicities were still evident, their amplitude was reduced, suggesting a considerable attenuation of the circadian rhythmicity. In contrast, in all rats, ultradian periodicities became predominant with periods ranging from 4.2 h to 6.6 h, with a mean period of 5.1 h. In the rats with intact circadian components the ultradian periods were harmonics of the circadian period; it hence could be speculated that the ultradian periodicities were by-products of the circadian process because of its nonsinusoidal form. However, the fact that ultradian rhythms were detected in rats whose circadian component was completely abolished, precludes the possibility that they were analytical artifacts.

Figure 3 illustrates the reestablishment of the circadian locomotor rhythm in a rat after removal from LL and reentrainment to a 24-h LD cycle. A circadian component seemed to appear immediately, and to gain stable entrainment via phase delay. The process was similar to the reentrainment of a free-running circadian rhythm to the LD cycle.

The mode of reestablishment of the circadian locomotor rhythm in LD as well as the splitting phenomenon in LL support the hypothesis that the emergence of a significant ultradian rhythm in LL is due to an unstable and weak coupling between two circadian oscillators. Under LL,

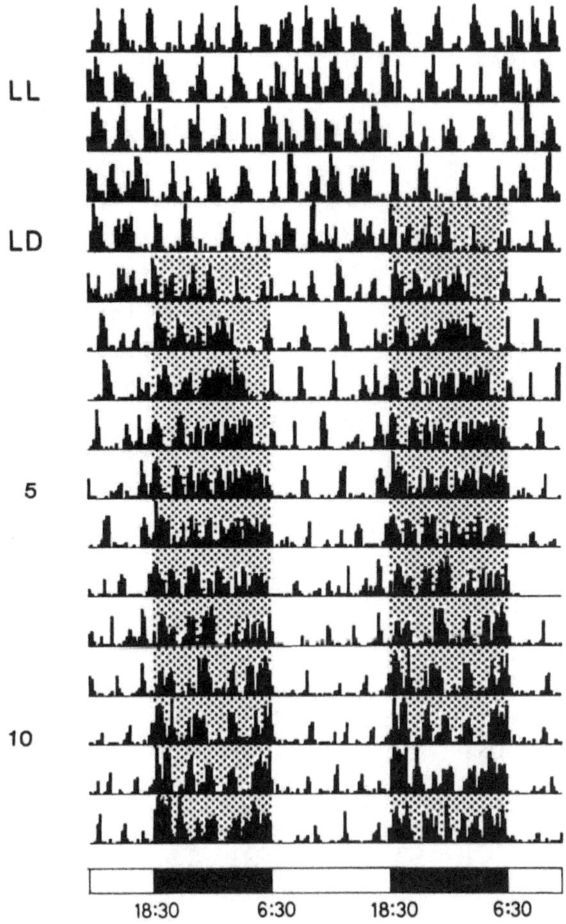

Fig. 3. The spontaneous locomotor activity of a rat returned to the 24-h LD cycle after the prolonged LL. The number on the ordinate indicates days after the return to the LD cycle. The dark period (1800-0600) is indicated by the shaded area

the phase-angle difference between the two postulated oscillators may increase, concomitantly with an increase in τ (SUMMER et al. 1984). This eventually resulted in a complete loss of coupling between the two circadian oscillators and the occurrence of ultradian locomotor bursts.

In addition to spontaneous locomotor activity, abdominal body temperature and plasma corticosterone levels were determined simultaneously and the phase relationships between these three parameters were examined under LL conditions. Body temperature was measured continuously using a telemetric system (HONMA and HIROSHIGE 1978a). Recently EASTMAN and RECHTSCHAFFEN (1983) reported that rats kept in constant dim light showed a dissociation between the circadian rhythms of wheel-running activity and body temperature; the activity rhythm dampened out, whereas the body temperature rhythm persisted in some animals. Previously, a similar finding was reported in the chicken (ASCHOFF and VON SAINT PAUL 1972). In the ultradian range such a dissociation was not detected in our rats (n = 5). Figure 4 depicts the phase relationship between locomotor bursts and abrupt rises in abdominal temperature. Generally, a locomotor burst preceded a rise in body temperature, but occasionally the rise in body temperature preceded the burst of locomotor activity. This reversed pattern seems to preclude the possibility that there is a causal relationship between the two parameters, even on the ultradian level (EASTMAN and RECHTSCHAFFEN 1983).

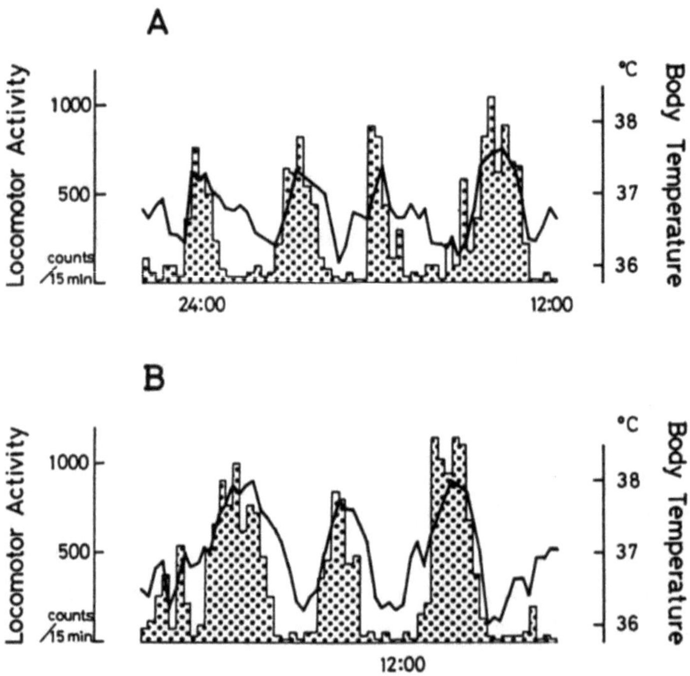

Fig. 4. Phase relation between bursts of locomotor activity and body temperature in rats exposed to prolonged LL. Spotted histograms indicate amounts of locomotor activity and solid lines fluctuation of body temperature. (HONMA and HIROSHIGE 1978c)

The phase relationship between the activity bursts and the plasma corticosterone level is shown in Fig. 5. To avoid stress due to continuous blood sampling, plasma hormone levels were taken at random intervals, spaced at least 4 h apart. Plasma hormone levels were plotted in relation to the onset of the prior activity burst. In Fig. 5, a standard sequence of two activity bursts is schematically illustrated for each rat. The superimposed plasma corticosterone level showed an initial rise anticipating the burst and a subsequent fall

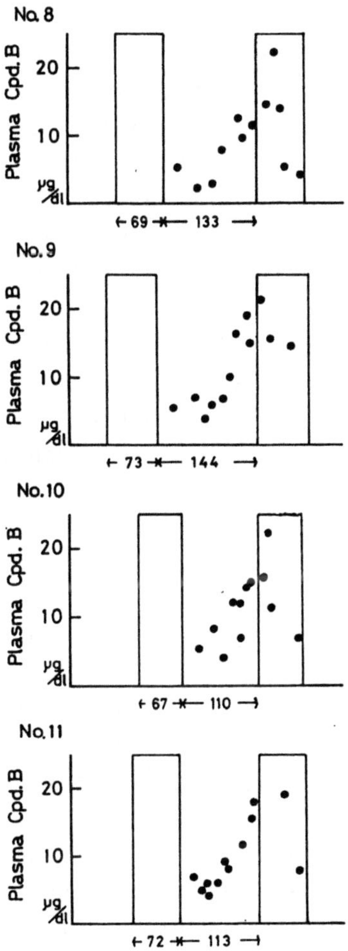

Fig. 5. Temporal relation between locomotor bursts and episodic variation of plasma corticosterone levels in four individual rats. Oblong columns indicate the standard locomotor burst in each rat. Mean duration of burst and mean interburst interval are given in min at the bottom of each panel. (HONMA and HIROSHIGE 1978c)

toward the end of the burst. This pattern suggests a fixed phase relationship between the locomotor burst and the episodic elevation of plasma corticosterone level. A similar temporal relationship was observed between the circadian rhythms of the two parameters (HONMA and HIROSHIGE 1978b).

ULTRADIAN RHYTHMS APPEARING AFTER THE DESTRUCTION, OR ISOLATION, OF THE SUPRACHIASMATIC NUCLEUS

Bilateral lesions of SCN abolished circadian rhythmicity in most, if not all, physiological and behavioral functions of rats (RUSAK and ZUCKER 1979). In SCN-lesioned rats, circadian rhythms were replaced by short-lasting bursts of activity (STEPHAN and ZUCKER 1972). Similarly, circadian rhythms disappeared after the isolation of SCN from the rest of the brain (WILSON and CRITCHLOW 1975; KAWAMURA and INOUYE 1979). Figure 6 displays double-plotted locomotor charts of rats under a 24-h LD cycle, whose SCNs were lesioned electrically, or isolated using the Halasz cut (HALASZ and PUPP 1965). The isolated island included SCN, parts of the paraventricular nucleus, the arcuate nucleus, and the ventromedial nucleus. Although the circadian rhythm was not detectable by visual inspection, period analysis revealed a significant peak at 24 h, which was probably due to the masking effect of light (ASCHOFF et al. 1982). The most predominant ultradian peak in this rat was 2.88 h. In the SCN-isolated rat, there was only a significant ultradian period (4.80 h). The distributions of ultradian periods in SCN-lesioned and SCN-isolated rats, together with those in rats under prolonged LL conditions, are shown in Fig. 7. Ultradian periods ranged from 1 to 4 h in the SCN-lesioned rats (mean, 2.6 h), and from 0.8 to 4.6 h in the SCN-isolated rats (mean, 2.5 h). These means are significantly different from the ultradian periods of the rats exposed to prolonged LL, which were about twice as long (mean, 5.1 h).

Reports of persistence of the circadian body temperature rhythm in SCN-lesioned rats and, therefore, claims that there is a separate circadian pacemaker for body temperature, located outside the SCN, have been made by several authors (DUNN et al. 1977; ABE et al. 1979; NAKAYAMA et al. 1979; HALBERG et al. 1979). In most of these studies, however, there were difficulties in measuring body temperature (manual measurement) or there was no histological confirmation of the extent of SCN destruction. Recently, telemetrically monitored body tempera-

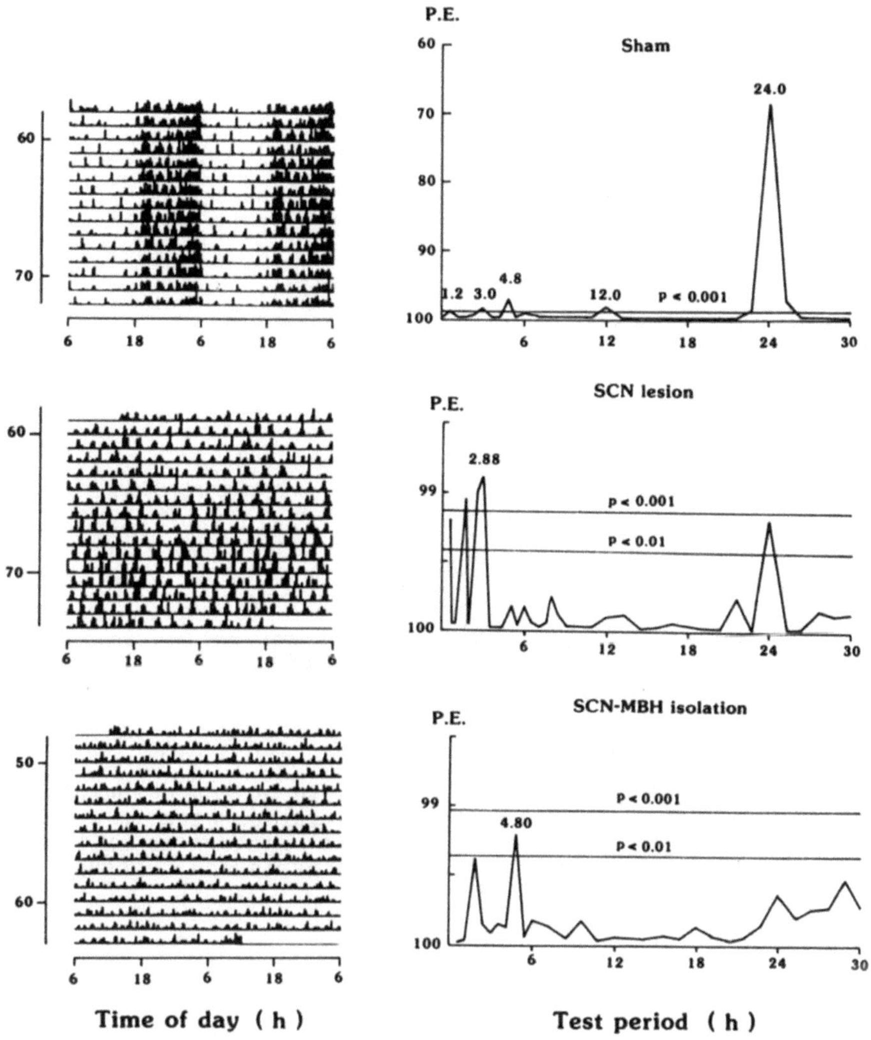

Fig. 6. The spontaneous locomotor activities in a sham-operated, an SCN-lesioned and an SCN-isolated rat, and results of a period analysis. The locomotor activity was measured under the 24-h LD cycle (light 0600-1800) and the records are double-plotted. PE indicates the percentage error. Significance of the peak in the spectrum was tested by analysis of variance. Horizontal bars in each panel indicate the levels of significance. Number on top of peak indicates period

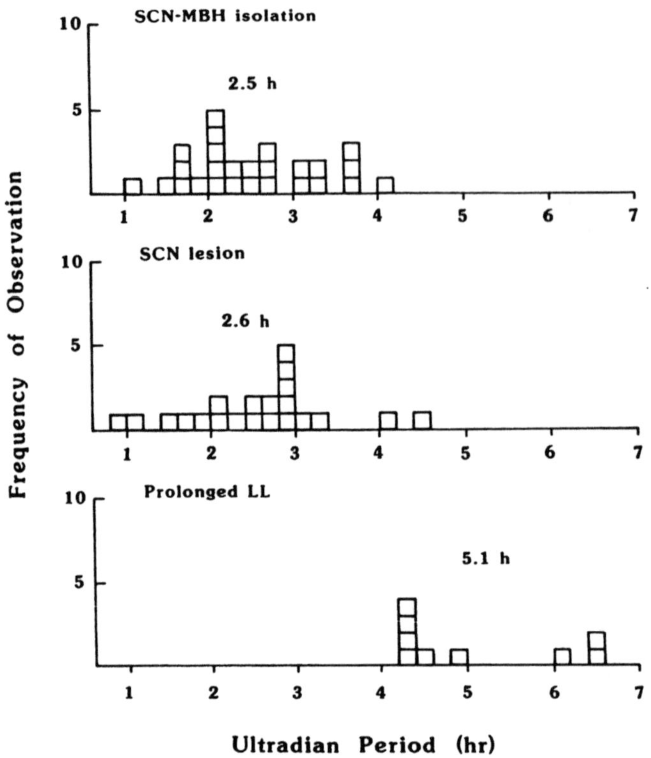

Fig. 7. Distributions of the ultradian period detected in three different conditions: after the SCN isolation (top, n = 26), SCN lesion (middle, n = 21) and the prolonged exposure to LL (bottom, n = 9). Time in each panel indicates the mean period of a group

ture and locomotor activity were obtained simultaneously from SCN-lesioned rats (EASTMAN et al. 1984). In that study there was no clear-cut dissociation between the circadian rhythms of locomotor activity and body temperature, an observation which does not support the hypothesis of a separate circadian oscillator for body temperature. However, this experiment was performed under constant dim light (1.5 lx), and as the authors themselves mentioned, arrhythmicity in body temperature could be due to the lighting condition rather than to the SCN lesions. Constant light was previously shown to have a disruptive effect on free-running rhythms in rats (EASTMAN and RECHTSCHAFFEN 1983). To exclude this possibility, we measured both locomotor activity and body temperature in a 24-h LD cycle. In four SCN-lesioned rats the circadian rhythms in both parameters disappeared, and short-lasting locomotor bursts and episodic elevations of body temperature became predominant.

The phase relationship between the locomotor bursts and the body temperature elevations in an SCN-lesioned rat is illustrated in Fig. 8. In this rat, as in most rats, the locomotor burst preceded the rise in body temperature. There were cases, however, where the temperature rise occurred prior to the locomotor burst. As mentioned before, similar phase relationships were observed in intact rats under continuous light.

The circadian rhythm in plasma corticosterone level is known to be abolished by SCN lesion (MOORE and EICHLER 1972) and to be replaced by episodic fluctuations in the level of the hormone (WATANABE and HIROSHIGE 1981). The phase relationship between the locomotor burst and plasma corticosterone level was examined in three SCN-lesioned rats (Fig. 9). In this experiment, blood samples were obtained every 15 min through a chronically indwelling catheter, from the jugular vein. In the SCN-lesioned rats, a locomotor burst always preceded the rise of plasma corticosterone by 15-30 min. Taking into account the time needed for the plasma corticosterone level to reach its maximum (15 min), it appears that the initiation of the locomotor burst and the secretion of the hormone occur almost simultaneously. This phase relationship in SCN-lesioned rats is different from that detected under prolonged LL (Fig. 5), where a rise in corticosterone generally preceded the locomotor burst.

In entrained adult rats, the circadian rise of plasma corticosterone level occurs just prior to the evening locomotor burst, but the hormone level decreases and remains low during the latter half of the subjective night. Thus, the special phase relationship observed under prolonged LL, is identical to the phase relationship between the hormone and the first nocturnal locomotor episode. This is different from infantile rats in which a rise of plasma corticosterone level is coupled to the morning component as well as to the evening component of the circadian locomotor rhythm (HIROSHIGE et al. 1982). Such a bimodality was observed in other pituitary hormones during the early ontogenic state (DOHLER and WUTTKE 1976). The similarity of the phase relationship between corticosterone and locomotor activity in the infantile rat and in rats under LL suggests that the rise of corticosterone in LL is coupled to both activity onset and activity offset. If this is the case, the probability of obtaining such a phase relationship between a locomotor burst and an episodic rise of plasma corticosterone level would be twice as high under prolonged LL as under

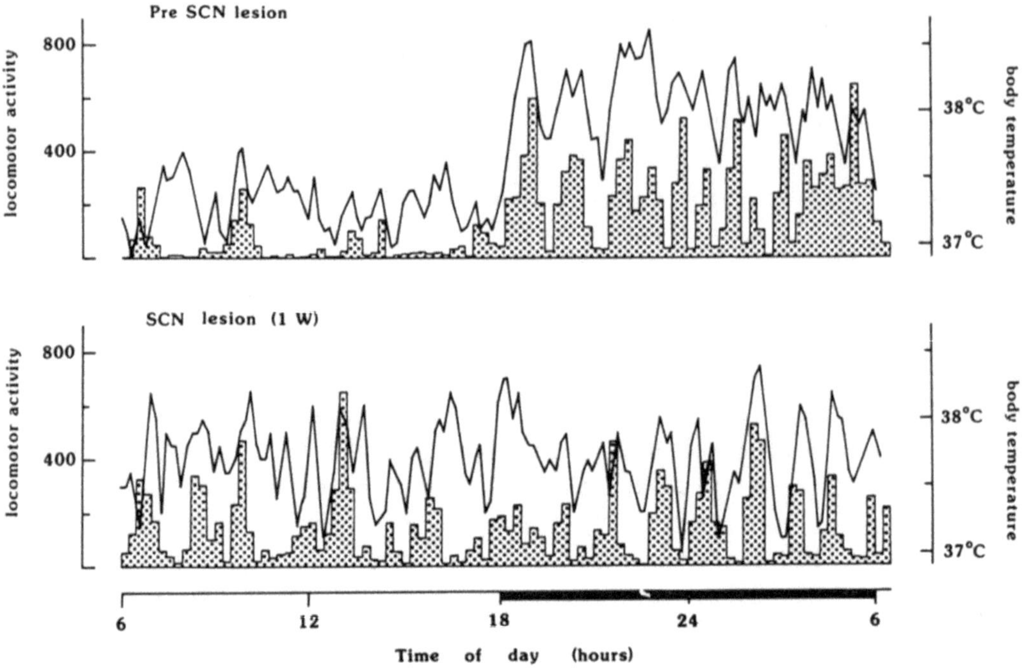

Fig. 8. Simultaneously determined locomotor activity and abdominal body temperature in an SCN-lesioned rat, before (upper) and 1 week after (lower) the lesion. Measurements were performed under the LD cycle. A black horizontal bar indicates the dark period

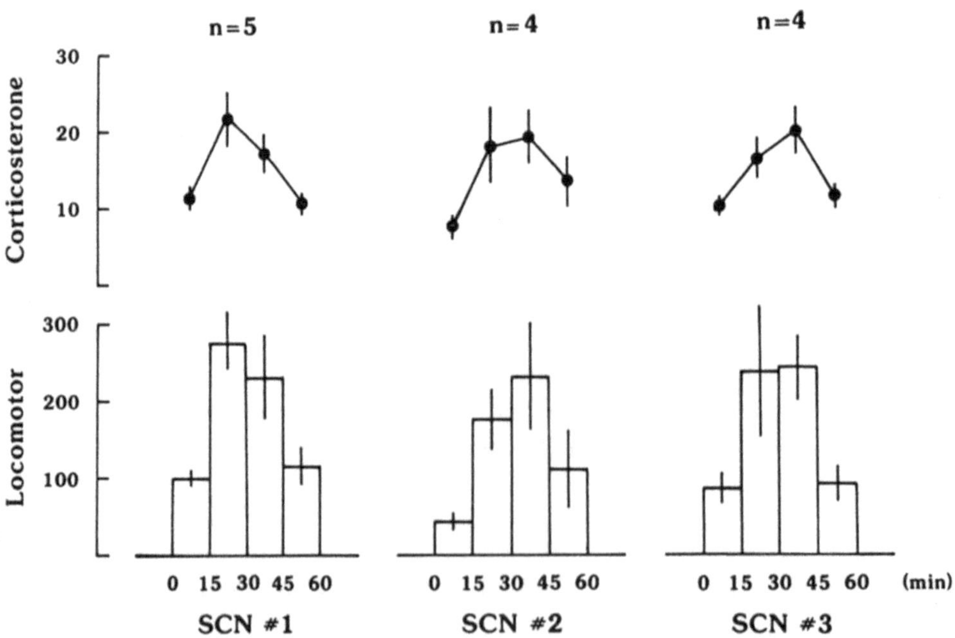

LD cycle. If the coupling of two circadian oscillators, which comprise the circadian pacemaker, is lost under prolonged LL, but the coupling between the oscillators and the overt locomotor bursts, or the hormone fluctuations is kept continuous, the specific phase relationship would be possible. Eliminating the circadian oscillators, for example, by SCN lesion, results in the disappearance of this special phase relationship.

HYPOTHESIS

There are two main differences between the ultradian rhythms observed under prolonged LL and those observed after lesioning, or isolating, the SCN, i.e., under two conditions in which circadian rhythmicity is either disintegrated (LL) or completely lost (lesioning or isolating of SCN). The first difference is in the ultradian period and the second is in the phase relationship between a locomotor burst and the plasma corticosterone level. The ultradian period is distinctly longer under prolonged LL than after SCN manipulation. The phase relationship under prolonged LL is similar to the phase relationship between the circadian rhythms in these two variables, which is different from the phase relationship after SCN lesion.

To explain these differences, the following hypothesis is advanced. We assume that the amplitude and period of each unit component which generates ultradian rhythmicity is influenced by the circadian rhythm (HONMA and HIROSHIGE 1978c; TANNENBAUM and MARTIN 1976; HONMA and HONMA, this volume, Chap. 6). There is evidence that the circadian pacemaking system in rats is comprised of two mutually coupled circadian oscillators. In extending this hypothesis, we suggest that the temporal arrangement of the locomotor burst or episodic hormone fluc-

◁ Fig. 9. The phase relation between a locomotor burst and a rise of plasma corticosterone level in 3 SCN-lesioned rats. Blood samples were taken sequentially, monitoring the spontaneous locomotor activity simultaneously. The locomotor activity was counted every 15 min. In the lower panel, the mean locomotor burst was constructed by superimposing several episodes of the burst in reference to the onset of burst (0 min), and the mean activity level (oblong column) and Standard Error (SE) (vertical bar) were calculated. The hormone levels were similarly superimposed, and expressed by the mean and SE (vertical bar) (upper panel); n indicates the number of episodes tuation, is influenced by the degree of coupling between the two circadian oscillators and the overt rhythms, on the one hand, and by the degree of the internal coupling between the two oscillators themselves, on the other hand. The strength of the former coupling primarily affects the amplitude of the unit component (amplitude modulation), and that of the internal coupling primarily affects the clustering of the units (frequency modulation). Under prolonged LL, the internal coupling between the two circadian oscillators is weakened and the phase-angle difference between them increases. This results in a more dispersed distribution of the unit components. Concomitantly, the coupling between each circadian oscillator and the overt rhythmic components is weakened. This causes a loss of amplification of unit components by circadian oscillators, resulting in the replacement of the circadian rhythm by the ultradian rhythm.

Since, however, the influence of the two circadian oscillators is not completely abolished, the distribution of the unit components is still affected to some extent by the degree of internal coupling. This may explain the relatively longer ultradian periods observed under LL, as compared with the SCN-lesioned or SCN-isolated rats. The phase relationship between a locomotor burst and the prior rise of plasma corticosterone level is probably maintained by the coupling between the circadian oscillators and the overt rhythms. In contrast, the ultradian rhythms in the SCN-lesioned rats, which are released from any influence of circadian oscillators, lose this phase relationship, and locomotor bursts and rises in corticosterone level occur almost simultaneously. Taking the cascade of the hormone secretion (CRF-ACTH-corticosterone) into account, it is difficult to understand such a temporal coincidence without assuming the existence of a single mechanism that initiates both the locomotor burst and the episodic corticosterone secretion.

REFERENCES

ABE K, KRONING J, GREER MA, CRITCHLOW V (1979) Effects of destruction of the suprachiasmatic nuclei on the circadian rhythms in plasma corticosterone, body temperature, feeding and plasma thyrotropin. Neuroendocrinology 29:119-131

ALBERS HE, GERALL AA, AXELSON JF (1981) Circadian rhythm dissociation in the rat: effects of long-term constant illumination. Neurosci Lett 25:89-94

ASCHOFF J (1979) Circadian rhythms: influences of internal and external factors on the period measured in constant conditions. Z Tierpsychol 49:225-249

ASCHOFF J, VON SAINT PAUL U (1972) Brain temperature as related to gross motor activity in the unanestetized chicken. Physiol Behav 10:529-533

ASCHOFF J, DAAN S, HONMA K (1982) Zeitgeber, entrainment, and masking: some unsettled questions. In: ASCHOFF J, DAAN S, GROOS G (eds) Vertebrate circadian system. Structure and physiology. Springer, Berlin Heidelberg New York, p 13

BÜNNING E (1973) The physiological clock. English University Press, London

DOHLER KD, WUTTKE W (1976) Circadian fluctuations of serum hormone levels in prepubertal male and female rats. Acta Endocrinol 83:269-279

DUNN JD, CASTRO AJ, McNULTY JA (1977) Effect of suprachiasmatic ablation on the daily temperature rhythm. Neurosci Lett 6:345-348

EASTMAN C, RECHTSCHAFFEN A (1983) Circadian temperature and wake rhythms of rats exposed to prolonged continuous illumination. Physiol Behav 31:417-427

EASTMAN C, MISTLBERGER RE, RECHTSCHAFFEN A (1984) Suprachiasmatic nuclei lesions eliminate circadian temperature and sleep rhythms in the rat. Physiol Behav 32:357-368

HALASZ B, PUPP L (1965) Hormone secretion of the anterior pituitary gland after physical interruption of all nervous pathways to the hypophysiotropic area. Endocrinology 77:553-562

HALBERG F, LUBANOVIC WA, SOTHERS RB, BROCKWAY B, POWELL EW, PASLEY JN, SCHEVING LE (1979) Nomifensine chronopharmacology, schedule-shifts and circadian temperature rhythms in di-suprachiasmatically lesioned rats - modeling emotional chronopathology and chronotherapy. Chronobiologia 6:405-424

HIROSHIGE T, HONMA K, WATANABE K (1982) Ontogeny of the circadian rhythm of plasma corticosterone in blind infantile rats. J Physiol (Lond) 325:493-506

HONMA K, HIROSHIGE T (1978a) Simultaneous determination of circadian rhythms in locomotor activity and body temperature of the rat. Jpn J Physiol 28:159-169

HONMA K, HIROSHIGE T (1978b) Internal synchronization among several circadian rhythms in rats under constant light. Am J Physiol 235: R243-R249

HONMA K, HIROSHIGE T (1978c) Endogenous ultradian rhythms in rats exposed to prolonged continuous light. Am J Physiol 135:R250-R256

HONMA S, HIROSHIGE T (1977) Pubertal manifestation of sex difference in circadian rhythm of corticotropin-releasing activity in the rat hypothalamus. Acta Endocrinol 86:225-234

HONMA S, HONMA K, HIROSHIGE T (1984) Dissociation of circadian rhythms in rats with a hypothalamic island. Am J Physiol 246:R949-R954

INOUYE ST, KAWAMURA H (1979) Persistence of circadian rhythmicity in a mammalian hypothalamic 'island' containing the suprachiasmatic nucleus. Proc Natl Acad Sci (USA) 76:5962-5966

KAWAMURA H, INOUYE ST (1979) Circadian rhythm in a hypothalamic island containing the suprachiasmatic nucleus. In: SUDA M, HAYAISHI O, NAKAGAWA H (eds) Biological rhythms and their central mechanism. Elsevier/North-Holland Biomedical, Amsterdam New York Oxford, pp 335-341

KLEITMAN N (1961) The nature of dreaming. In: WOLSTENHOLME GEW, O'CONNOR M (eds) The nature of sleep. Churchill, London, p 349

MOORE RY, EICHLER VB (1972) Loss of a circadian adrenal corticosterone rhythm following suprachiasmatic lesions in the rat. Brain Res 42:201-206

NAKAYAMA T, ARAI S, YAMAMOTO K (1979) Body temperature rhythm and its central mechanism. In: SUDA M, HAYAISHI O, NAKAGAWA H (eds) Biological rhythms and their central mechanism. Elsevier/North-Holland Biomedical, Amsterdam New York Oxford, p 395

PITTENDRIGH CS (1960) Circadian rhythms and the circadian organization of living system. Cold Spring Harbor Symp Quant Biol 25:159-184

PITTENDRIGH CS, DAAN S (1976) A functional analysis of circadian pacemakers in nocturnal rodents. V. Pacemaker structure: a clock for all seasons. J Comp Physiol 106:333-355

RUSAK B, ZUCKER I (1979) Neural regulation of circadian rhythms. Physiol Rev 59:449-526

STEPHAN FK, ZUCKER I (1972) Circadian rhythms in drinking behaviors and locomotor activity of rats are eliminated by hypothalamic lesions. Proc Natl Acad Sci (USA) 69:1583-1586

SUMMER TL, FERRADO JS, McCORMACK CE (1984) Phase-response and Aschoff illuminance curves for locomotor activity rhythm of the rat. Am J Physiol 246:R299-R304

TANNENBAUM GS, MARTIN JB (1976) Evidence for an endogenous ultradian rhythm governing growth hormone secretion in the rat. Endocrinology 98:562-570

WATANABE K, HIROSHIGE T (1981) Phase relation between episodic fluctuations of spontaneous locomotor activity and plasma corticosterone in rats with suprachiasmatic nuclei lesions. Neuroendocrinology 33:52-59

WILSON MM, CRITCHLOW V (1975) Absence of a circadian rhythm in persisting corticosterone fluctuations following surgical isolation of the medial basal hypothalamus. Neuroendocrinology 19:185-192

Interaction Between Circadian and Ultradian Rhythms of Spontaneous Locomotor Activity in Rats During the Early Developmental Period

S. Honma and K. Honma
Department of Physiology, Hokkaido University, School of Medicine, Sapporo 060, Japan

INTRODUCTION

Most of the physiological functions in newborn rats lack circadian rhythmicity until a certain postnatal age is reached. Different circadian rhythms appear at different developmental stages, according, most probably, to the maturation of the coupling system between the overt rhythm and the underlying oscillators one of which is located in the suprachiasmatic nucleus (SCN) of the hypothalamus (MOORE 1979). There is evidence that the oscillator becomes functional earlier (DAVIS 1981), even at a prenatal stage. A day-night difference in the deoxyglucose uptake was reported in SCN of fetal rats, which was coordinated by the rhythms of the mother (REPPART and SCHWARTZ 1983). We showed maternal phase-setting of fetal circadian oscillations (HONMA S et al. 1985), a phenomenon difficult to explain without assuming the existence of a fetal oscillator.

In contrast to the circadian system, it is not clear whether a specific oscillator regulates ultradian rhythms. Before the appearance of overt circadian rhythms, many functions show continuously low levels of activity or random fluctuations. Period analyses, however, often reveal that these apparently arrythmic fluctuations contain short-term periodicities whose periods are within the ultradian range. Ultradian phenomena during the ontogeny of spontaneous locomotor activity of laboratory rats have already been observed half a century ago (RICHTER 1927).

In this context, changes in the 24-h pattern of activity during the course of postnatal development may provide a good example of the interaction between circadian and ultradian rhythms. In the present experiments, spontaneous locomotor activity was recorded from infantile

Experimental Brain Research, Suppl. 12

rats using an Animex apparatus (HONMA and HIROSHIGE 1978), and their periodicities were analyzed using a modified chi-square periodogram (SOKOLOVE and BUSHELL 1978), or a least-square spectrum method (HONMA and HIROSHIGE 1978). The purpose of these experiments was to understand the interaction between ultradian and circadian rhythms from an ontogenic point of view.

CIRCADIAN ORGANIZATION OF ULTRADIAN LOCOMOTOR RHYTHMS IN INFANTILE RATS

Spontaneous locomotor activity of the infantile rat shows short-lasting activity bouts throughout the day until about the third postnatal week, when the circadian locomotor rhythm appears (HONMA and HIROSHIGE 1977). The process of the circadian organization has been explained in two different ways (DAVIS 1981). The first is the so-called amplitude modulation explanation. It assumes that an increase in the amplitude of the ultradian bouts during the active phase, and/or a decrease in the amplitude of the bouts during the rest phase, are responsible for the emergence of the circadian rhythm. According to this explanation, the period of the ultradian rhythm remains unchanged. The second explanation is the so-called frequency modulation. It assumes changes in the period of ultradian bouts, or progressive clustering of the bouts, occurring predominantly in the active phase. Thus, ultradian periods during the active phase become progressively shorter than those during the rest phase.

In order to study the process of the circadian organization in entrainment and free-running conditions, sighted and blinded pups were investigated. Sighted pups were weaned at 21 days after birth, and their spontaneous locomotor activity was recorded in a 24-h light-dark cycle (LD 12:12) continuously for at least 4 weeks. A period analysis was performed in two 10-day intervals: one obtained immediately after weaning and the second 3 weeks later. The level of significance of a spectral peak was determined using an analysis of variance. The spontaneous locomotor activity of two male pups is shown in Fig. 1. Soon after weaning, locomotor activity appeared predominantly in the dark period. This asymmetric distribution increased markedly during the post-weaning development. A large peak in the circadian range was detected using the period analysis even in the first 10-day period. In addition, several statistically significant periodicities within the

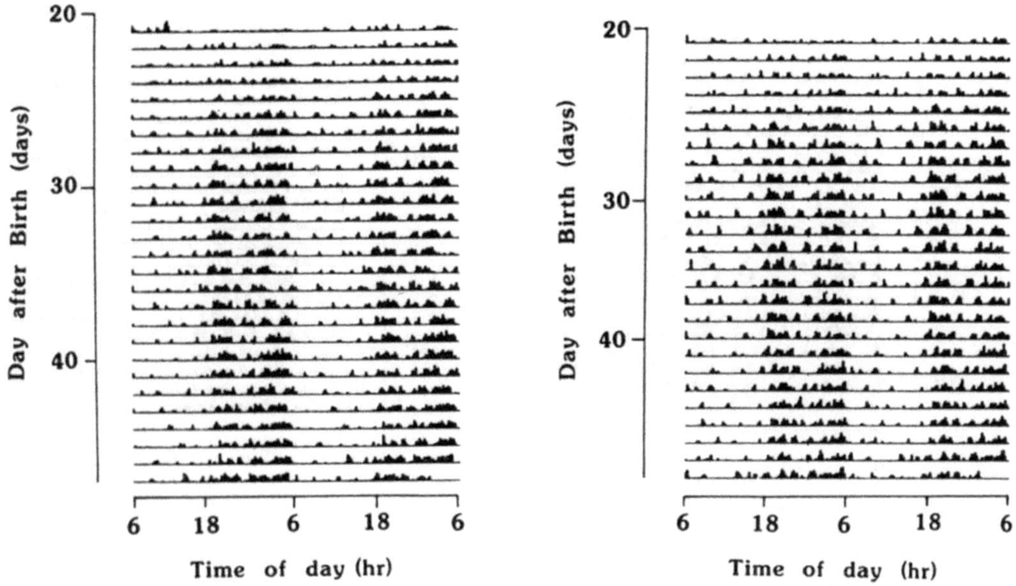

Fig. 1. Double-plotted spontaneous locomotor activity of intact male rats under 24-h LD cycle (light on, 0600-1800). Locomotor activity expressed in counts per 15 min

ultradian range were also observed. In the second 10-day period, the circadian peak was higher, although some ultradian periodicities, similar to those in the earlier period, were still detectable.

In the next stage, the development of the circadian organization in locomotor activity was studied in blinded rats. Rat pups were blinded by ocular enucleation under hypothermic anesthesia, immediately after birth. Locomotor activity was recorded from individual pups, weaned at 21 days of age, for at least 3 weeks. Period analysis was performed from activity records of the first and last 10-day periods. Actographs of two representative pups are shown in Fig. 2. The free-running circadian locomotor rhythms were manifest immediately after weaning. Generally, around the day of weaning, activity offset was more clearly expressed than activity onset. Activity bouts representing an offset component were usually larger than those representing an onset component. These bouts enlarged during the post-weaning development. The period analysis revealed in most rats a predominant circadian peak in the first 10-day record. The amplitude of the circadian peak increased markedly in the last 10-day record. Several peaks were observed in the ultradian range. Most of them were harmonic components of the dominant

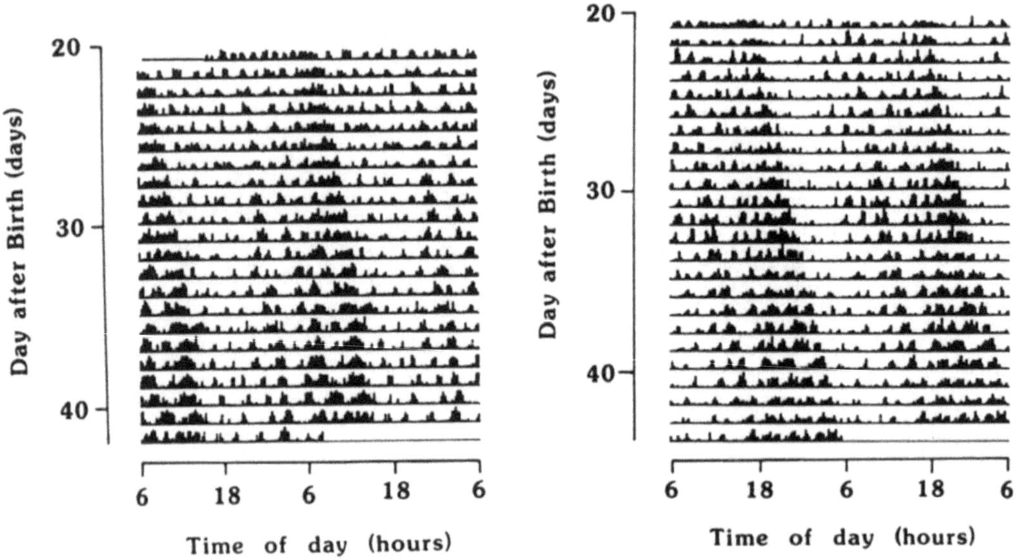

Fig. 2. Double-plotted spontaneous locomotor activity of neonatally blinded male rats, weaned at 21 days after birth

circadian period. Besides the harmonics, significant peaks of non-harmonic components were also evident. These results favor the amplitude modulation over the frequency modulation explanation for the development of the circadian locomotor rhythms in weaned rats.

DIRECT EFFECT OF LIGHT ON THE LOCOMOTOR ACTIVITY OF INFANTILE RATS

As has been shown in different animals, including rodents (ASCHOFF and MEYER-LOHMANN 1954; ASCHOFF 1955), a light-dark cycle is not necessary for the manifestation of circadian rhythmicity in spontaneous locomotor activity. A day-night difference in locomotor activity was reported in 15-day-old sighted rats (TEICHER and FLAUM 1979), whereas the circadian locomotor rhythm was delayed by about a week in neonatally blinded pups (HONMA S et al. 1985). Furthermore, the level of the locomotor bursts of sighted pups during the light period was lower in comparison with that of blinded pups (Figs. 1 and 2). Therefore, one possible explanation for the difference in the time of appearance of the circadian rhythm between sighted and blinded pups is a direct masking effect of light on locomotor activity (ASCHOFF et al. 1982).

To examine this possibility, well-nourished sighted and blinded pups were weaned on the 14th postnatal day, and reared under LD cycles. Their body weight, measured in the sixth postnatal week, was not significantly different from that of pups weaned at 21 days in our previous study. Figure 3 illustrates representative records of sighted and blinded pups, and the results of the period analysis. In sighted pups, the 24-h locomotor rhythm already appeared in the second postnatal week. This finding confirms a previous report by TEICHER and FLAUM (1979). In contrast, neonatally blinded pups did not show a circadian locomotor rhythm in the second postnatal week. Similar to the results mentioned previously, the circadian locomotor rhythm in these rats appeared only after the third week. The earlier appearance of the circadian locomotor rhythm in sighted pups than in blinded pups was confirmed several times without exception. In blinded pups, ultradian periods of circa 2.5 h were observed consistently before the emergence of the circadian rhythm (Fig. 3). This periodicity is of special interest since it was also observed in SCN-lesioned rats (HONMA and HONMA, this volume, Chap. 5). It appears that the same mechanism is regulating the 2.5-h ultradian rhythms observed during the early ontogenic stage, and those after the destruction or isolation of SCN. This periodicity was not detected once the circadian rhythmicity had fully emerged.

The present results may support the idea that a direct effect of light (masking) causes suppression of the locomotor activity during the light period, which results in a clearer appearance of the 24-h rhythm in sighted rats. As an alternative explanation one could hypothesize that the LD cycle amplifies the oscillations of an already present but weak circadian oscillator.

INTERACTIONS AMONG MULTI-OSCILLATORS UNDERLYING THE DEVELOPMENT OF THE CIRCADIAN LOCOMOTOR RHYTHM

It has long been noticed that the circadian locomotor rhythm of nocturnal rodents is comprised of onset and offset components (ASCHOFF 1954). These two components are sometimes split under constant illumination (PITTENDRIGH 1967; BOULOS and TERMAN 1979). Recently, we demonstrated that the onset and offset of the circadian locomotor rhythm in rats responded to a single light pulse differently. This suggests that the two components are driven by different mechanisms

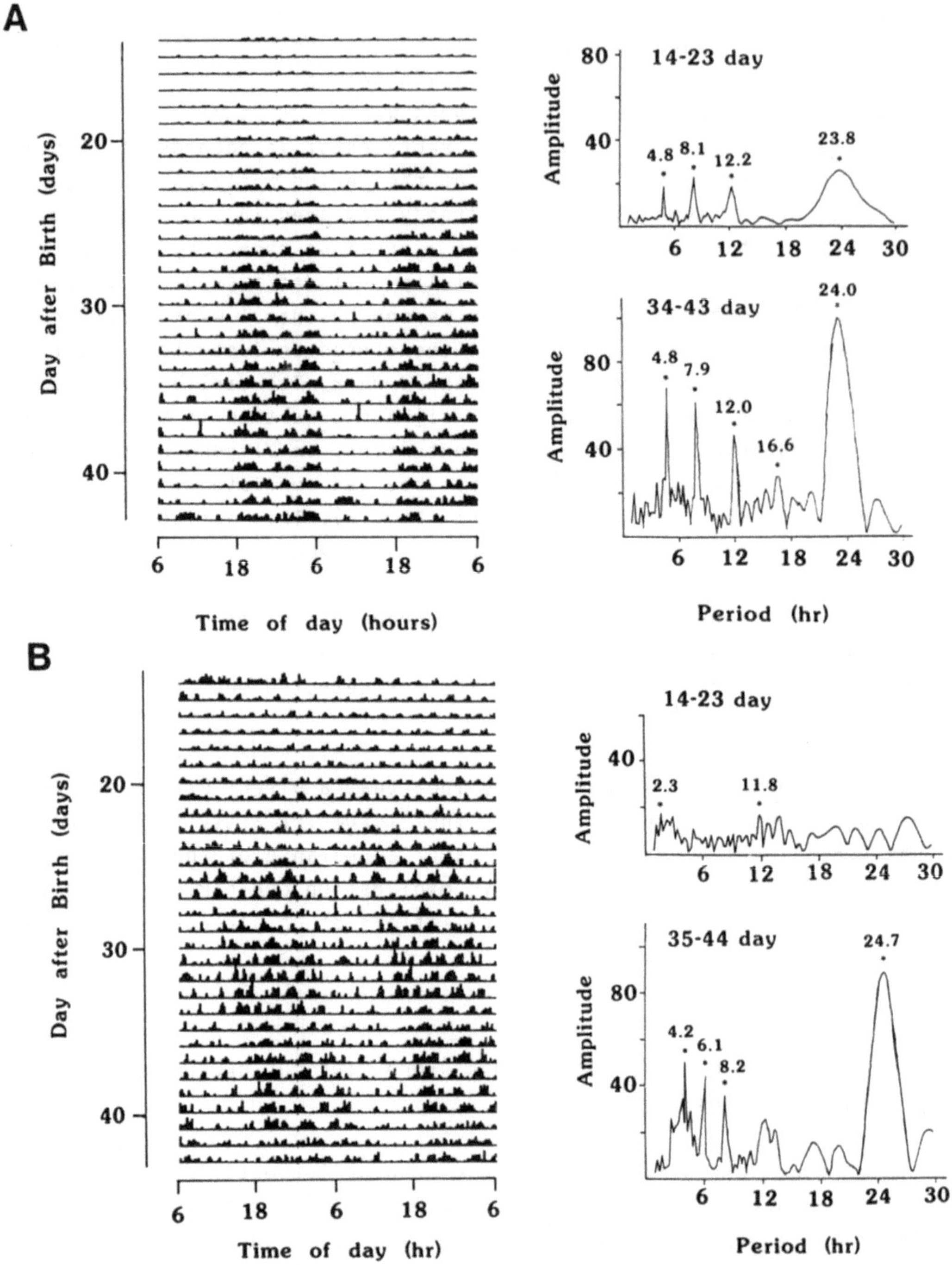

Fig. 3. Spontaneous locomotor activity (left panels) and least-square spectra (right panels) of an intact pup (A) and a neonatally blinded pup (B) weaned at 14 days after birth. Spectral analysis was done for the first 10 days (upper right panel) and the last 10 days (lower right panel) of each rat. Significant periods (h) are indicated (light on, 0600-1800)

(HONMA K et al. 1985). The findings are in good agreement with the two-oscillator model of PITTENDRIGH and DAAN (1976), which assumes that two mutually coupled oscillators regulate the onset and offset components. Therefore, it was of interest to us to examine the phase relationship between the activity onset and offset during the developmental course of the circadian rhythm. During this period, the formation of coupling between the two oscillators was hypothesized. Some examples in which the phase relationship between activity onset and offset progressively changed during the early post-weaning period can be seen in Fig. 4. For several days after weaning, the circadian locomotor rhythms of these blinded rats appeared as if there were two activity components per day. Interestingly enough, at this stage of development the circadian organization of plasma corticosterone level is also characterized by bimodality (HIROSHIGE et al. 1982). In locomotor activity (see left panel, Fig. 4), short locomotor bursts during the subjective day were clustered into longer bursts which were consolidated into the onset component of the circadian rhythm. Such a coalescence of locomotor bouts is not clear in the record of the rat presented in the middle panel. On the other hand, the rat shown in the right panel showed significant increase of activity time (α), measured from activity onset to offset.

The locomotor activity described above was recorded from blinded pups nursed previously by sighted mothers, under a light-dark cycle. Since it was shown that restriction of daily food and water supply to the nursing mothers had phase-setting effects on the pups' circadian rhythms (HONMA S et al. 1985), we investigated for these effects on the phase relationship between activity onset and offset in blinded pups. The development of circadian locomotor rhythms in blinded pups whose mothers were subjected to restricted feeding (RF) is shown in Fig. 5. Remarkable changes in α were observed during the early post-weaning period. The daily phase shift of activity onset was relatively small and constant, whereas the phase shift of activity offset was larger, particularly during the first 2 weeks after weaning. Thereafter, the rate of the phase shift of activity offset was abruptly slowed down and synchronized with that of activity onset. αs were calculated on the day of weaning and 3 weeks later for the control blinded pups, whose mothers were fed ad libitum, and for the experimental blinded rats whose mothers were subjected to RF (Table 1). As shown in Table 1, in both groups α increased significantly in the sixth week, but the increment was larger in the RF group (2.3 h vs 0.4 h).

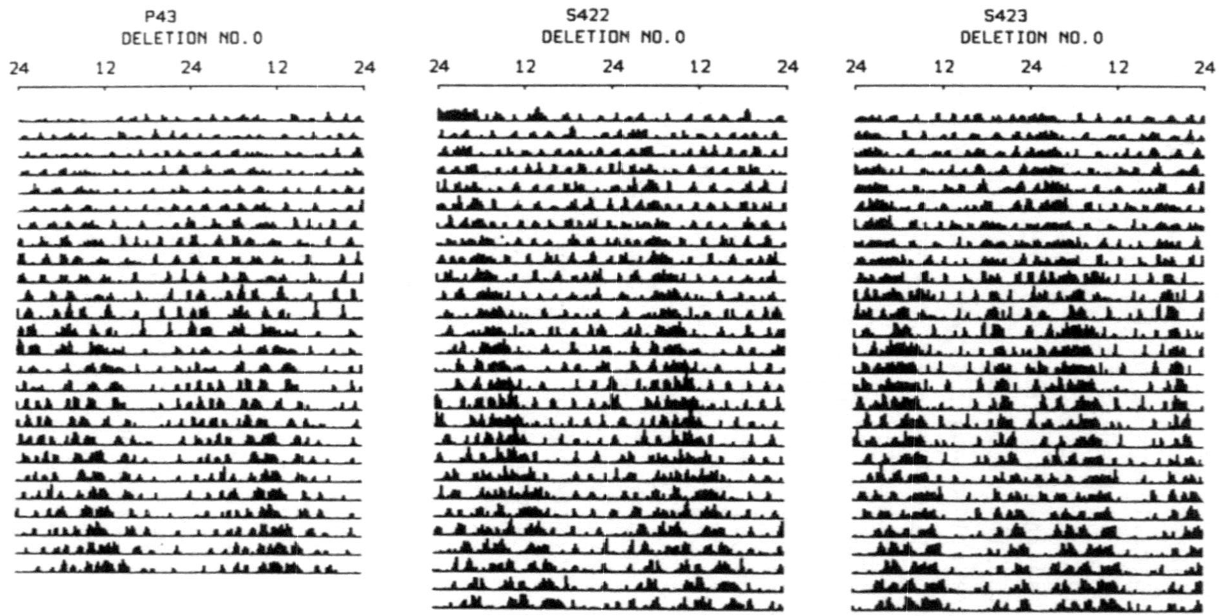

Fig. 4. Double-plotted spontaneous locomotor activity of neonatally blinded rats. The records begin on the day of weaning, 21 days after birth. Scales on top of the record indicate time of day (h)

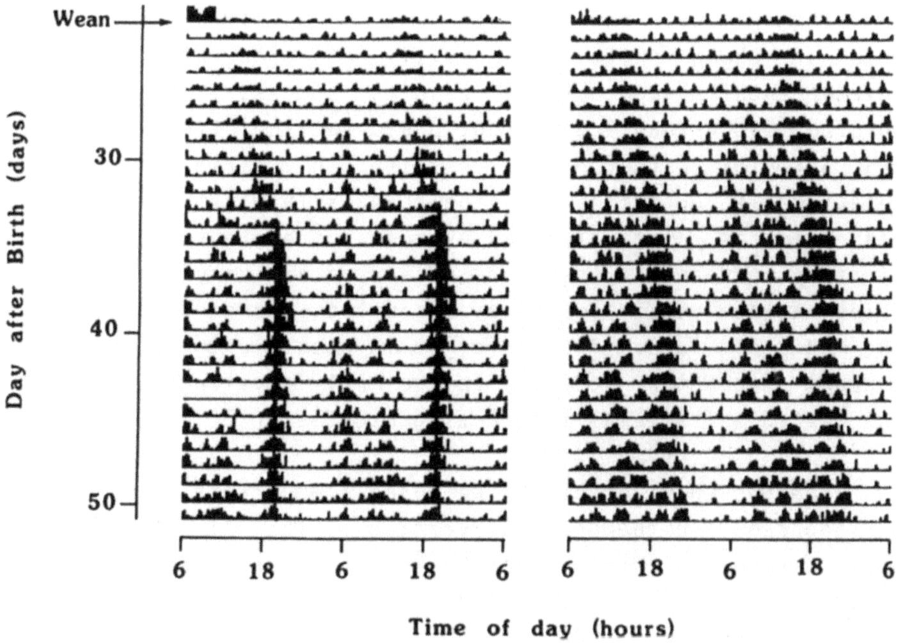

Fig. 5. Spontaneous locomotor activity of neonatally blinded rats, whose nursing mothers were subjected to restricted daily feeding (feeding, 0600-1000) until weaning (left panel), or until 2 weeks after delivery (right panel). Pups were weaned at 21 days after birth (light on, 0600-1800)

Table 1. Activity time (α), mean ± SE, measured at weaning (21 days after birth) and at the sixth postnatal week in neonatally blinded pups whose nursing mothers were fed at libitum (control), or were under restricted daily feeding (RF). Statistical significance was tested using one-way analysis of variance (*, $p < 0.01$; RF vs control)

Condition	Sample size	α (h)	α (h) at weaning	Difference (h) at 6th week
Control	18	13.6 ± 0.3	14.0 ± 0.3	0.4 ± 0.1
RF	14	12.4 ± 0.3*	14.7 ± 0.3	2.3 ± 0.4*

In both groups, the increase in α was mainly due to a change in the rate of phase shift of activity offset occurring during the early developmental stage. These results suggest that imposing RF on the nursing mother had a different phase-setting effect on activity onset and activity offset. This resulted in a marked change in the phase relationship between them. The changes in α after weaning can be interpreted as a transient adjustment to retain the stable phase relationship between the onset and offset of activity. The phase relationship between activity onset and offset changed, though slightly, in the control rats. Therefore, it is possible to speculate that spontaneous locomotor activity in the early developmental stage is under the control of several interacting oscillators.

To examine this problem, we used a modification of the chi-square periodogram. In the original version of the chi-square periodogram, a weak period is often hidden by a strong period and not detected even though it exists. The activity record and the process of the period analysis is shown in Fig. 6. An original locomotor record is analyzed using a chi-square periodogram. Then values which represent the activity component whose period has the highest Qp value are eliminated from the original data, and newly constructed time series data are reanalyzed using the periodogram. This procedure is repeated until the new period is discovered. In the case of Fig. 6, the period with the highest Qp value was 24.5 h. Then the values consisting of the onset and offset components were deleted from the original data. After this procedure, the period analysis revealed a new period, 25.0 h, which was hidden by the period of 24.5 h in the first analysis.

Figure 7 illustrates a case in which two oscillatory components appeared to be involved in the regulation of locomotor activity. The chi-square periodogram revealed two periodicities, 24.5 h and 24.0 h. The former period seems to be that of activity onset and offset, and the latter represents a very regular component occurring at around 2000. Though we cannot exclude the possibility that the 24-h component was an artifact, the fact that other pups studied simultaneously in the same room did not show this periodicity allows speculation on the existence of an independent oscillator. After filtering out the 24.5 h component, the chi-square periodogram showed three periodicities: 24.0, 24.5 and 25.0 h. These results suggest that interacting of multi-oscillators which participate in the process of the circadian organization are responsible for the short-lasting locomotor bouts.

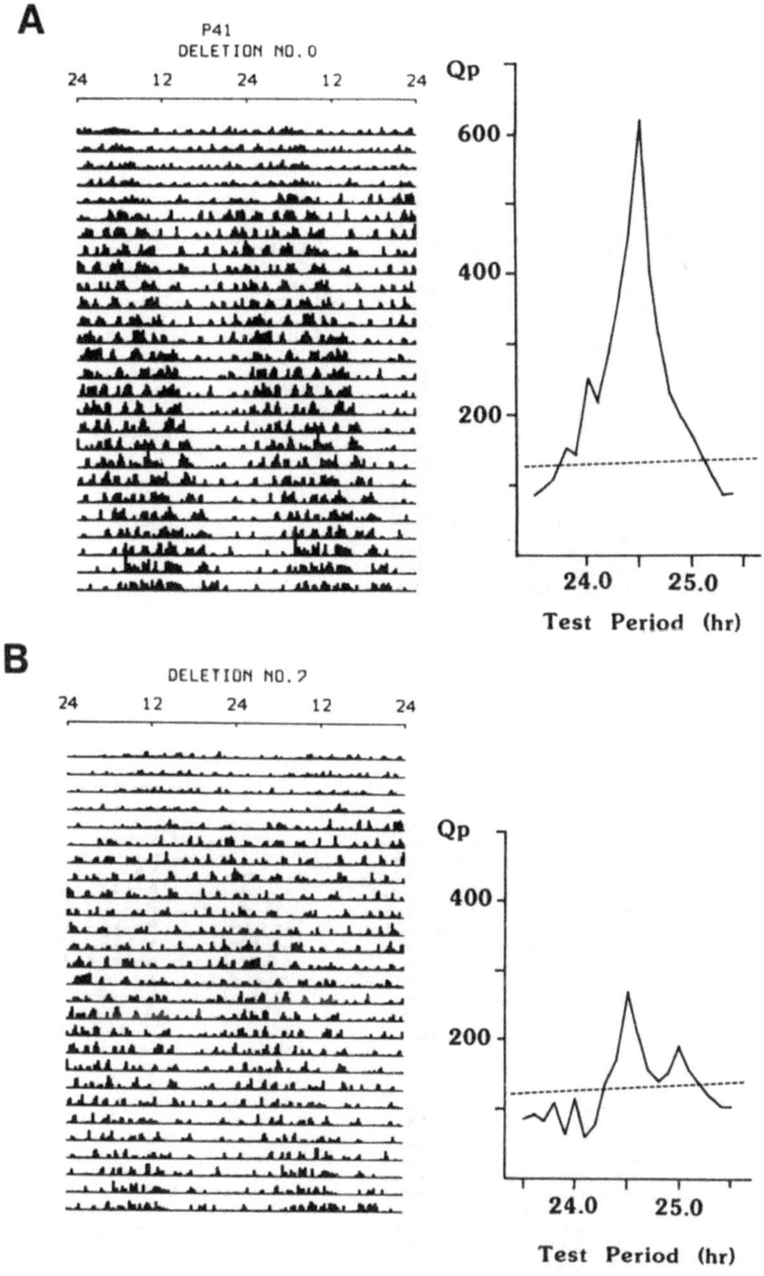

Fig. 6. Locomotor activity (left panels) and chi-square periodogram (right panels) of original data (A) and filtered data (B). Broken lines in the left panel indicate the level of significance ($p \leqq 0.01$)

A

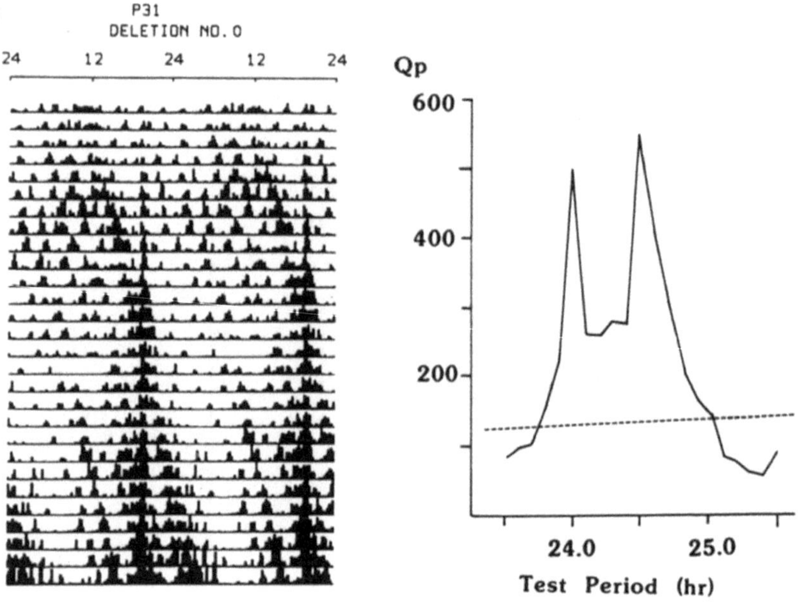
P31
DELETION NO. 0
24
12
24
12
24
Qp
600
400
200
24.0
25.0
Test Period (hr)

B

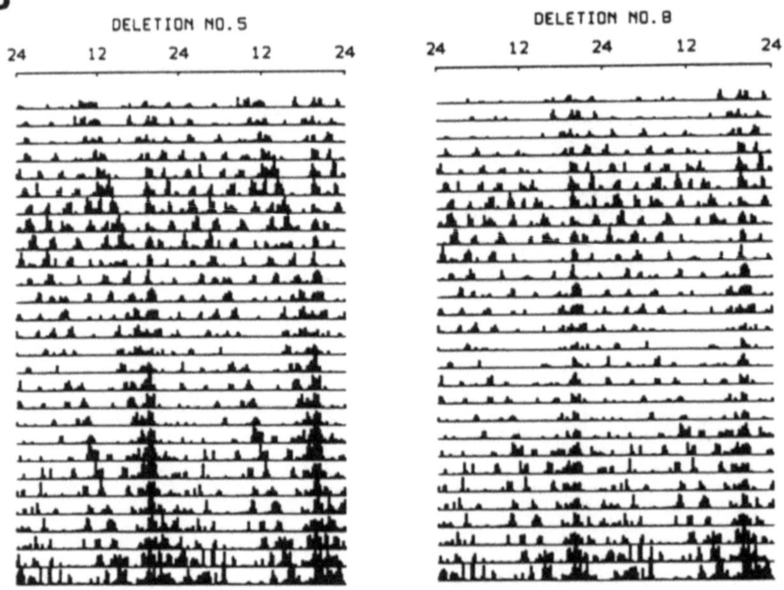
DELETION NO. 5
24
12
24
12
24
DELETION NO. 8
24
12
24
12
24

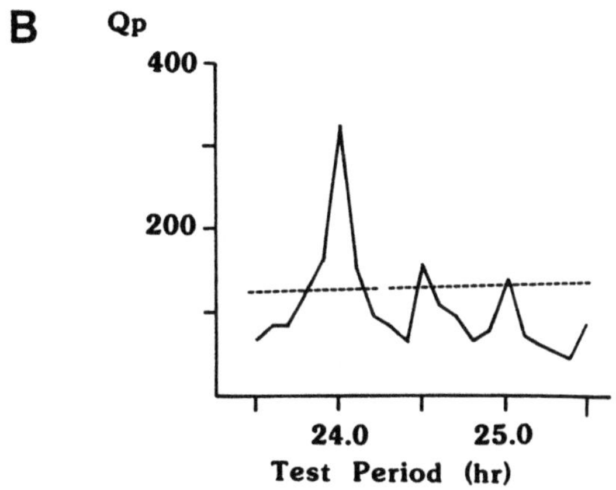

◁Fig. 7. Locomotor activity and modified chi-square periodogram in a neonatally blinded rat; the original data (A) and filtered data (B). The period analysis in B was done in the record of the deletion 8. See Fig. 6 and the text for details

CONCLUSION

The data presented above lead to the following conclusions. The day-night difference in locomotor activity appears earlier in sighted rats than in blinded rats, because short-lasting locomotor bursts of blinded pups were not attenuated during the subjective day, while the burst amplitude of sighted pups was reduced during the light period. Therefore, the earlier manifestation of the circadian rhythm in sighted pups is due, at least in part, to the suppressive effect of light on the ultradian locomotor bouts (negative masking). Besides amplitude modulation there was also frequency modulation. During the early ontogenetic stage, before the appearance of the circadian locomotor rhythm, a 2.5-h ultradian rhythm predominated, while after the appearance of the circadian rhythm, ultradian rhythms of longer periodicities emerged. Therefore, the circadian rhythm develops not only by amplitude modification of locomotor bouts, but also by a frequency modification of the bouts. In addition, the phase relation between activity onset and offset changes during the early developmental stage. These findings can be interpreted in a context of a two-oscillator model, whose coupling is unstable in the early postnatal period, but becomes more stable later on. It appears that the restricted daily feeding imposed on the nursing mother modifies the internal coupling of pups' circadian system. A statistical analysis revealed more than two oscillatory components in the early postnatal period. It is surmised that an interaction of multi-oscillators is involved in the circadian organization of locomotor bouts.

In summmary, the circadian locomotor rhythm of rats under an LD cycle may develop in two different modes. The first is a modulation of the amplitude of each ultradian component, which is carried out mainly by the circadian oscillator; but the suppression of amplitude by direct effect of light is superimposed in sighted rats under LD cycle. The second is the clustering of each activity component which is caused by changes in the internal coupling of multi-oscillators underlying the circadian pacemaking system of rats.

REFERENCES

ASCHOFF J (1954) Zeitgeber der tierischen Tagesperiodik. Naturwissenschaften 41:49-56

ASCHOFF J (1955) Tagesperiodik bei Mäusestämmen unter konstanten Umgebungsbedingungen. Pflügers Arch 262:51-59

ASCHOFF J, MEYER-LOHMANN J (1954) Angeborene 24-Stunden-Periodik beim Kücken. Pflügers Arch 260:170-176

ASCHOFF J, DAAN S, HONMA K (1982) Zeitgebers, entrainment, and masking: some unsettled questions. In: ASCHOFF J, DAAN S, GROOS GA (eds) Vertebrate circadian systems. Springer, Berlin Heidelberg New York, p 3

BOULOS Z, TERMAN M (1979) Splitting of circadian rhythms in the rats. J Comp Physiol 134:75-83

DAVIS FC (1981) Ontogeny of circadian rhythms. In: ASCHOFF J (ed) Handbook of behavioral biology, vol 4. Biological rhythms. Plenum, New York, p 257

HIROSHIGE T, HONMA K, WATANABE K (1982) Ontogeny of the circadian rhythm of plasma corticosterone in blind infantile rats. J Physiol (Lond) 325:493-506

HONMA K, HIROSHIGE T (1978) Internal synchronization among several circadian rhythms in rats under constant light. Am J Physiol 235: R243-R249

HONMA K, HONMA S, HIROSHIGE T (1985) Phase response curves of circadian locomotor rhythm in rats: analysis of the pacemaker structure. In: HIROSHIGE T, HONMA K (eds) Circadian rhythms (in print)

HONMA S, HIROSHIGE T (1977) Pubertal manifestation of sex difference in circadian rhythm of corticotrophine-releasing activity in the rat hypothalamus. Acta Endocrinol 86:225-234

HONMA S, HONMA K, SHIRAKAWA T, HIROSHIGE T (1984a) Effects of elimination of maternal circadian rhythms during pregnancy on the postnatal development of circadian corticosterone rhythm in blinded infantile rats. Endocrinology 114:44-50

HONMA S, HONMA K, SHIRAKAWA T, HIROSHIGE T (1984b) Maternal phase setting of fetal circadian oscillation underlying the plasma corticosterone rhythm in rats. Endocrinology 114:1791-1796

HONMA S, HONMA K, HIROSHIGE T (1985) Ontogeny of circadian corticosterone and locomotor rhythms: effects of maternal rhythms and restricted daily feeding. In: HIROSHIGE T, HONMA K (eds) Circadian rhythms (in print)

MOORE RY (1979) The anatomy of central neural mechanisms regulating endocrine rhythms. In: KRIEGER DT (ed) Endocrine rhythms. Raven, New York, p 63

PITTENDRIGH CS (1967) Circadian rhythms, space research and manned space flight. Life Sci Space Res 5:122-134

PITTENDRIGH CS (1974) Circadian oscillations in cells and the circadian organization of multicellular systems. In: SCHMITT FO, WORDEN FG (eds) The neuroscience: third study program. MIT Press, Cambridge, p 437

PITTENDRIGH CS, DAAN S (1976) A functional analysis of circadian pacemakers in nocturnal rodents. V. Pacemaker structure: a clock for all seasons. J Comp Physiol 106:333-355

REPPART SM, SCHWARTZ WZ (1983) Maternal coordination of the fetal biological clock in utero. Science 220:969-971

RICHTER CP (1927) Animal behavior and internal drives. Q Rev Biol 2:307-343

SOKOLOVE PG, BUSHELL WN (1978) The Chi square periodogram: its utility for analysis of circadian rhythms. J Theor Biol 72:131-160

TEICHER MH, FLAUM LE (1979) Ontogeny of ultradian and nocturnal activity rhythms in the isolated albino rat. Dev Psychobiol 12:441-454

Urinary Ultradian Rhythms in Dogs

C. R. Gordon and P. Lavie

Sleep Laboratory, Unit of Behavioral Biology, Faculty of Medicine, Gutwirth Building, Technion – Israel Institute of Technology, Haifa 32000, Israel

INTRODUCTION

In searching for changes in physiologic parameters associated with the rapid eye movement (REM) sleep, MANDELL et al. (1966) described ultradian fluctuations in urinary excretions of sleeping catheterized patients. Urine flow decreased and osmolality increased every 90 min, coinciding with REM periods. Rhythmic secretion of antidiuretic hormone (ADH) was suggested as the regulating mechanism, but the hormone itself was not measured. Similar ultradian rhythms in urine flow, osmolality, and electrolyte concentrations were later demonstrated in waking subjects who consumed a constant amount of fluid and voluntarily urinated every 10 min (LAVIE and KRIPKE 1977). Subsequent studies in our laboratory showed that the reported ultradian rhythms were dependent on hydration level and posture but were not correlated with plasma ADH levels (LUBOSHITZKY et al. 1978; LAVIE et al. 1980).

In addition, ultradian fluctuations have been described in various physiologic parameters in animals. Gastric motility in rabbits, sheep, and dogs displays rhythmic changes with periods approximating 90 min (GRIVEL and RUCKEBUSCH 1972). Ultradian rhythms of congestion and decongestion of the nasal mucosa were shown in domestic pigs, rabbits, and cats and were attributed to periodic fluctuations in autonomic activity (BAMFORD and ECCLES 1982). Also reported were 90-min variations for plasma noradrenaline (LEVIN et al. 1978) and cortisol (HOLADAY et al. 1977) in monkeys. Surprisingly, in spite of a priori assumptions, many of these rhythmic fluctuations in animals are not associated with the REM-NREM cycle.

Experimental Brain Research, Suppl. 12

In recent years, we have investigated possible ultradian fluctuations in urinary variables in the laboratory dog. The present report summarizes our findings of urinary ultradian rhythms and describes the relationships between urine parameters, sleep-wake stages, and hormones regulating fluid balance. The role of sympathetic innervation in the regulation of the urinary rhythms was investigated using pharmacologic and surgical procedures.

MATERIAL AND METHODS

Six adult female mongrel dogs (18-30 kg) were trained to spend prolonged periods under laboratory conditions (GORDON and LAVIE 1982, 1984a,b). Urine samples for determination of urinary flow, osmolality, and sodium and potassium concentrations were collected through an indwelling bladder catheter every 10 min. Sleep-wake stages were monitored continuously. Blood samples for hormone determinations were obtained via a catheter inserted into a leg vein. All experimental sessions began at 0800, lasted for 24 h, and were conducted under controlled light-dark conditions (L, 0700-1900; D, 1900-0700). In each experiment dogs were tested under two hydratory conditions: (a) hydration (H), where the dogs drank voluntarily a 1:4 milk/water solution (20 ml/kg body weight) at 0800 and at 1900, and (b) progressive dehydration (PD), where water was allowed ad libitum up to the time the experiment began and then was withheld during the entire experimental session. The study included: (a) a control experiment providing baseline parameters including plasma level determinations of ADH, renin activity (PRA), cortisol, and aldosterone; (b) an adrenergic β receptors blockade experiment using systemic administration of propranolol hydrochloride; (c) an adrenergic α receptors blockade experiment using systemic administration of phentolamine; (d) a kidney denervation experiment after surgical renal denervation. All experiments were separated by at least a 2-week recovery period.

RESULTS

Control Experiment

Urine flow and osmolality displayed clear ultradian fluctuations in all dogs, during at least one of the experimental sessions. Figure 1 shows urine volume, osmolality, electrolyte concentrations (raw data), and sleep-wake patterns of one dog during a 24-h period in the PD condition. Figures 2a and 2b show the same variables for two different dogs during a day and a night period in the H condition.

Appreciable fluctuations every 150-200 min in the flow of urine, out of phase with fluctuations in osmolality, are clearly visible during the day in the PD condition. The magnitude of the fluctuations in both variables progressively diminished and almost disappeared during the night.

The amplitude of the diurnal ultradian variations is noteworthy in both hydratory conditions. In several cases osmolality varies from 300-400 mosm/kg to 1000 mOsm/kg (Fig. 2a). Similar variations, albeit less consistent, were found in electrolyte concentrations.

Visual examination of the data also revealed that ultradian variations in urine flow and osmolality were related to the dogs' sleep-wake cycles. Urine volume increased (and osmolality decreased) during waking episodes, and conversely, urine volume decreased (and osmolality increased) during sleep episodes. This relationship was consistent regardless of the dogs' levels of motor activity during the wake periods.

Time-series analysis techniques were used to determine the rhythmic structure of the data (JENKINS and WATTS 1968). Spectral analysis confirmed the visual impression. Table 1 summarizes the significant spectral peaks for each urine variable during the different light and hydratory conditions.

The average spectra for urine flow during the day showed ultradian frequencies which peaked at 7.2 c/d (circa 200 min) in the H condition, and at 14.4 c/d (circa 100 min) in the PD condition. These peaks significantly ($p \leq 0.05$) exceeded both the spectral variance of the

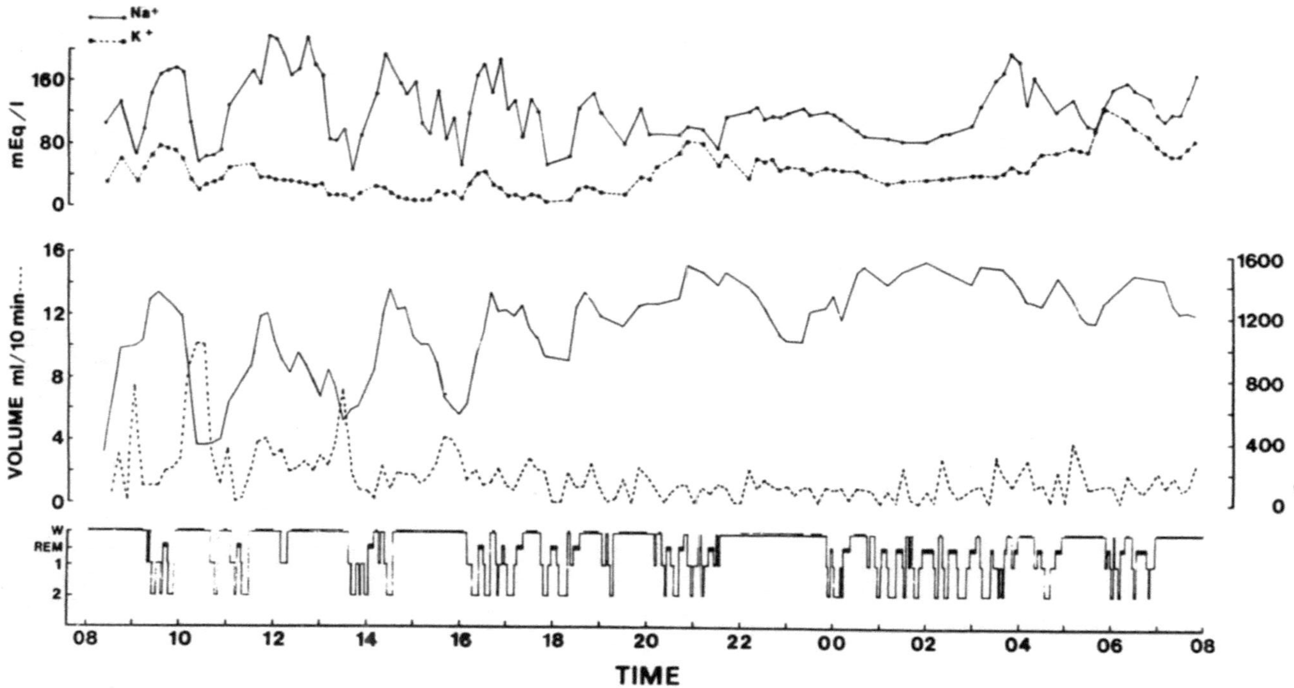

Fig. 1. Raw data of urine volume, osmolality, sodium and potassium concentrations, and sleep-wake stages for one dog in progressive dehydration condition during 24 h. The lines above the wake periods indicate locomotor activity (from GORDON and LAVIE 1982)

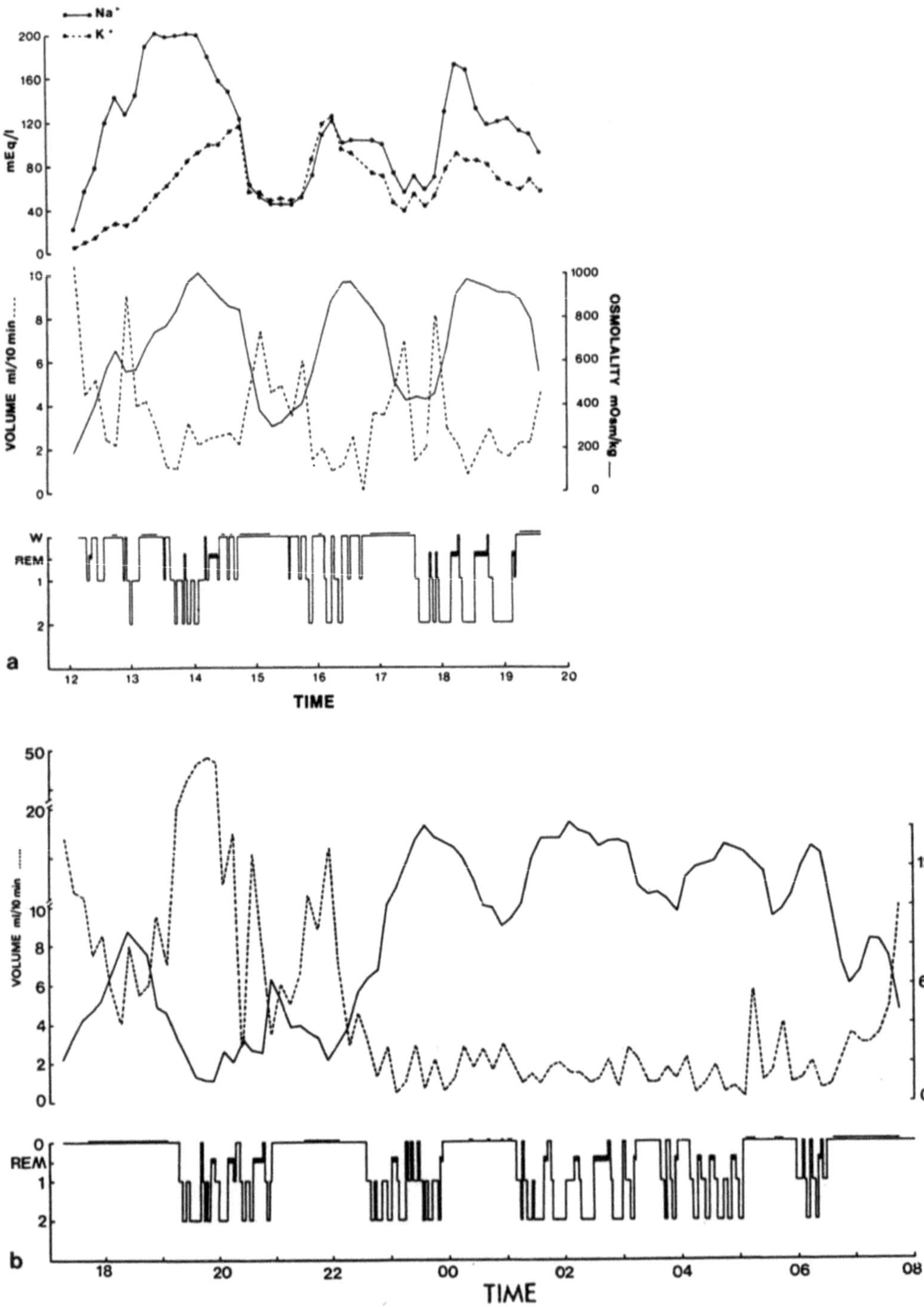

Fig. 2. Raw data of urine variables and sleep-wake stages for two dogs in hydration condition during the day (a) and night (b) periods. The lines above the wake periods indicate locomotor activity. (from GORDON and LAVIE 1982)

Table 1. Summary of spectral peaks at ultradian frequencies in each experiment, for each urine variable, and for the day and night periods

		Volume	Osmolality	Na^+	K^+
CONTROL					
Day	H	7.2 c/d**	7.2 c/d**	-	7.2-10.8**[a]
	PD	14.4 c/d**	7.2 c/d**	3.6-7.2 c/d[ns]	7.2-10.8[ns]
Night	H	-	-	-	-
	PD	-	-	3.6-7.2 c/d[ns]	-
PROPRANOLOL					
Day	H	-	4.8-9.6 c/d**[a]	4.8 c/d*	4.8 c/d*
Night	H	-		3.6-7.2 c/d**[a]	7.2 c/d*
PHENTOLAMINE					
Day	H	-	9.6 c/d[ns]	4.8 c/d*	9.6 c/d*
Night	H	-	7.2 c/d**	7.2 c/d*	7.2 c/d**
DENERVATION					
Day	H	-	14.4-19.2 c/d[ns]	4.8-9.6 c/d*	9.6 c/d*
Night	PD	-	9.6 c/d[ns]	-	-
Night	H	-	-	-	-
	PD	-	-	-	-

*$p < 0.1$ (bordered statistical significance); **$p < 0.05$

[ns]nonsignificant

[a]spectral peak was shared by two adjacent frequencies

adjacent frequencies, and the mean spectral variance of the remaining spectral frequencies, as contrasted by two-tailed t tests.

The average spectra for urine osmolality during the day peaked significantly at 7.2 c/d in both hydratory conditions. The average spectra revealed rhythms with similar characteristics for urinary potassium concentration during the day, only in the H condition. Urinary sodium did not show consistent rhythmic patterns.

Spectral analysis did not show significant peaks at ultradian frequencies for any of the urine variables during the night period.

Hormone Patterns

Minimal variations were found in levels of ADH, PRA, cortisol, and aldosterone, with no evidence of ultradian or episodic fluctuations. Neither visual examination nor time-series analysis techniques revealed evidence of hormonal ultradian patterns. In spite of the marked ultradian variations in urine flow and osmolality, and the variations in locomotor activity associated with the sleep-wake cycle, ADH levels were constant and near the lowest sensitivity limit of our assay (0.25 pg/ml).

To investigate the possibility of an episodic pattern in hormonal secretion, we arbitrarily defined a plasma hormone level 1.5 SD above the mean level as a secretory peak. Since none of the hormones exceeded its mean level by 1.5 SD, the possibility of episodic fluctuations in these data was refuted.

Sleep-Wake Parameters

Dogs spent 45.6% of the 24 h asleep. The sleep period occupied 29.6% of the light period and 61.4% of the dark period. The major portion of sleep occurred between 2200 and 0400.

The mean sleep-wake cycle was 103.3 min (SD = 27.8), or 13.9 sleep-wake c/d. Sleep episodes were significantly longer during the dark period (66.8 min vs 29.4 min), and, conversely, waking episodes were longer during the light period (71.6 min vs 42.1 min). Consequently, there was no difference in the length of the average sleep-wake cycle between the light and dark period (101 min and 108 min, respectively). The average length of the REM-NREM sleep cycle was 20.6 min with REM sleep episodes averaging 6.6 min. The average number of REM episodes in each sleep-wake cycle was 2.1; there were 1.2 cycles in the light period and 3.0 in the dark period.

During waking episodes in the light period, dogs were active and walked around the laboratory room. During waking episodes in the dark period, animals were quiet and generally lying down with eyes open and occasional movements.

Table 2. Summary of average correlations between variables in each experimental condition

		Osm/Vol	Vol/Wake	Osm/Wake	Osm/Stage 2	K^+/Na^+
CONTROL						
Day	H	-0.40****	0.20***	-0.11****	0.11**	0.34**
	PD	-0.29***	0.23**	-0.09	0.05	0.09
Night	H	-0.44**	0.27*	-0.38****	0.30****	0.33****
	PD	-0.12	0.11	-0.29***	0.22***	-0.18
PROPRANOLOL						
Day	H	-0.33	0.47*	-0.30	0.21	-0.04
Night	H	-0.59****	0.16	-0.05	0.03	0.21
PHENTOLAMINE						
Day	H	-0.16	-0.02	0.05	0.03	0.01
Night	H	-0.39*	0.07	-0.20	0.20***	0.58
DENERVATION						
Day	H	-0.13	0.15	0.09	0.02	0.18
	PD	-0.25*	0.16*	-0.06	0.08	-0.20
Night	H	-0.47***	0.35***	0.22*	0.13	0.24
	PD	-0.24	0.20***	0.22	-0.21**	-0.27

*$p < 0.05$; **$p < 0.02$; ***$p < 0.01$; ****$p < 0.001$

Correlations

Table 2 summarizes the average Pearson correlation coefficients between urinary variables, and between these variables and behavioral states.

Correlations were calculated separately for the day and night periods, and for the H and PD conditions, after the subtraction of the linear trends from each variable. The linear trends were subtracted because we were interested in the covariations between the variables within the fast ultradian range, and not in the common circadian trends.

The correlational analyses confirmed the visual impression. The average z-transformed correlation between volume and osmolality was significantly negative, except during the night period in the PD condition.

Volume was significantly positively correlated with the minutes of wakefulness, except during the night period in the PD condition, and significantly negatively correlated with minutes of stage 2. Conversely, osmolality was significantly negatively correlated with minutes of wakefulness, except during the day in the PD condition, and positively correlated with minutes of stage 2 sleep.

Average correlations between electrolyte concentrations and sleep-wake stages were not consistent. In addition, correlations between hormones and urine variables as well as between hormones and behavioral states were not statistically different from zero.

Adrenoceptors Blockade and Kidney Denervation Experiments

Figure 3 shows the temporal changes in urine flow and osmolality for the different experiments under H condition, during the day period. Propranolol, phentolamine, and surgical denervation of the kidneys had similar effects on the ultradian patterns in urine variables. In both pharmacologic and surgical treatments, urine flow fluctuated episodically with no apparent rhythmicity or regularity. In contrast, urine osmolality fluctuated at ultradian frequencies associated with the sleep-wake cycle.

The amplitude of the ultradian fluctuations in osmolality was lower in the drug and denervation experiments than the corresponding fluctuations during the control experiment. Also, oscillations were more variable, covering several ultradian frequencies (see below).

Time-series analyses confirmed the visual examination (see Table 1). While the average spectrum for urine flow was flat, the average spectrum for osmolality peaked at frequencies within the ultradian range. In general, the mean spectral peaks for osmolality were broad, covering several spectral frequencies, and did not reach statistical significance. The reason for the nonsignificance becomes apparent once the individual spectra are inspected. In all experiments, almost every dog showed a clear spectral peak within the wide frequency range of from 7.2 c/d to 19.2 c/d. However, because of this variability, averaging across dogs resulted in nonsignificant peaks.

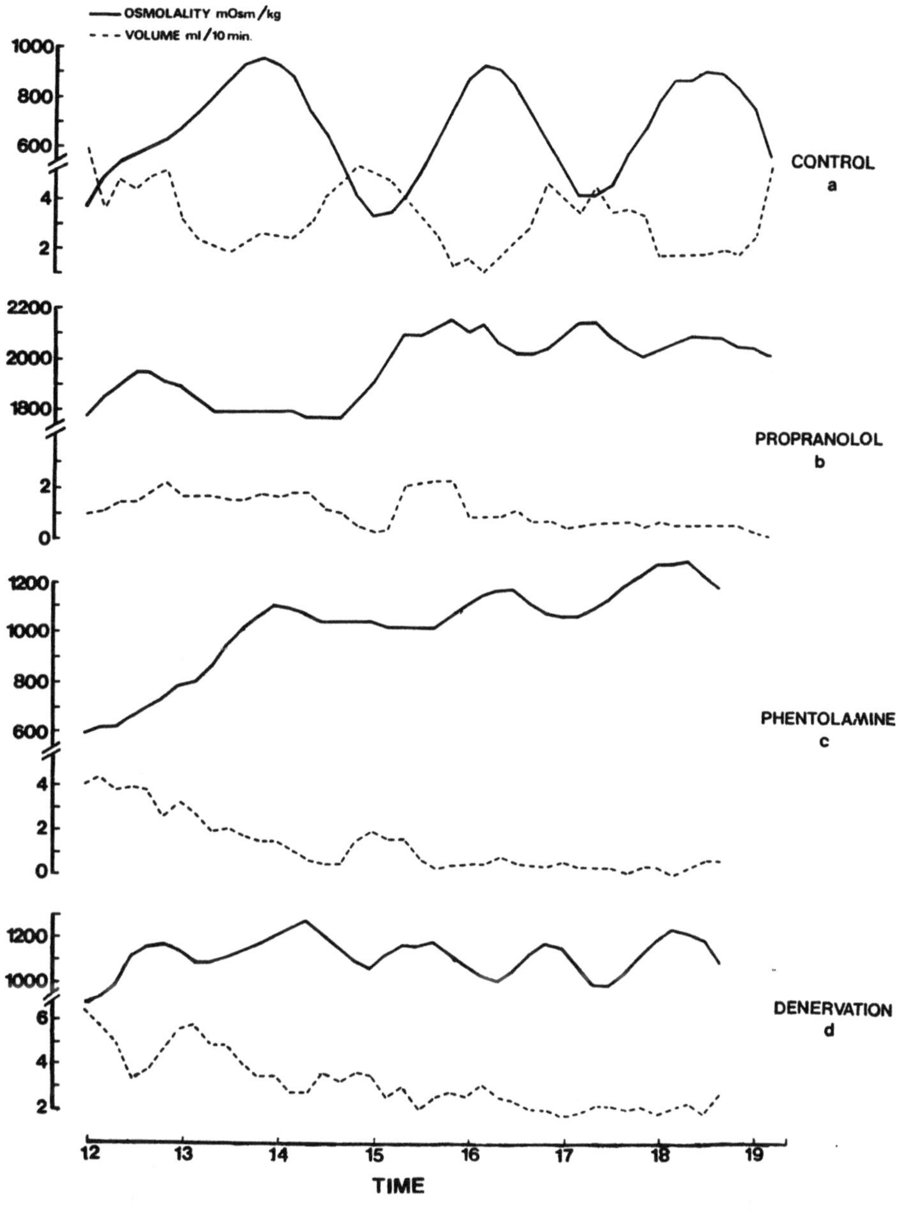

Fig. 3. Smoothed data of urine flow and osmolality, in control (a), propranolol (b), phentolamine (c) and denervation (d); experiments under hydration condition during the light period

Peaks at ultradian frequencies, bordering on statistical significance, were found in electrolyte concentrations after both pharmacologic ahd surgical treatments (see Table 1).

In light of the fact that ultradian fluctuations in urine flow and osmolality were related to the dog's sleep-wake cycle, it should be noted that the length of the sleep-wake cycle was unaffected by either experimental treatment.

Correlations

In general, average correlations in the drug and denervation experiments maintained trends similar to those of the control experiment (see Table 2). Volume was consistently negatively correlated with osmolality, and positively correlated with the minutes of wakefulness. Conversely, osmolality was consistently negatively correlated with wakefulness. The average correlations were statistically significant in a few cases. Potassium and sodium were not significantly correlated with each other in any experiment. Also, none of the average correlations between electrolyte concentration and sleep-wake stages were consistent in any experiment.

DISCUSSION

Rhythmic Characteristics

The present study describes impressive ultradian rhythms of about 200 min/c in urine flow and osmolality, which were synchronized with the diurnal (lights on) sleep-wake cycle of the dogs. Urine flow increased and osmolality decreased during diurnal waking episodes, and conversely, flow decreased and osmolality increased during the diurnal sleep episodes. The amplitude of the rhythms was strikingly higher than the corresponding amplitude in humans (MANDELL et al. 1966; LAVIE and KRIPKE 1977; LUBOSHITZKY et al. 1978). Also, the period of these rhythms was about twice as long as in humans. In addition, the period was not in the frequency range of the dog's REM-NREM cycle, but about twice as long as the sleep-wake cycle.

Ultradian rhythmicities in urine parameters were affected by both the photo period and by the level of hydration. The light-dark differences in the ultradian rhythmicity may be explained by at least two factors: (a) circadian modulation, i.e., fast ultradian rhythmicity which is superimposed upon the circadian rhythm, appears only during the light period; (b) the existence of day-night differences in other physiologic subsystems which indirectly influence the ultradian rhythmicity in urine excretions. The significant relationship between the dog's sleep-wake stages and the fluctuations in flow and osmolality suggests that the sleep-wake cycle itself, or physiologic variables related to this cycle, may affect the rhythms in urine excretion. These may also account for the day-night differences. Despite the fact that the length of the sleep-wake cycle during the day and night was identical, the content of the cycles was different. Waking episodes during the day were longer and more active than the corresponding nocturnal waking episodes. Posture and postural changes have previously been shown to influence the ultradian rhythmicity in urine excretion in humans. While well-hydrated subjects in an upright posture showed 90-min rhythms in volume and osmolality, which were mutually synchronized, subjects in a supine posture showed 90-min rhythms only in osmolality (LUBOSHITZKY et al. 1978). It is possible, then, that some level of motor activity or changes in body posture are necessary for the appearance of ultradian rhythmicity. This can also account for the fact that the period length of the urine cycle was twice as long as the sleep-wake cycle.

Urinary ultradian rhythms were also affected by the level of hydration. The amplitude of the rhythms in urine flow and osmolality depends on the level of hydration (Fig. 1). It is interesting to note that the changes in the level of hydration influence differently the cycle duration in flow and osmolality. In the control experiment osmolality showed a significant spectral peak at 7.2 c/d in both hydratory conditions, while the ultradian rhythms in urine flow in the PD condition were faster than in the H condition (14.4 c/d vs 7.2 c/d). A certain level of hydration is possibly a prerequisite for the appearance of the ultradian rhythmicity in renal excretion.

Possible Regulatory Mechanisms

Although we do not yet know the nature of the regulatory mechanisms, the inverse phases of the rhythms in volume and osmolality could suggest that they are regulated by rhythmic fluctuations in ADH levels. However, in spite of the marked ultradian variations in urine flow, osmolality, and locomotor activity, ADH levels were constant and near the lowest sensitivity limit of the assay. Similarly, in humans there were no consistent phase relationships between episodic secretion of ADH and ultradian rhythms in urine flow and osmolality (LUBOSHITZKY et al. 1978; LAVIE et al. 1980). Furthermore, the lack of consistent ultradian rhythms in the concentration of sodium and potassium and the lack of a consistent relationship between these two electrolytes make it unlikely that water reabsorption mechanisms alone are involved. The fact that neither levels of PRA nor aldosterone nor cortisol were correlated with urine variables also suggests that plasma hormone levels are not associated with ultradian urinary fluctuations. However, the possibility cannot be excluded that intrarenal hormonal activity changes, not detected in peripheral blood, are involved in the regulation of urinary ultradian rhythms.

Both α- and β-adrenoreceptor blockage and surgical denervation of the kidneys suppress the ultradian rhythmicity in the flow of urine, but do not suppress (or suppress to a lesser degree) the ultradian rhythm in osmolality. The fact that α and β blockage had the same effect as kidney sympathectomy suggests that an adequate balance between the two types of sympathetic fibers is necessary for the emergence of these rhythms.

The loss of ultradian rhythmicity in the flow of urine can be explained in at least two ways: (a) kidney sympathectomy eliminates rhythmic input generated by an extrarenal ultradian oscillator, or by oscillators residing within the autonomic nervous system, whose activities are independent of the ultradian oscillators that regulate solute excretion and the sleep-wake cycle; (b) kidney sympathectomy suppresses the coupling mechanism between the "slave" water excretion oscillator and an ultradian oscillator in other subsystems such as solute excretion mechanisms or the sleep-wake cycle oscillator.

Our findings support the assertion that in a similar way to the urinary circadian system, the ultradian rhythms in urine excretions are organized as a "multi-oscillatory" system, comprised of separate and potentially independent oscillators for flow, osmolality, and sodium and potassium concentrations, rather than a single "master oscillator" with several "slaves" (WEVER 1979; MOORE-EDE et al. 1976). The persistence of ultradian rhythmicity in osmolality, dissociated from the random and episodic pattern of urine flow, in the drugs and denervation experiments, indicates a separate and independent control mechanism for these rhythms. The finding that sympathectomy modified the frequency of the ultradian osmolality rhythm from a single dominant frequency of 7.2 c/d toward several frequencies with a rather wide range (7.2 c/d - 19.2 c/d) raises the possibility that the ultradian rhythms in flow entrained the osmolality oscillator whose frequency is less stable. After renal denervation the unstable frequency of the ultradian rhythms in osmolality resembles the wide range of frequencies seen in "free-running" sleep-wake rhythms of humans isolated from time cues (WEVER 1979). A clear separation between the rhythms in osmolality and changes in the flow of urine were described also in humans (LUBOSHITZKY et al. 1978). In well-hydrated supine waking subjects there were rhythms of about 80-130 min in urinary osmolality and in sodium and potassium concentrations, while urine flow was unrelated, fluctuating randomly and much more rapidly.

The relationship between the ultradian changes in urine excretion and the sleep-wake stages in the normal dog points to the possibility that sympathetic oscillations may be concomitants of central nervous changes associated with the sleep-wake transitions.

We do not yet know how the sympathetic nervous system regulates the ultradian rhythms in the flow of urine. The automatic nervous system has both direct and indirect effects on renal water and salt regulation (SCHRIER et al. 1977; DIBONA 1978). The numerous effects of renal nerve activity on vascular, hormonal, and tubular responses within the kidney could play a major role in the regulation and appearance of urinal ultradian rhythms.

ACKNOWLEDGEMENTS. We gratefully acknowledge the technical help of Mr. S. Vinocur and Ms S. Medan. We also thank Prof. S. Glick and Dr. L. Kahana for the hormone assays, and Dr. S. Keidar for the surgical preparation of the dogs.

REFERENCES

BAMFORD OS, ECCLES R (1982) The central reciprocal control of nasal vasomotor oscillations. Pflügers Arch 394:139-143

DIBONA GF (1978) Neural control of renal function. Fed Proc 37: 1191-1221

GORDON C, LAVIE P (1982) Ultradian rhythms in renal excretion in dogs. Life Sci 31:2727-2734

GORDON C, LAVIE P (1984a) Effect of adrenergic blockers on the dog's sleep-wake pattern. Physiol Behav 32:345-350

GORDON C, LAVIE P (1984b) Ultradian rhythms in renal excretions in dogs: lack of relationship with hormones influencing fluid balance. Neuroendocrinol Lett 6:101-107

GRIVEL ML, RUCKEBUSCH Y (1972) The propagation of segmental contractions along the small intestines. J Physiol 227:611-625

HOLADAY JW, MARTINEZ HM, NATELSON BM (1977) Synchronized ultradian cortisol rhythms in monkeys: persistence during corticotropin infusion. Science 198:56-58

JENKINS GM, WATTS DG (1968) Spectral analysis and its applications. Holdenday, New York

LAVIE P, KRIPKE DF (1977) Ultradian rhythms in urine flow in waking humans. Nature 269:142-144

LAVIE P, LUBOSHITZKY R, KLEINHOUSE N, SHEN-ORR Z, BARZILAI D, GLICK SM, LEROITH D, LEVY J (1980) Rhythms in urine flow are not correlated with rhythmic secretion of antidiuretic hormone (ADH) in well hydrated men. Horm Metab Res 12:66-70

LEVIN BE, GOLDSTEIN A, NATELSON BH (1978) Ultradian rhythm of plasma noradrenaline in rhesus monkeys. Nature 272:164-166

LUBOSHITZKY R, LAVIE P, SOK Y, GLICK SM, LEROITH D, SHEN-ORR Z, BARZILAI D (1978) Antidiuretic hormone secretion and urine flow in aged catheterized patients. T-I-T J Life Sci 8:99-103

MANDELL AJ, CHAFFEY B, BRILL P, MANDELL MP, RODNICK J, RUBIN RT, SCHEFF R (1966) Dreaming sleep in man: changes in urine volume and osmolality. Science 151:1558-1560

MOORE-EDE MC, SCHMELZER WS, KASS DA, HERD JA (1976) Internal organization of the circadian timing system in multicellular animals. Fed Proc 25:2333-2338

SCHRIER RW, BERL T, ANDERSON RJ, McDONALD KM (1977) Nonosmolar control of water renal excretion. In: ANDREALI TE, GRANTHAN JJ, RECTOR FC Jr (eds) Disturbance in body fluid osmolality. American Physiological Society, Bethesda, pp 149-178

WEVER RA (1979) The circadian system of man. Springer, Berlin Heidelberg New York

Timing Function of the Dorsal Raphe Nucleus and the Temporal Organization of the Ultradian Sleep Cycle

R. Lydic, R. W. McCarley, and J. A. Hobson
Loboratory of Neurophysiology, Department of Psychiatry, Harvard Medical School, Boston, MA 02115, USA

INTRODUCTION

With exception perhaps of the echidna (ALLISON and VAN TWYVER 1972) all mammals studied to date display two electroencephalographic (EEG) phases of sleep. Desynchronized sleep is characterized by fast frequency, low voltage EEG activity, rapid eye movements (REM) and, in humans, dreaming (DEMENT and KLEITMAN 1957). During synchronized sleep the EEG displays slow frequency, high voltage waves and there are no rapid eye movements (NREM). Within a sleep interval, REM sleep occurs with cyclic regularity and exhibits species-specific period lengths which are always in the ultradian (i.e., much less than 24 h) range (ZEPELIN and RECHTSCHAFFEN 1974). These characteristics support the concept that REM sleep, like respiration, is a fundamental biological rhythm.

REM sleep has long been known to have an essential role in maintaining the temporal aspects of homeostasis. Consequently, the structural and functional substrates influencing sleep have been the subject of many integrative reviews (JOUVET 1962, 1972; MORGANE and STERN 1974; McGINTY et al. 1974; STERIADE and HOBSON 1976; HOBSON and McCARLEY 1977; McCARLEY 1978; JACOBS and JONES 1978; McGINTY and DRUCKER-COLIN 1982; HOBSON and STERIADE, in press), and there have been numerous efforts to systematically describe the temporal organization of the ultradian sleep cycle (KRIPKE 1974; CZEISLER and GUILLEMINAULT 1980; SCHULZ et al. 1980; LAVIE and KRIPKE 1981; URSIN and STERMAN 1981; URSIN et al. 1983; LAVIE 1982; KRAEMER et al. 1984).

Experimental Brain Research, Suppl. 12

This chapter focuses upon the dorsal raphe nucleus (DRN) and reviews recent evidence concerning the time course of DRN discharge relative to the temporal organization of sleep. These time course data provide a new form of correlative evidence strongly supporting the hypothesis that DRN is part of an anatomically distributed central pattern generator involved in physiologically regulating the ultradian sleep cycle.

DOES DRN REGULATE ULTRADIAN OSCILLATIONS OF BEHAVIORAL STATE?

Discharge Rate Data

It should be acknowledged from the beginning that there are no data which unequivocally prove that discharge levels of DRN are necessary and sufficient for generating the REM sleep state or any REM-specific physiological trait. However, it is fair to conclude that the cessation of DRN discharge during physiological REM sleep is one of the most consistent findings in behavioral neurobiology (McGINTY and HARPER 1976; TRULSON and JACOBS 1979; HOBSON et al. 1983; LYDIC et al. 1984b). To date, there have been no exceptions to this rule: in the intact, undrugged, adult cat the rate of DRN discharge is always inversely related to the onset and maintenance of REM sleep.

This inverse relationship between the cessation of DRN firing and the onset of REM sleep is referred to as a REM-off discharge (Fig. 1). Single-cell activity recorded from the DRN of intact, undrugged cats displays a slow and regular firing rate during wakefulness (about 2 spikes/s), a decreased firing rate during NREM sleep and transition (0.5-1.0 spike/s), and a cessation of regular discharge during REM sleep (0.05-0.2 spikes/s). It should be noted that the slow and reg-

Fig. 1. Dorsal raphe nucleus (DRN) discharge rate relative to the behavioral states comprising the ultradian sleep cycle. (A) Ending of wakefulness and the onset of synchronized (NREM) sleep; (B) ending of synchronized sleep and the onset of a behavioral state referred to as transition. Note the increased frequency of ponto-geniculo-occipital (PGO) waves and the slowing of DRN discharge; (C) ending of transition and the onset of REM sleep; (D) ending of REM sleep and the resumption of the subsequent sleep cycle. Electromyogram (EMG), electroencephalogram (EEG), Schmitt trigger output of DRN single-cell activity (Unit), Schmitt trigger output divided by 10 (Unit/10), electrooculogram (EOG). The black bar above the EOG trace is a time mark, the length of which indicates 1 min ▷

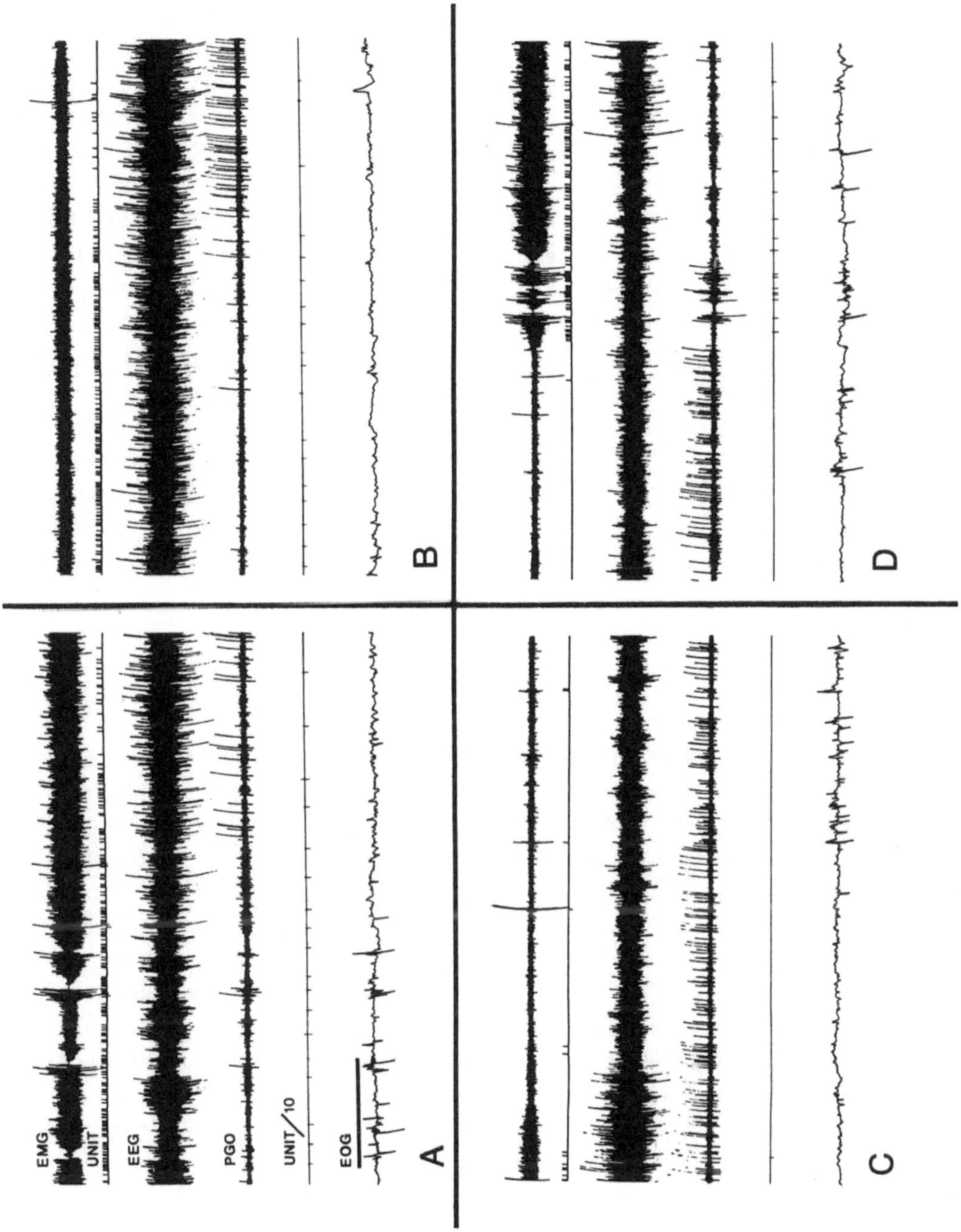
EMG
UNIT
EEG
PGO
UNIT/10
EOG
A
B
C
D

ular DRN discharge is strikingly different from the discharge of nearby reticular neurons which can fire at rates approaching 300 spikes/s (SZERB 1967). The REM-off quality of DRN cells also differs from surrounding reticular neurons, many of which are more active during REM sleep than during wakefulness (STERIADE and HOBSON 1976).

Global classifications of behavioral state (i.e., wakefulness, NREM, REM) are traditional and offer a productive scheme for studying cellular elements or entire organ systems across the ultradian sleep cycle. For example, such classification schemes were the first to show that average DRN discharge rates during NREM sleep were 46% lower than in waking, and that during REM sleep DRN discharge was 96% less than in wakefulness (McGINTY and HARPER 1976). However, the precise phase relationships between DRN discharge and specific components of the ultradian sleep cycle are not readily apparent when DRN discharge rate is averaged within global and arbitrarily discontinuous classifications of behavioral state.

Time Course Data

The period, amplitude, and phase relationships between cellular discharge and the behavioral states comprising the ultradian sleep cycle can be evaluated using time course analyses. Time course data preserve the temporal continuity across the complete sleep cycle while quantifying physiological and behavioral ultradian oscillations.

The technique of time-normalized sleep cycle averaging transforms the ultradian period length in minutes into a standardized duration expressed as percentage of the cycle completed. It thus becomes possible to average any dependent variable across multiple ultradian oscillations in spite of the noted variability of REM sleep duration (URSIN and STERMAN 1981; URSIN et al. 1983). Time-normalized sleep cycle averaging (Fig. 2) has been used (LYDIC et al. 1984b) to characterize the time course of DRN discharge and involved four analytic steps: (a) Simultaneous polygraphic and single-cell recordings were scored for behavioral state and DRN discharge rate across successive 1-min bins. (b) A sleep cycle was operationally defined from the end of one REM sleep episode through the end of the subsequent REM sleep episode. Specific REM sleep cycles to be normalized and averaged were chosen from a series (Fig. 2A) on the basis of period length. Single-cell DRN

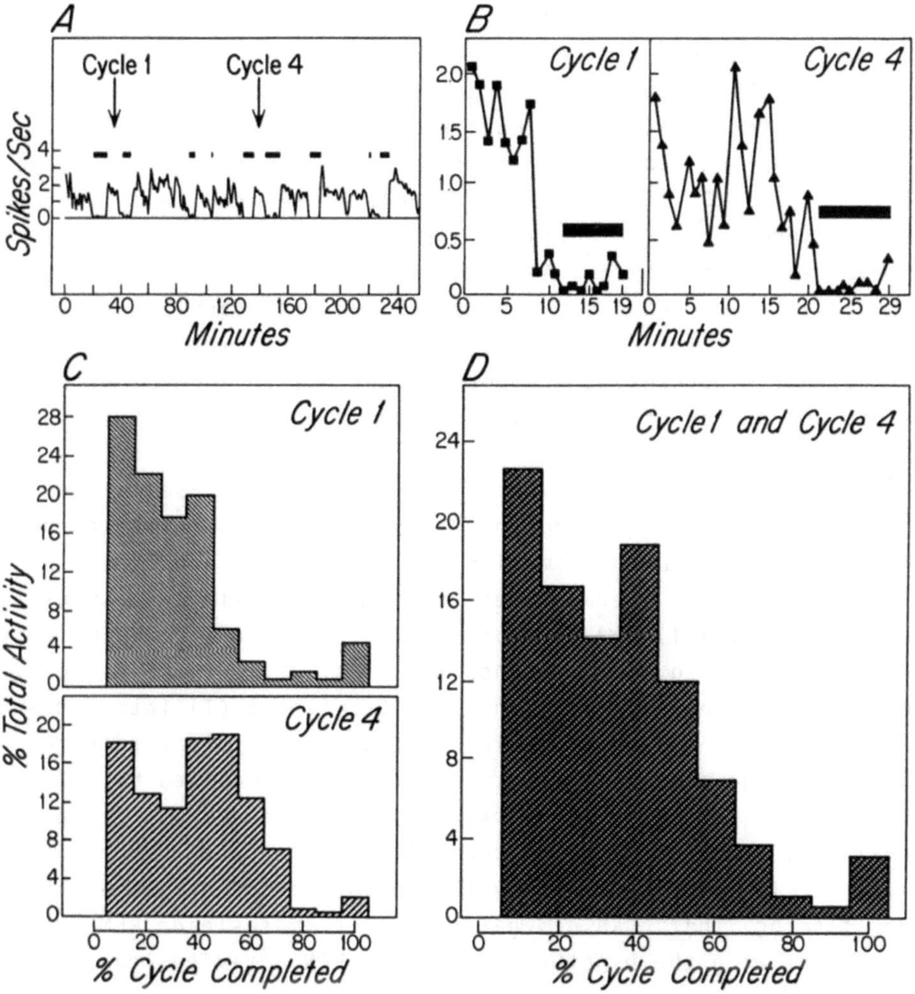

Fig. 2. Schematic illustration of the four analytic steps used for time-normalized sleep cycle averaging. (A) DRN discharge profile in spikes/s across 240 min of recording. Black bars indicate REM sleep epochs. (B) Enlarged view of two cycles with period lengths in the 15-30 min range. (C) Conversion to the time-normalized format. (D) Multiple cycle averaging of time-normalized profiles. Text provides additional details

discharge across a sleep cycle (Fig. 2B) was expressed as spikes/s during each minute of the recording. (c) These data were entered into a computer which converted the period length of the cycle from time in minutes into 10 standardized bins. These time-normalized bins ranged from 0-100 and correspond to the percentage of the sleep cycle completed (Fig. 2C). (d) At each of these time-normalized bins, the computer expressed DRN activity as a percentage of the total DRN discharge within the cycle. Although the ultradian period lengths varied (Fig. 2B), the DRN discharge associated with each bin in the normalized format (Fig. 2C) had the same arithmetic weighting. Thus, time-normalization preserves temporal continuity and permits averaging across any number of sleep cycles (Fig. 2D).

The results of these analyses (Fig. 3) provide phase-aligned comparisons between DRN discharge profiles and the oscillating behavioral state profiles which comprise the ultradian sleep cycle. During the first half of this average sleep cycle, little time was spent in wakefulness (Fig. 3A1) and most of the time was spent in NREM or synchronized sleep (Fig. 3A2). Across the first half of the sleep cycle the amount of time spent during the transition (T) into REM increased (Fig. 3A3). As early as 20% into the average cycle (ordinate, bin 20) DRN discharge level (abscissa of Fig. 3A5) declined by 20% from initial values. Thus, the progression towards the REM-off firing pattern, consistently described by discharge rate analyses, began early in the sleep cycle and well before the appearance of REM sleep (Fig. 3A4). The finding that DRN discharge offset phase-leads REM onset is not consistent with the hypothesis that the DRN activity levels are merely

Fig. 3. Time course histograms derived from the time-normalized ▷ average of multiple sleep cycles. Regular cycles represent the average of 11 individual cycles with durations ranging from 15-30 min and averaging 25 min. Frames A1 A4 illustrate the average time course of the behavioral state profiles. For the top four histograms, note that the % state values (ordinates) sum to 100 at each portion of the % cycle completed (abscissa), e. g., during the first 10% of the average cycle wakefulness (W) accounted for about 20% of the state distribution, NREM or synchronized (S) sleep for about 70%, transition (T) for about 10%, and 0% of the behavioral state was accounted for by REM or desynchronized (D) sleep. The bottom frame (A5) illustrates the time-normalized average of DRN discharge. At each portion of cycle completed (abscissa) DRN discharge is expressed as % total spikes. Thus, the DRN profile sums to 100 across the 10 bins of the average cycle. The time-normalized profiles for the long cycles were derived from 15 individual sleep cycles with period lengths ranging from 31 to 46 min, and averaging 38 min

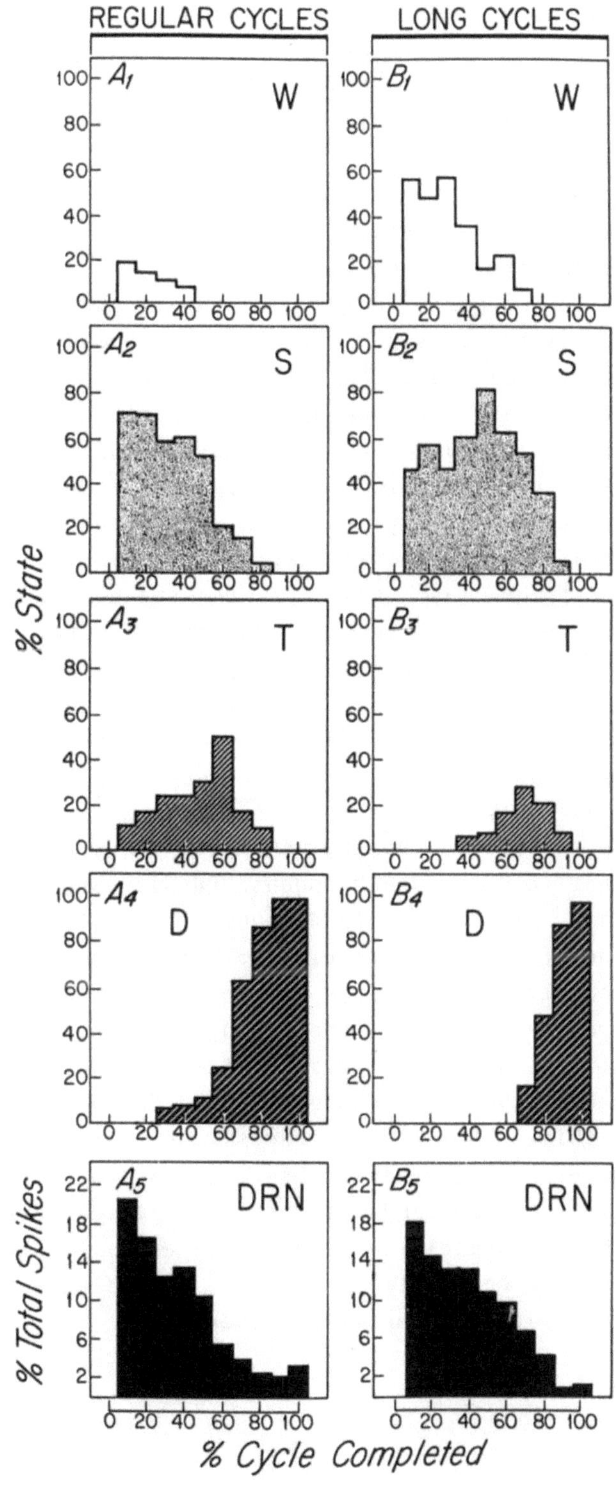
REGULAR CYCLES
LONG CYCLES
A1
W
B1
W
A2
S
B2
S
A3
T
B3
T
A4
D
B4
D
A5
DRN
B5
DRN
% State
% Total Spikes
% Cycle Completed

following REM sleep. Alternatively, these time course data provide additional correlative evidence supporting the hypothesis (McGINTY and HARPER 1976) that DRN exerts a permissive, disinhibitory role which actively influences the onset and maintenance of REM sleep.

Furthermore, when DRN discharge is evaluated within NREM sleep or within REM sleep, one may observe statistically significant phase-dependent firing rates (LYDIC et al. 1984a,b). The within sleep, phase-dependent nature of DRN discharge has been described using regression analyses (Fig. 4). Time in minutes within a single behavioral state was taken as the independent variable (x) and DRN discharge served as the dependent variable (y). In addition, the average DRN discharge rate was calculated during the first half and during the last half of individual epochs of wakefulness, NREM, and REM sleep. Split-half comparisons then were used to calculate the number of cases within a behavioral state where DRN discharge was greatest during the first or last half of the state. The binomial theorem was used to calculate the exact probability of obtaining the ordering observed by split-half analyses vs the null hypothesis of random ordering. The results revealed that DRN discharge rates were greatest during the first half of wakefulness (Fig. 4W) in 9 of 11 epochs analyzed ($p \leq 0.03$). During NREM sleep (Fig. 4S) split-half analyses revealed lower DRN firing rates during the later half of 14 of 15 NREM epochs ($p \leq 0.0005$). During REM sleep (Fig. 4D) split-half analyses revealed that in 9 of 10 REM epochs DRN discharge rates were greater during the last half of REM ($p \leq 0.009$), prior to the resumption of wakefulness. Therefore, the results of regression analyses (Fig. 4) and the probability calculations demonstrate that changes in DRN discharge rate precede in time the changes in behavioral state which comprise the ultradian sleep cycle. Declining DRN discharge within NREM sleep precedes the appearance of REM. Likewise, the increasing DRN discharge within REM sleep precedes the offset of REM and the resumption of wakefulness. These observations concerning phase (derived from both time course and regression analyses) are consistent with the notion that DRN activity levels are causally involved in regulating the temporal organization of the ultradian REM cycle and are not simply a correlate of the state sequence.

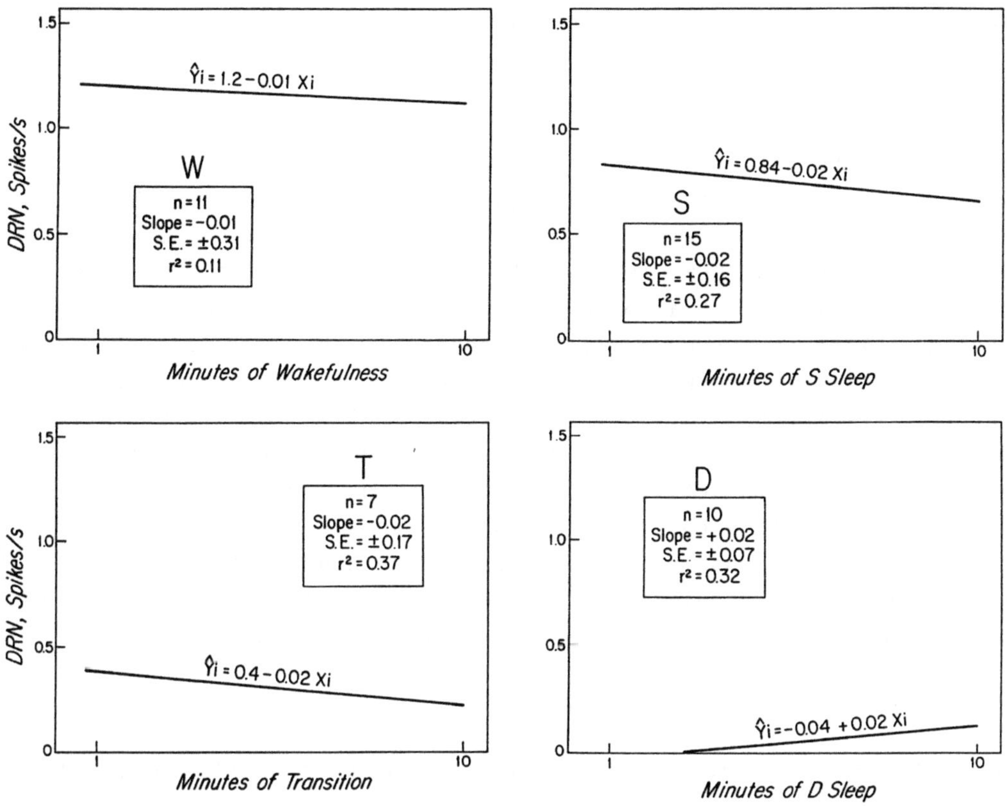

Fig. 4. Simple linear regression lines illustrating DRN discharge rate (spikes/s) as a function of time within behavioral states comprising the ultradian sleep cycle. The regression lines represent the average, least-squares best fit of DRN discharge (y) across time (x) spent in a behavioral state. The number (n) of behavioral state epochs used in these analyses is given for wakefulness (W), NREM or synchronized (S) sleep, a transition (T) state operationally defined by the presence of eight or more PGO waves per min, and REM or desynchronized (D) sleep. The calculated slopes refer to spikes/s within each min of a given behavioral state (W, S, T, D) and SE indicates the average standard error of the estimate for the fitted regression function. The proportion (or percentage) of the total variation in DRN discharge that was accounted for by the fitted regression is given by the coefficient of determination (r squared)

Effects of Forced Activity on Behavioral State

The temporal organization of mammalian sleep has often been studied by exposing animals to periods of forced locomotor activity (TAKAHASHI et al. 1978; BORBELY et al. 1980; TOBLER et al. 1983). Manipulation of the sleep cycle with forced locomotor activity recently was used to examine the null hypothesis that there is no consistent relationship between the temporal organization of sleep and the DRN discharge rhythm (LYDIC et al. 1984b).

The results revealed that the ultradian sleep cycle and the DRN discharge rhythm always exhibited similar period lengths (τ), yet both were altered significantly and with a high degree of coherence by forced activity. Following forced activity, τ was reduced during wakefulness, NREM and REM sleep and the DRN discharge profiles were less noise dominated and also tended to have shorter τs (Fig. 3, regular cycles). With no forced activity sleep cycle period lengths tended to be longer and there was more time spent in wakefulness and less time spent in REM sleep (Fig. 3, long cycles). Analyses of 177 sleep cycles with τs ranging from 8.9-78.0 min revealed that in no case was it possible for forced activity to dissociate the time course of DRN discharge from the time course of the ultradian sleep cycle. These faithful phase relationships also support the hypothesis that DRN has a causal role in regulating the temporal organization of behavioral states which comprise the ultradian sleep cycle.

DOES DRN REGULATE THE ULTRADIAN OSCILLATIONS OF SPECIFIC PHYSIOLOGICAL TRAITS?

The technique of time-normalized averaging also has been used to characterize the time course of DRN discharge relative to the time course of simultaneously recorded physiological traits, such as ponto-geniculo-occipital (PGO) waves and somatic muscle tone (LYDIC et al. 1983a).

Time Course of PGO Waves and DRN Discharge

Although PGO waves were first described in the late 1950s (JOUVET and MICHEL 1959), their functional significance remains unknown. These bipolar waves (amplitude, 200-300 microvolts; duration, 100 ms) always accompany the onset and maintenance of REM sleep. The original firing rate analyses of DRN activity during REM sleep (McGINTY and HARPER 1972) noted that DRN discharge ended before the appearance of PGO waves. This led to the hypotheses (McGINTY and HARPER 1976) that (a) DRN discharge during wakefulness inhibits PGO waves and (b) loss of

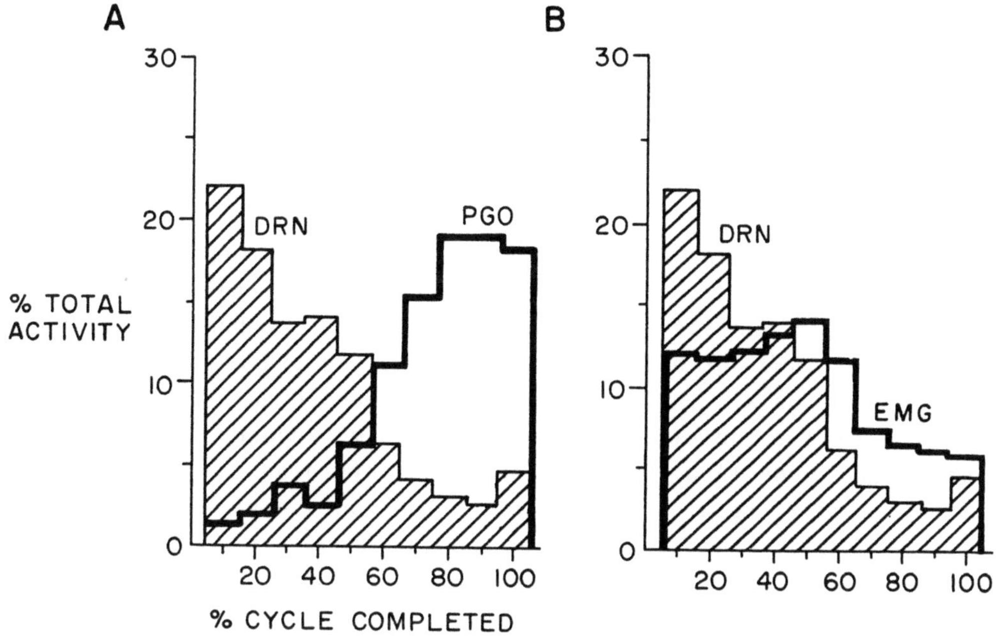

Fig. 5. Histograms illustrating the average time-course of three physiological traits simultaneously recorded across 11 sleep cycles. Frame A gives the time course of single-cell discharge recorded from the DRN and PGO wave activity recorded from the lateral geniculate nucleus. Frame B contrasts the activity profile of the same DRN cell with the profile of EMG activity simultaneously recorded from neck muscles

regular DRN activity during REM sleep might serve as a permissive, disinhibitory signal to putative PGO generating mechanisms.

The time-normalized averaging technique has been used to begin characterizing (Fig. 5A) the period, amplitude, and phase properties of PGO waves and DRN discharge. The time course profile of PGO waves begins to rise at the same time (20% cycle completed) that the DRN discharge profile begins to decline. These inversely related activity profiles intersect one another near midcycle (50%-60% cycle completed) and when DRN activity is minimal PGO activity is maximal (80%-90% cycle completed). At cycle end, PGO waves begin to decline as DRN discharge starts to increase. At each bin of the time-normalized cycle, these negative correlations are statistically significant. These data support and extend the findings obtained from discharge rate analyses: across the ultradian sleep cycle, the time course of DRN discharge is negatively correlated with the time course of PGO waves. These data also support the hypothesis that DRN discharge levels regulate, through disinhibition, the timing of PGO wave activity.

Time Course of Muscle Tone and DRN Discharge

With exception of the extraocular and some respiratory muscles, there is a rhythmic loss of muscle tone which always parallels the ultradian oscillation of REM sleep. Are the physiological mechanisms generating atonia the same or different from the mechanisms which regulate the timing of atonia? The atonia of REM has been hypothesized to result from the active inhibition of spinal alpha motoneurons by supraspinal mechanisms (POMPEIANO 1967). The precise physiological mechanisms mediating atonia remain to be determined but spinal motoneurons are known to receive inhibitory input from the medullary nucleus reticularis magnocellularis which is regulated, in part, by the mediodorsal pontine tegmentum (SAKAI et al. 1981).

Several lines of evidence associate DRN activity levels with muscle tone. DRN discharge declines to the lowest levels during the atonia of physiological REM sleep (McGINTY and HARPER 1976). DRN firing levels also have been grossly correlated with levels of tonic motor activity during behavioral arousal (TRULSON and JACOBS 1979). During the REM sleep-like state associated with lesions of the pontine tegmentum and the absence of atonia, the offset of DRN discharge is greatly reduced (TRULSON et al. 1981). Atonia induced by microinjections of carbachol into the medial pontine reticular formation (mPRF) also has been associated with significant reductions in DRN discharge, an effect which could not be replicated by systemic administration of peripherally acting muscle relaxants (STEINFELS et al. 1983). Although direct relationships between muscle tone and DRN discharge have not been demonstrated, these data indicate that dissociations can be surgically and pharmacologically induced between muscle tone, a REM sleep-like state, and DRN discharge levels. Considered together, these findings suggest that any involvement which DRN may have in generating atonia is likely to be of a modulatory and nonspecific nature (JACOBS et al. 1981; STEINFELS et al. 1983).

Data obtained from the intact, undrugged cat do not reveal a simple, linear relationship between DRN discharge and the atonia of REM sleep (LYDIC et al. 1983a). During the first half of the sleep cycle, time course analyses (Fig. 5B) showed increased nuchal muscle tone which was negatively correlated ($r = -0.73$, $p \leq 0.001$) with declining DRN discharge. During the last half of the time-normalized cycle (50%-100% cycle completed) there was a positively correlated ($r = +0.97$, $p \leq 0.001$) decline in both muscle tone and DRN discharge. NREM sleep accounts for most of the behavioral state during the first half of the ultradian sleep cycle (Fig. 3A2) and during NREM others (DICK et al. 1982) have noted an increased muscle tone in the respiratory-related activity of the external intercostal musculature.

Thus, in the intact, undrugged cat the relationship between muscle tone and DRN discharge is not monotonically linear and instead is highly dependent upon the temporal phase of the ultradian sleep cycle. The phase-dependent nature of this relationship may mean that DRN discharge plays a key role in timing the ultradian bouts of muscular atonia.

Effects of Forced Activity on DRN Discharge and PGO Waves

As previously noted (Fig. 3), a high degree of coherence always was observed between the period length of the sleep cycle and the period length of DRN discharge. The results of cosinor spectral analyses (Fig. 6) illustrate in further detail the ability of forced activity to alter the periodic properties of DRN discharge and PGO waves. With no forced activity, a 6-h recording of DRN discharge revealed a dominant period length in the range of 35-40 min; 5 h of DRN discharge recorded from the same cell after the cat had been exposed to 12 or more h of forced locomotor activity in a treadmill revealed a dominant τ in the 15-20 min range.

Similar analyses applied to simultaneously recorded PGO waves also revealed significant shifts in PGO period length as a function of forced activity (Fig. 6). With no forced activity, PGO waves had a dominant ultradian rhythm in the 35-40 min range. Following 12 h of forced activity, recordings from the same animal showed that the dominant rhythm in the PGO wave activity spectrum was shortened and ranged from 17 to 22 min. Thus, time series analyses demonstrate that major shifts in the period length of DRN discharge are accompanied by similar periodic changes in the temporal distribution of PGO waves.

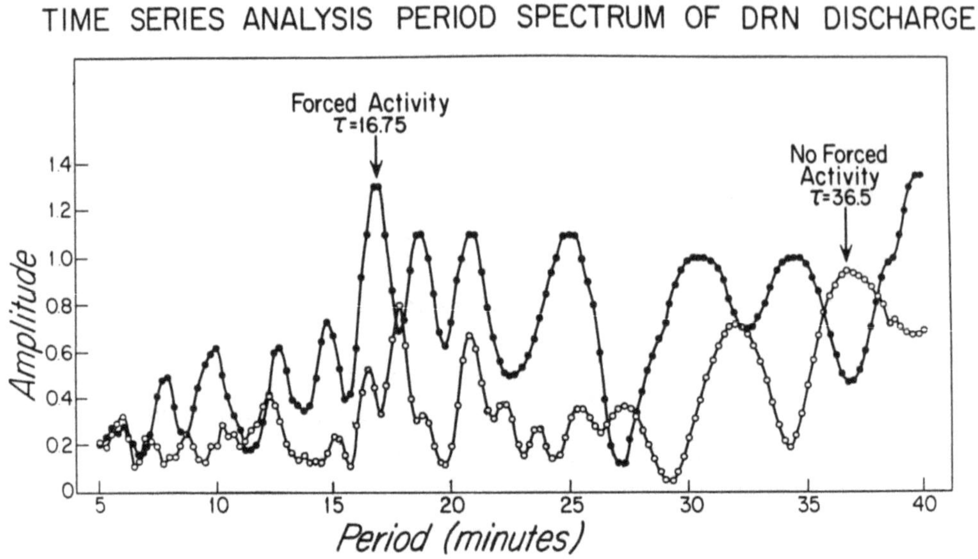

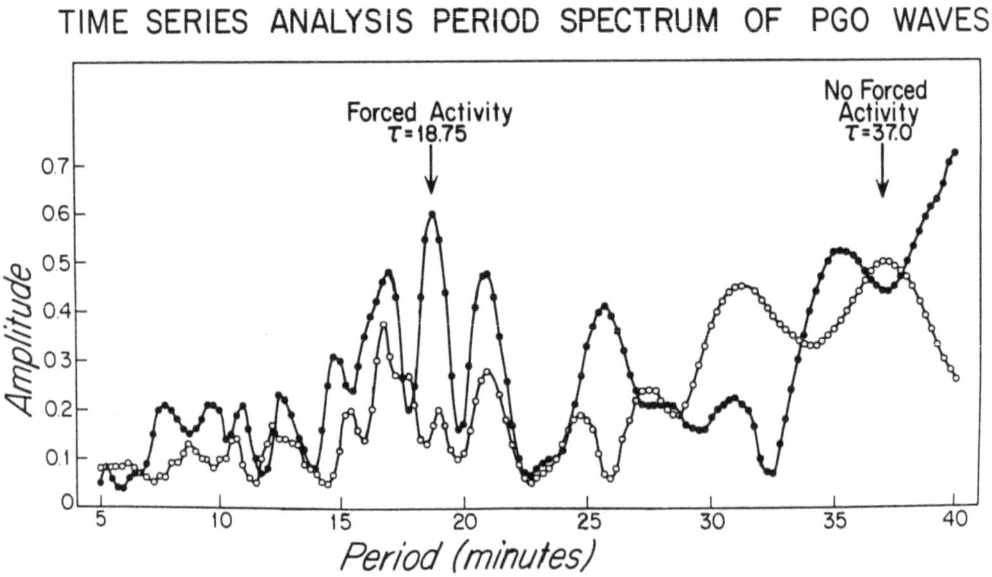

Fig. 6. Time series period spectrum for DRN single unit activity and PGO wave activity recorded before (open circles) and after (closed circles) forced locomotor activity. Peaks of the dominant periods (τ) are indicated by arrows for each variable and condition. These peaks represent period lengths significantly different at the 0.001 level, by the least-squares method of wave fitting (RUMMEL et al. 1974), from the null hypothesis that the amplitude of the rhythm was zero. When these data were detrended, similar results concerning period length of these ultradian rhythms were obtained from the Fourier transform method (DIXON 1983) of time series analyses. Forced activity produced significant reductions in the ultradian rhythms of both DRN discharge and PGO wave activity

CONCLUSIONS: CORRELATION AND CAUSATION

Many of the reviews cited in the introduction to this chapter note that REM sleep can occur following total destruction of DRN. Such findings do not prove a lack of causal involvement by DRN in sleep cycle regulation. Instead, the persistence of REM sleep following DRN destruction exemplifies a property characteristic of diverse neuronal systems putatively regarded as pattern generators. For example, the respiratory rhythm is believed to be generated by neurons concentrated in the brain stem dorsal and ventral respiratory groups, but respiration persists following destruction of these medullary nuclei (SPECK and FELDMAN 1982). Similarly, the medial pontine reticular formation (mPRF) is believed to be involved in controlling REM sleep and waking eye movements, yet neither REM sleep nor waking eye movements are abolished by lesions of the mPRF (SASTRE et al. 1981; DRUCKER-COLIN and BERNAL-PEDRAZA 1983). The hypothalamic suprachiasmatic nuclei (SCN) are known to be involved in generating rhythmic physiological and behavioral rhythms but these rhythms are not uniformly abolished by lesions of the SCN (FULLER et al. 1981; ALBERS et al. 1984). Therefore, persisting ultradian or circadian rhythms following destruction of a putative generator do not prove that a nuclear group is not causally involved in the generation or temporal organization of a rhythm. These numerous examples illustrate the complexity of physiological mechanisms involved in rhythm generation, the need for detailed temporal analyses of persisting rhythms, and the limitations of present concepts regarding correlation and causation.

Clearly, one must be cautious not to infer that levels of DRN discharge cause REM sleep simply because of their consistent and statistically reliable correlations. Such correlational evidence is, however, powerfully revealing of the relationship between DRN discharge, behavioral states, and physiological traits. Furthermore, the correlational approach represents an essential step which can neither be bypassed nor devalued in the search for causality. The period (Fig. 6), amplitude, and phase relationships (Figs. 1, 3, 5) observed in intact and undrugged cats are consistent with the notion that DRN is causally involved in sleep cycle regulation.

Four lines of evidence suggest that DRN exerts an anatomically distributed timing function.

1. Since a cause cannot follow the event it causes, the results reviewed here strongly support the hypothesis that the temporal distribution of atonia, PGO waves, and REM sleep may be timed, at least in part, by DRN discharge levels.

2. A large body of literature, briefly reviewed elsewhere (LYDIC et al. 1984b), clearly demonstrates that DRN is involved in regulating multiple physiological systems. With regard to the timing concept, it may be noted that DRN interacts with physiological systems which are predominantly oscillatory. In addition to the ultradian sleep cycle, these include the neuroendocrine system controlling ovulation, temperature regulation, the cardiovascular, the respiratory, and the circadian systems.

3. At the synaptic level, the ability of the putatively serotonergic DRN to provide a timing function would require DRN-mediated enhancement or suppression of a postsynaptic response to some presynaptic input. This characteristic constitutes a definition of synaptic modulation and monoaminergic neurotransmission that has been postulated (JACOBS and JONES 1978) to be of a modulatory nature. The regularity of DRN firing during wakefulness and the rhythmic DRN discharge across the ultradian sleep cycle are consistent with the notion that DRN may subserve a timing function which modulates the temporal organization of multiple physiological systems. This pacemaking notion is supported by recent neurophysiological evidence. For example, intracellular DRN recordings reveal pacemaker potentials in the form of calcium-dependent alterations in potassium conductance (VANDERMAELEN and AGHAJANIAN 1983; AGHAJANIAN and LAKOSKI 1984). DRN discharge also is believed to be subject to automodulation by recurrent inhibitory axon collaterals (WANG and AGHAJANIAN 1977). Recently it was shown that the facilitatory response of DRN cells to electrical stimulation of the medial pontine reticular formation was, in part, dependent on the temporal phase of the ultradian sleep cycle (LYDIC et al. 1983b). Cellular tests of the DRN timing hypothesis, however, require demonstration of the fact that the rate and/or time course of DRN discharge can alter the response of another cell to some parallel synaptic input.

4. The timing concept is a specific corollary of a broader working hypothesis (HOBSON et al. 1975; McCARLEY and HOBSON 1975; McCarley and MASSAQUOI, this volume, Chap. 17), namely, that DRN may be part of an anatomically distributed neuronal system which generates the ultradian sleep cycle.

ACKNOWLEDGEMENTS. We thank Dr. H. Baghdoyan for critical reading of the manuscript. This work was supported by National Institute of Mental Health grants MH-13923 (J.A.H.), MH-280 (R.W.M.), National Research Service Award MH-14275 (R.L.), and by an award to R.L. from the William F. Milton Fund of Harvard Medical School.

REFERENCES

AGHAJANIAN GK, LAKOSKI JM (1984) Hyperpolarization of serotonergic neurons by serotonin and LSD: studies in brain slices showing increased potassium conductance. Brain Res 305:181-185

ALBERS HE, LYDIC R, GANDER PH, MOORE-EDE MC (1984) Role of the suprachiasmatic nuclei in the circadian timing system of the squirrel monkey. I: The generation of rhythmicity. Brain Res 300:275-284

ALLISON T, VAN TWYVER H (1972) Electrophysiological studies of the Echidna Tachyglossus aculeatus. II: Dormancy and hibernation. Arch Ital Biol 110:185-194

BORBELY AA, STEIGRAD P, TOBLER I (1980) Effect of sleep deprivation on brain serotonin in the rat. Behav Brain Res 1:205-210

CZEISLER CA, GUILLEMINAULT C (1980) REM sleep: a workshop on its temporal distribution. Sleep 2:285-286

DEMENT WC, KLEITMAN N (1957) Cyclic variations in EEG during sleep and their relation to eye movements, body motility, and dreaming. Electroencephalogr Clin Neurophysiol 9:673-690

DICK TE, PARMEGGIANI PL, OREM J (1982) Intercostal muscle activity during sleep in the cat: an augmentation of expiratory activity. Resp Physiol 50:255-265

DIXON WJ (ed) (1983) BMDP statistical software. University of California Press, Berkeley

DRUCKER-COLIN R, BERNAL-PEDRAZA JG (1983) Kainic acid lesions of gigantocellular tegmental field (FTG) neurons does not abolish REM sleep. Brain Res 272:387-391

FULLER CA, LYDIC R, SULZMAN FM, ALBERS HE, TEPPER B, MOORE-EDE MC (1981) Circadian rhythm of body temperature persists after suprachiasmatic lesions in the squirrel monkey. Am J Physiol 241:R385-R391

HOBSON JA, McCARLEY RW (eds) (1977) Neuronal activity in sleep: 1969-1984. An annotated bibliography. Brain Information Service, University of California, Los Angeles

HOBSON JA, STERIADE M (in press) The neuronal basis of behavioral state control. In: BLOOM F, MOUNTCASTLE V (eds) Internal regulatory systems of the brain. American Physiological Society (Handbook of physiology)

HOBSON JA, McCARLEY RW, WYZINSKY PW (1975) Sleep cycle oscillation: reciprocal discharge by two brainstem neuronal groups. Science 189: 55-58

HOBSON JA, McCARLEY RW, NELSON JP (1983) Location and spike-train characteristics of cells in anterodorsal pons having selective decreases in firing rate during desynchronized sleep. J Neurophysiol 50:770-783

JACOBS BL, JONES BE (1978) The role of central monoamine and acetycholine systems in sleep-wakefulness states: mediation or modulation? In: BUTCHER LL (ed) Cholinergic-monoaminergic interactions in the brain. Academic, New York, pp 271-290

JACOBS BL, HEYM J, TRULSON ME (1981) Behavioral and physiological correlates of brain serotonergic unit activity. J Physiol (Paris) 77: 431-436

JOUVET M (1962) Recherches sur les structures nerveuses et les méchanismes responsables des différentes phases du sommeil physiologique. Arch Ital Biol 100:125-206

JOUVET M (1972) The role of monoamines and acetylcholine-containing neurons in the regulation of the sleep-waking cycle. Ergebn Physiol 64:166-307

JOUVET M, MICHEL F (1959) Corrélations électromyographiques du sommeil chez le chat décortiqué et mésencéphalique chronique. C R Soc Biol (Paris) 153:422-425

KRAEMER HC, HOLE WT, ANDERS TF (1984) The detection of behavioral state cycles and classification of temporal structure in behavioral states. Sleep 7:3-17

KRIPKE DF (1974) Ultradian rhythms in sleep and wakefulness. In: WEITZMAN ED (ed) Advances in sleep research, vol 1. Spectrum, Flushing, pp 305-325

LAVIE P (1982) Ultradian rhythms in human sleep and wakefulness. In: WEBB WB (ed) Biological rhythms, sleep, and performance. Wiley, New York, pp 239-272

LAVIE P, KRIPKE DF (1981) Ultradian circa 1 1/2 hour rhythms: a multioscillatory system. Life Sci 29:2445-2450

LYDIC R, McCARLEY RW, HOBSON JA (1983a) The time-course of dorsal raphe discharge, PGO waves, and muscle tone averaged across multiple sleep cycles. Brain Res 274:365-370

LYDIC R, McCARLEY RW, HOBSON JA (1983b) Enhancement of dorsal raphe discharge by medial pontine reticular formation stimulation depends on behavioral state. Neurosci Lett 38:35-40

LYDIC R, McCARLEY RW, HOBSON JA (1984a) Dorsal raphe discharge is phase-dependent within sleep states. Sleep Res 13:18

LYDIC R, McCARLEY RW, HOBSON JA (1984a) Dorsal raphe discharge is phase-dependent within sleep states. Sleep Res 13:18

LYDIC R, McCARLEY RW, HOBSON JA (1984b) Forced activity alters sleep cycle periodicity and dorsal raphe discharge rhythm. Am J Physiol 247:R135-R145

McCARLEY RW (1978) Control of sleep-waking state alterations in Felix domesticus. In: FERRENDELLI JA (ed) Neuroscience symposia, vol 3. Aspects of behavioral neurobiology. Society for Neuroscience, Bethesda, pp 90-128

McCARLEY RW, HOBSON JA (1975) Neuronal excitability over the sleep cycle: a structural and mathematical model. Science 189:58-60

McGINTY DJ, DRUCKER-COLIN RR (1982) Sleep mechanisms: biology and the control of REM sleep. Int Rev Neurobiol 101:569-575

McGINTY DJ, HARPER RM (1972) 5-HT-containing neurons: unit activity during sleep. Sleep Res 1:27

McGINTY DJ, HARPER RM (1976) Dorsal raphe neurons: depression of firing during sleep in cats. Brain Res 101:569-575

McGINTY DJ, HARPER RM, FAIRBANKS MK (1974) Neuronal activity and the control of sleep states. In: WEITZMAN ED (ed) Advances in sleep research, vol 1. Spectrum, Flushing, pp 173-216

MORGANE PJ, STERN WC (1974) Chemical anatomy of brain circuits in relation to sleep and wakefulness. In: WEITZMAN ED (ed) Advances in sleep research, vol 1. Spectrum, Flushing, pp 1-131

POMPEIANO O (1967) The neurophysiological mechanisms of the postural and motor events during desychronized sleep. Proc Assoc Res Nervous Mental Dis 45:351-423

RUMMEL J, LEE JK, HALBERG F (1974) Combined linear-nonlinear chronobiologic windows by least squares resolve neighboring components in a physiologic rhythm spectrum. In: FERIN M, HALBERG F, RICHART RM, VAN DE WIELE RL (eds) Biorhythms and human reproduction. Wiley, New York, pp 53-82

SAKAI K, SASTRE JP, KANAMORI N, JOUVET M (1981) State-specific neurons in the ponto-medullary reticular formation with special reference to the postural atonia during paradoxical sleep in the cat. In: POMPEIANO O, AJMONE MARSAN C (eds) Brain mechanisms and perceptual awareness. Raven, New York, pp 405-429

SASTRE JP, SAKAI K, JOUVET M (1981) Are the gigantocellular tegmental field neurons responsible for paradoxical sleep? Brain Res 229:147-161

SCHULZ H, DIRLICH G, BALTESKONIS S, ZULLEY J (1980) The REM-NREM sleep cycle: renewal process or periodically driven process? Sleep 2:319-328

SPECK DF, FELDMAN JL (1982) The effects of microstimulation and microlesions in the ventral and dorsal respiratory groups in medulla of cat. J Neurosci 2:744-757

STEINFELS GF, HEYM J, STRECKER RE, JACOBS BL (1983) Raphe unit activity in freely moving cats is altered by manipulations of central but not peripheral motor systems. Brain Res 279:77-84

STERIADE M, HOBSON JA (1976) Neuronal activity during the sleep-waking cycle. Prog Neurobiol 6:155-376

SZERB JC (1967) Cortical acetycholine release and electroencephalographic arousal. J Physiol (Lond) 192:329-343

TAKAHASHI Y, EBIHARA S, NAKAMURA Y, TAKAHASHI K (1978) Temporal distributions of delta wave sleep and REM sleep during recovery sleep after 12 hours forced wakefulness in dogs: similarity to human sleep. Neurosci Lett 10:329-334

TOBLER I, MURISON R, URSIN R, BORBELY AA (1983) The effect of sleep deprivation and recovery sleep on plasma corticosterone in the rat. Neurosci Lett 35:297-300

TRULSON ME, JACOBS BL (1979) Raphe unit activity in freely moving cats: correlation with level of behavioral arousal. Brain Res 163: 135-150

TRULSON ME, JACOBS BL, MORRISON AR (1981) Raphe unit activity during REM sleep in normal cats and in pontine cats displaying REM sleep without atonia. Brain Res 226:75-91

URSIN R, STERMAN MB (1981) A manual for standardized scoring of sleep waking states in the adult cat. Brain Information Service/Brain Research Institute, University of California, Los Angeles

URSIN R, MOSES J, NAITOH P, JOHNSON LC (1983) REM-NREM cycle in the cat may be sleep dependent. Sleep 6:1-9

VANDERMAELEN CP, AGHAJANIAN GK (1983) Electrophysiological and pharmacological characterization of serotonergic dorsal raphe neurons recorded extracellularly and intracellularly in rat brain slices. Brain Res 289:109-119

WANG RY, AGHAJANIAN GK (1977) Antidromically identified serotonergic neurons in the rat midbrain raphe: evidence for collateral inhibition. Brain Res 132:186-193

ZEPELIN H, RECHTSCHAFFEN A (1974) Mammalian sleep, longevity, and energy metabolism. Brain Behav Evol 10:425-470

Section 2
Sleep and Waking Rhythms

The search for ultradian rhythms during waking, and the question whether they are a counterpart to the ultradian REM-NREM cycles, as suggested by Kleitman's BRAC hypothesis, are topics of the second part of the book. In Chapter 9 Lavie summarizes the evidence supporting the existence of multifrequency ultradian cycles in sleepiness. He suggests that the waking cycles and the sleep REM-NREM cycles serve a similar function, to provide sleep and wakefulness with multiple gates during which the transitions from one state to the other are greatly facilitated.

In Chapter 10 Schulz investigates the temporal structure of the occurrence of diurnal sleep episodes in narcoleptic patients in light of the BRAC hypothesis. He demonstrates that at least in some patients, sleep episodes do not occur randomly, but are spaced approximately 1.5 h apart, possibly in phase with the REM cycle during the preceding nocturnal sleep period. Sterman, in Chapter 11, first summarizes his pioneering original experiments which provided experimental support for the BRAC hypothesis, and then describes his current study on ultradian rhythms in electroencephalographic activity of waking humans who performed repetitively a flight simulation task.

The instability of the ultradian rhythms in waking behaviors, and their sensitivity to external conditions, are demonstrated in Chapter 12. Kripke, Mullaney and Fleck describe their search for ultradian rhythms in 42 h of sustained performance under highly demanding conditions. No ultradian rhythms could be found in performance under such conditions. The effects of various experimental conditions on the manifestation of the 1.5-h waking rhythms in human performance are also discussed by Broughton in the next chapter (13). He concludes that such cycles are accentuated under conditions of increased sleep need, lowered motivation, and lowered performance demand. He also discusses the evidence supporting the existence of 12-h rhythms in sleep tendency and the temporal relationships between the various ultradian rhythms.

In the last chapter of this section (14) Campbell and Zulley provide further evidence for a 12-h rhythm in sleep tendency. Subjects in a "disentrained" environment with reduced behavioral alternatives to sleep showed a bimodal distribution of sleep propensity with a primary nocturnal and a secondary mid-afternoon sleepiness crest.

Ultradian Rhythms: Gates of Sleep and Wakefulness

P. Lavie

Sleep Laboratory, Unit of Behavioral Biology, Faculty of Medicine, Technion – Israel Institute of Technology, Haifa 32000, Israel

INTRODUCTION

The study of the 1.5-h ultradian rhythms is a relatively recent by-product of the rapidly expanding field of sleep research. The discovery of the 1.5-h REM-NREM cycles by ASERINSKY and KLEITMAN in 1953, and KLEITMAN's (1961, 1963) theoretical speculation that these cycles are only the nocturnal fragment of a 24-h basic rest-activity cycle (BRAC), have captured the imagination of many investigators who have begun searching for waking counterparts of the sleep REM-NREM cycles. In a recent evaluation of the current status of his BRAC hypothesis, KLEITMAN (1983) cited more than 50 studies that investigated the possible existence of a waking counterpart of the sleep cycles. Since all these studies were motivated by his hypothesis, it is not surprising that any finding of a waking cyclicity with a dominant period of about 1.5 h was parsimoniously interpreted as directly supporting BRAC. Only in recent years has it become clear that 1.5-h ultradian rhythms represent a multi-oscillatory phenomenon rather than the activity of a single oscillator.

At least three separate physiological systems were shown to have 1.5-h rhythms: renal secretion, gastric motility, and cortical alertness. There is convincing evidence that although the rhythms in renal secretion and the rhythms in gastric motility persist during sleep, they are independent of the sleep REM-NREM rhythms (for a recent review see LAVIE and KRIPKE 1981). Thus, the rhythms in alertness remain as candidates for a test of KLEITMAN's BRAC hypothesis.

The rhythms in alertness, the focus of the present chapter, fit closest to KLEITMAN's original BRAC hypothesis. This chapter will summarize the evidence supporting the existence of ultradian rhythms in

Experimental Brain Research, Suppl. 12

alertness during the habitual waking day. It will be suggested that these rhythms and the REM-NREM sleep cycles reflect the activity of related mechanisms. It will be argued that the waking and sleeping portions of BRAC provide distinct "gates" during which the transitions from one state to the other are greatly facilitated.

EVIDENCE FOR ULTRADIAN RHYTHMS IN ALERTNESS - EEG MEASURES

A number of studies have shown ultradian variations in EEG measures of alertness. In isolated subjects under partial sensory deprivation, KRIPKE (1972) reported significant ultradian rhythms in EEG activity in various frequencies that were most prominent in the delta frequency band (0.5-0.3 cps). KRIPKE and SONNENSCHEIN (1978) studied the relationships between waking REMs, EEG alpha activity, and daydreaming in isolated subjects and in subjects in their natural social environment. Daydreaming, characterized by vivid sensory imagery resembling the hypnagogic hallucinations of sleep onset, occurred at 90-min intervals and was correlated with continuous alpha activity and diminution of eye movements. The effects were most pronounced in the isolated condition. In a later replication of this study, somewhat longer periodicities in daydreaming were evident, in the order of 3 h (see KRIPKE et al., this volume, Chap. 12). As we shall see, there are additional findings of ultradian rhythms in alertness with periodicities longer than 1.5 h.

LAVIE and SCHERSON (1981) postulated that ultradian rhythms, if present, should be reflected in subjects' ability to fall asleep at different times during the day. They instructed subjects to close their eyes and to fall asleep during 5-min periods of darkness at 20-min intervals for 12 h. Sleep and wakefulness were defined using EEG. Significant 1.5-h ultradian rhythms in sleep stage 1 were evident. As one might expect, ability to fall asleep was also modulated by circadian rhythmicity, with increased wake time toward the evening hours, and increased sleepiness around midafternoon. Consequently, the ultradian variations in alertness levels were much more prominent during the morning.

The rhythmic occurrence of naps during the day was also observed in patients suffering from pathologic somnolence. In a recent study, VOLK et al. (1984) investigated the temporal structure of the diurnal sleep attacks in narcoleptic patients. These patients suffer from uncontrollable naps during the day, sometimes in combination with cataplexy. For some narcoleptics, napping occurred cyclically, with a periodicity of about 1.5 h. Similar results were previously reported by PASSOUANT et al. (1969).

LAVIE and ZOMER (1983) investigated the relationship between the daytime rhythms in the ability to fall asleep and the sleep REM-NREM cycles. They awakened subjects from either REM or NREM periods and instructed them, as in the study by LAVIE and SCHERSON (1981), to attempt to fall asleep during 5-min periods of darkness at 15-min intervals for 8 h. The results of this study showed that the ultradian variations in the ability to fall asleep during the day show two different ultradian components: a 1.5-h ultradian component which appeared synchronized, albeit weakly, with the REM-NREM cycle, and a second morning component of a much slower ultradian cycle of about 3.5-4 h. Similar experiments were also carried out, without prior sleep, between 1600 and 2400. There were no significant ultradian components during the afternoon-evening experiments.

Evidence for the existence of a second ultradian component slower than 1.5 h was provided by other studies as well. In the second part of the study by LAVIE and SCHERSON (1981) mentioned previously, the effects of sleep deprivation on the ability of subjects to fall asleep at different times of the day were investigated. Under these conditions, the 1.5-h ultradian rhythmicity was modified toward a much slower rhythmicity of about 3.5-4 h. These rhythms appeared synchronized across subjects. It should be mentioned that the subjects in the study by LAVIE and ZOMER (1983) were also partially sleep deprived because of the early REM and NREM awakenings which can account for the slow ultradian component detected in that study. Two studies that investigated sleep-wake cycles under the conditions of prolonged continuous bed rest also provided evidence for the existence of slow ultradian cyclicity in alertness (NAKAGAWA 1980; CAMPBELL 1984). In both studies a sleep-wake cycle of about 3-4 h was evident, although its statistical reliability had not been determined.

GERTZ and LAVIE (1983) investigated the possible existence of ultradian rhythms in EEG alertness in yet a different way. They recorded

5 min EEG activity from 11 subjects at 10-min intervals for 7.5 h under baseline and biofeedback conditions. In the latter condition subjects attempted to either raise or lower the frequency of their EEG with the aid of biofeedback. Both in the baseline and in the feedback conditions there were ultradian rhythms of 7.2-14.4 cycles per day (c/d) (corresponding to periodicities of about 100 min/c and 200 min/c) in the mean frequency of the EEG and in the mean amplitude and integrated amplitude of the EEG. The biofeedback task had little effect upon the ultradian rhythms with respect to baseline. High resolution spectral analysis revealed that the dominant frequency was closer to 7.2 c/d than to 14.4 c/d.

OKAWA et al. (1984) investigated the occurrence of spontaneous fluctuations in vigilance, measured by continuous electrographic recordings in subjects who were encouraged to continue their normal daytime activities. Their results showed that vigilance fluctuations occurred cyclically with periodicities within the wide range of 60-100 min. The authors indicated, however, that the fluctuations were more rapid and pronounced during the first half of the day and that there was evidence for even longer periodicities in the data. As reported by others, the ultradian rhythms were unstable when intrasubject comparisons were made between experiments conducted several weeks apart. They concluded that the daytime ultradian rhythms may be explained by the simultaneous occurrence of several frequencies.

MANSEAU and BROUGHTON (1984) also investigated the occurrence of ultradian rhythms in EEG activity of awake individuals during the habitual waking day. They recorded frontal and parietal EEGs every 15 min for a duration of 100 s for 8 h. Spectral analyses showed the presence of significant 16 c/d rhythms (corresponding to periodicities of 72-100 min/c) in the total power of the EEG, of each hemisphere. The ultradian rhythms in electrocortical activity were synchronized between the two hemispheres. For individual EEG bandwidths, only the frontal theta showed a marked, though not statistically significant, ultradian rhythm at the expected 16 c/d frequency. In agreement with other studies, a substantial number of spectral peaks were at the slower frequency of 8 c/d corresponding to periodicities centered at 3 h/c.

Several animal studies also showed ultradian variations in EEG measures of arousal. KRIPKE et al. (1976) made continuous 24-h electrophysiological recordings in rhesus monkeys maintained on 12:12 h LD cycles. They observed ultradian rhythms in several EEG frequency bands during both periods. During the lights-on period, the rhythms were considerably slower, 120 min/c vs 60 min/c during the lights-off period, suggesting a circadian modulation of the ultradian frequency. Since the phase relationships among the physiological variables were preserved from lights-on to lights-off, KRIPKE et al. (1976) suggested a joint mechanism for the sleep and wakefulness rhythms. In cats that were sleep deprived by brain-stem electrical stimulation, LUCAS and HARPER (1976) reported 24-h, 3-h, and 20-min periods in the rate of brain stimulation required to maintain arousal. The 20-min periodicity is identical to the cat's REM-NREM periodicity during sleep.

OTHER PHYSIOLOGICAL VARIABLES

A number of studies have demonstrated ultradian rhythms in other psychophysiological indices which reflect alertness levels. OTHMER et al. (1969) reported cyclic changes in REM activity in waking subjects which resemble REM cycles in sleep. However, no statistical analysis accompanied their conclusion. HORNE and WHITEHEAD (1976) reported approximately 90-min rhythms in respiratory rate of isolated subjects. LAVIE (1979) demonstrated significant 70 to 90-min ultradian rhythms in pupillary diameter and in an index of pupillary stability in subjects investigated every 15 min for 10 h. The ultradian cycles in pupillary measures were superimposed on much slower trends, which probably reflected circadian variations. Recently, PRESSMAN et al. (1984) reported on similar cyclicity in pupillary diameter in two of three narcoleptics investigated every 30 min in a similar way.

ULTRADIAN RHYTHMS IN ALERTNESS - ONE OR MORE FREQUENCIES?

The above data, coming from studies employing widely different methodological approaches, indicate that there are short-term rhythmic variations in the level of alertness during the habitual waking day.

Although admittedly these cycles are unstable, having large intra- and intersubject variability, at certain portions of the habitual waking day they appear remarkably prominent. In most studies there was evidence for more than a single ultradian component predicted by the

BRAC hypothesis. The most preponderant component has been a periodicity of about 3-4 h.

The existence of a "slow" ultradian rhythm in alertness with a period of about 3-4 h is not unexpected, even though it was not explicitly predicted by the BRAC hypothesis. KLEITMAN (1963) viewed the structure of the infant's 3-4 h sleep-wake cycle as a coalescence of shorter cycles with a 50 to 60-min periodicity, which correspond to the length of the infant's REM-NREM cycles. Two things occur during ontogenesis: (a) the 50 to 60-min cycles are lengthened to 80 to 100-min cycles, and (b) the 3-4 h sleep-wake cycles are consolidated into the adult form of the 24-h sleep-wake cycles. Although the disappearance of the 3-4 h sleep-wake cycles may imply that the oscillator regulating these cycles ceases to function during maturation, it is also possible that its activity is only dampened. It may be released, however, under certain conditions, such as continuous bed rest (CAMPBELL 1984; NAKAGAWA 1980) or increased "sleep pressure" due to sleep deprivation (LAVIE and SCHERSON 1981). It may be noted that BROUGHTON (1975) suggested that the post-lunch dip in arousal, which was noted in most studies investigating ultradian rhythmicity in alertness, is a part of a "slow" cycle with a period of about half a day.

ULTRADIAN RHYTHMS IN ALERTNESS - NONSTATIONARITY OF FREQUENCY AND AMPLITUDE

The finding that the ultradian rhythms in alertness are more rapid and pronounced during the first half of the day than during the second half, has not received much attention so far. This finding has important theoretical and practical implications with respect to the understanding of the daytime structure of alertness. It suggests that the habitual waking day is comprised of two zones which are different from each other with respect to the frequency characteristics of the ultradian rhythms in alertness. The first half of the day is characterized by fast and pronounced rhythms and the second half of the day is characterized by slow and dampened rhythms. The dividing point is the mid-afternoon dip in alertness.

We utilized a new set of experimental data from our laboratory to demonstrate these differences. Figure 1 depicts the mean percentages of sleep stage 1 obtained by 16 subjects who attempted to fall asleep

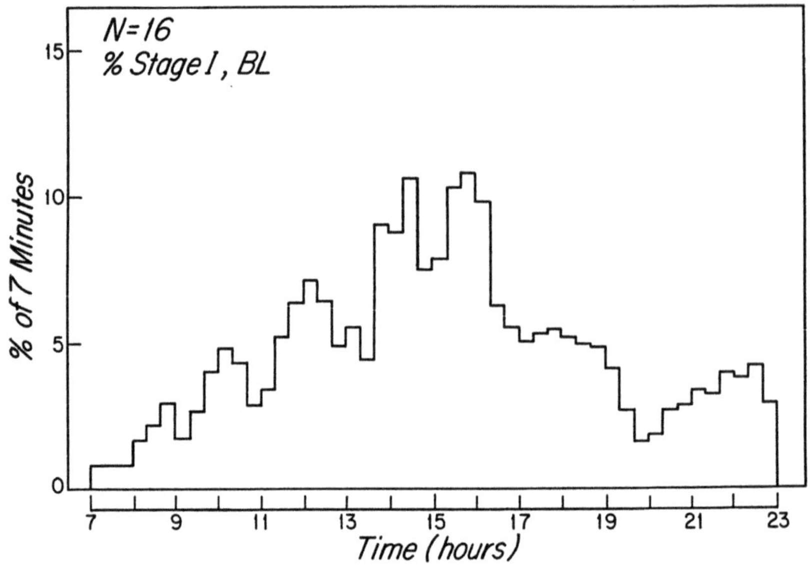

Fig. 1. Mean histogram of sleep stage 1 based on 16 subjects who attempted to fall asleep for 7-min periods at 20-min intervals, from 0700 until 2300. All had a normal night's sleep before testing

for 7-min periods in a darkened room at 20-min intervals, from 0700 until 2300. All had a normal night of sleep prior to testing (for further details of this experimental paradigm see LAVIE and SCHERSON 1981; LAVIE and ZOMER 1983). It can be clearly seen from the figure that the variations in the amount of sleep stage 1 for the period 0700-1600 were widely different from the variations during the period 1600-2300. During the morning-midafternoon period there were rapid fluctuations in the amount of sleep stage 1, superimposed on an ascending trend of sleepiness, reaching a peak at 1500-1600. The interpeak interval for that period was about 80-100 min. In contrast with this pattern, there was a pronounced decrease in sleepiness from 1600-2300, reaching a nadir at 2000, without any evidence of rapid fluctuations. This structure of rapid and pronounced variations in alertness during the first half of the day, followed by much slower and less pronounced variations during the second half of the day, is in good agreement with the results reported by LAVIE and SCHERSON (1981), LAVIE and ZOMER (1983) and OKAWA et al. (1984).

The nonstationarity of the variations in alertness levels across the complete 24-h cycle in individual subjects is shown in Fig. 2. It presents the data of four subjects who attempted to fall asleep for 7-

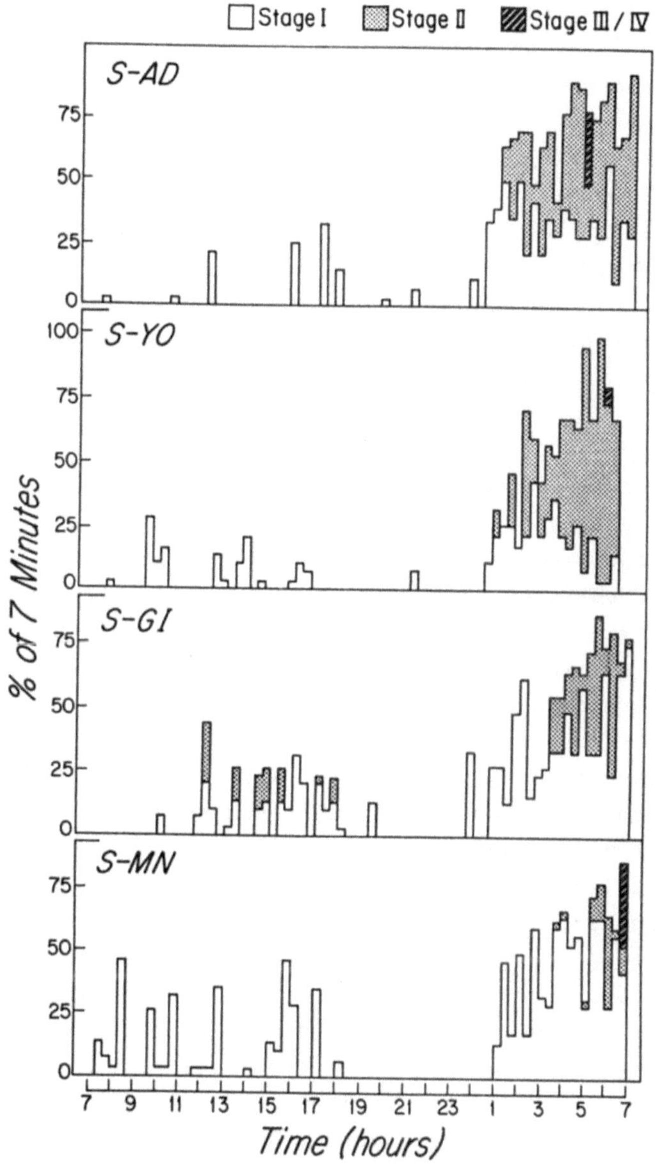

Fig. 2. Sleep histograms of four subjects who attempted to fall asleep for 7-min periods at 20-min intervals, from 0700 until 0700 the next day. All had a normal night of sleep before testing

min periods in a darkened room at 20-min intervals, from 0700 until 0700 the next day. As before, all subjects had a normal night of sleep prior to testing. In agreement with the mean data (Fig. 1), three of the four subjects (YO, GE, and MN) had rapid fluctuations in sleep stage 1 during the period 0700-1700. The fourth subject, AD, had fewer sleep episodes which mostly occurred around 1600-1700. All four subjects had almost no sleep episodes during the period 1800-2400, but showed increasing amounts of sleep from 0100-0200 onward. Interestingly enough, ultradian variations in the ability to fall asleep were evident even during the night when subjects easily fell asleep on every trial (subjects MN and AD, in particular).

The nonstationarity of the rhythms in alertness was further demonstrated in our laboratory in a sleep deprivation experiment. After one night of sleep deprivation, eight subjects began an experimental protocol in which they were required to resist sleep for 7-min periods in a darkened room, spaced at 20-min intervals, from 0700 until 0700 the next day. Subjects spent the 13-min intertrial intervals awake, outside the bedroom, under supervision. Electrophysiological recordings were conducted during the 7-min resisting sleep trials.

The frequency of "awake" and "asleep" trials for each hour of the experiment for the eight subjects can be seen in Fig. 3. An "awake" trial was defined as any trial with no more than 1 min of sleep and an "asleep" trial was defined as at least 75% sleep of any stage. In good agreement with our previous findings, the occurrence of irresistible sleep episodes during the day occurred only during the period 0700-1700. Furthermore, in spite of the progressive sleep deprivation, 50% of all the trials between 1800 and 2300 were "awake" trials. A secondary daytime peak in the ability to resist sleep (40%) occurred at 1200-1600. Also in agreement with the baseline data, the largest increment in the number of "asleep" trials occurred at 2400 (from nine "asleep" trials at 2300 to 19 at 2400).

It can therefore be concluded that the midafternoon dip in alertness divides the habitual waking day into two zones: the morning-midafternoon zone, characterized by rapid rhythms in alertness, with a dominant frequency of about 14.4-16 c/d (a periodicity of about 90-100 min/c), and the midafternoon-evening zone, characterized by an overall decrease in the level of sleepiness and by much slower fluctuations in alertness. This latter zone is ended by a rather abrupt change in

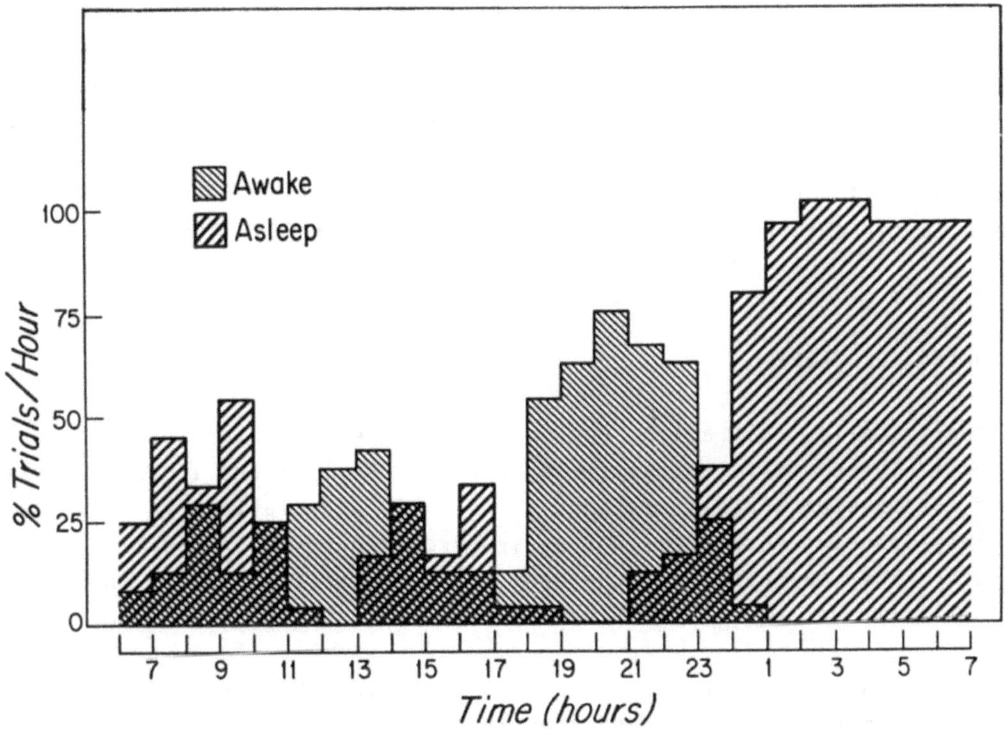

Fig. 3. The frequency of "awake" and "asleep" trials for eight subjects who attempted to resist sleep after a night of deprivation during 7-min periods at 20 min-intervals, from 0700 until 0700 the next day. See text for definitions of "awake" and "asleep" trials

alertness which corresponds to the onset of the "rest" phase of the circadian cycle.

METHODOLOGICAL IMPLICATIONS

It should be pointed out that the nonstationarity of the ultradian rhythms in alertness has important methodological implications. It suggests that analyses of alertness data collected before and after the midafternoon dip in alertness as a single time series, may produce artifactual ultradian frequencies. Possibly, at least some of the unexplained variability and frequency instability reported before can be explained by the nonstationary nature of the data. Except for a single study (LAVIE and ZOMER 1983) that compared ultradian rhythms in alertness before and after the "dip" and demonstrated morning-evening differences in frequency, none of the other studies took this nonstationarity into account. A possible way to overcome the bias introduced by the abrupt changes in the frequency and amplitude of the rhythms is to divide the time series into two stationary parts. This approach has been used before (LAVIE 1976) to demonstrate that the ultradian rhythms in the perception of the spiral after-effect apparent motion are faster during the first half of the day than the second.

"GATES" OF SLEEP

Theoretically, ultradian rhythms in alertness can be explained in at least three different ways: as rhythmic oscillations in the activity of brain-stem arousal systems, as rhythmic oscillations in brain-stem sleep-inducing mechanisms, or as synchronized rhythmic oscillations in both. The majority of the data collected so far support the second possibility of periodic activation of sleep-inducing mechanisms.

The following evidence supports this assertion. KRIPKE (1972) reported that under partial sleep and sensory deprivation, ultradian rhythms in electrocortical activity were most pronounced in the delta bandwidth, which is characteristic of slow-wave sleep. MANSEAU and BROUGHTON (1984), who sampled EEG activity in non-sleep deprived subjects, reported that the most pronounced ultradian rhythms were in theta activity, which is characteristic of light sleep. KRIPKE and SONNENSCHEIN (1978) demonstrated in isolated and non-isolated subjects periodic occurrence of fantasies which resembles sleep onset hypnagogic hallucinations. These fantasies were associated with diminution of eye movements and increased intensity of EEG alpha activity. Similar rhythms were shown by GERTZ and LAVIE (1983). During waking, increased alpha activity was associated with increased sleepiness. LAVIE and SCHERSON (1981) and LAVIE and ZOMER (1983) reported that subjects attempting to fall asleep every 20 min for 5 min showed ultradian rhythmicity in their ability to do so. Both CAMPBELL (1984) and NAKAGAWA (1980) reported on slow ultradian periodicity in spontaneous occurrence of sleep episodes under continuous bed rest conditions. Slow rhythmicity in the ability to fall asleep also predominated under sleep deprivation conditions (LAVIE and SCHERSON 1981; LAVIE and ZOMER 1983). It is therefore more appropriate to call the daytime rhythms ultradian rhythms in sleepiness rather than ultradian rhythms in alertness.

It could be argued that the rhythms in sleepiness may reflect periodic deactivation of arousal systems, rather than periodic activation of sleep-inducing mechanisms. This possibility, however, seems less likely in light of the findings by PASSOUANT et al. (1969) and VOLK et al. (1984) who showed that sleep attacks in narcoleptic patients occurred cyclically with periodicities within the ultradian range, suggesting periodic activation of sleep-inducing mechanisms. These may reflect pathologic exacerbation of the normally occurring 1.5-h sleepiness cycles.

The accumulated findings reviewed in the present chapter necessitate a revision of the prevailing views regarding the structure of the diurnal vigilance levels. It has been generally agreed in the literature that in subjects living under normal sleep-wake schedules cycles of arousal which were synchronized with the circadian cycle in body temperature reach a peak during the second half of the day, and a nadir during the second half of the night. A more complex function is emerging from our data. Figure 4 depicts a schematic representation of the 24-h variations in the probability of a wake-sleep transition ($P_r(W \rightarrow S)$).

The reason for representing the sleepiness rhythms as a probability function is that it provides a quantitative measure of sleepiness which is conceptually easy to understand and convenient to measure.

Evidently, the changes in sleepiness across the 24 h are comprised of both circadian and ultradian components. The ultradian rhythmicity which is superimposed on slow circadian trends reflects the periodic activation of sleep-inducing mechanisms. Sleepiness "gates" occur more frequently during the first half of the day, with periodicity close to the sleep REM-NREM cycles, than during the second half, when they are less frequent and less pronounced. The two most distinct sleepiness gates are the midafternoon and the nocturnal "gates", which are about 7-8 h apart. It should also be pointed out that ultradian rhythms in sleepiness are superimposed on the nocturnal crest of sleepiness as well.

Our preliminary investigations of the effects of sleep deprivation on the ultradian rhythms in sleepiness indicate that while the general 24-h structure remains, the frequency of the rhythms is considerably slower, in the order of 4-6 h, than the rhythms observed under normal sleep-wake conditions.

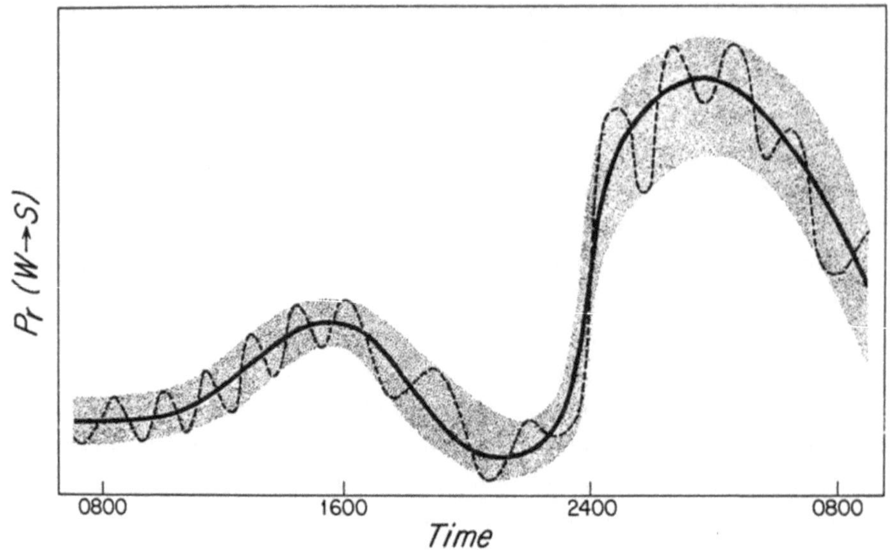

Fig. 4. A schematic representation of the 24 variations in the probability of a wake-sleep transition

"GATES" OF WAKEFULNESS

The nocturnal counterpart of the periodic activation of hypnogenic mechanisms during the waking state is the periodic activation of arousal mechanisms in sleep. These are manifested as the REM and NREM cycles. There is abundant evidence that REM sleep is a state of increased central arousal. It is sufficient to mention that brain temperature, oxygen consumption, and the rate of firing of brain cells are greatly increased during REM sleep.

The notion that the transition from sleep to wakefulness is greatly facilitated during the REM period is not new. As early as 1966 SNYDER suggested that one of the functions of REM sleep is to provide the sleeping organism with periodic "sentinal" arousals during which it can attend to environmental stimuli and be prepared for fight or flight. My only modification of SNYDER's hypothesis is that I suggest that the evolvement of REM sleep as a "gating mechanism" is not necessarily linked to defensive purposes, but to the more general purpose of providing the sleeping organism with a greater control over the length of sleep and over its phase with respect to the geophysical zeitgebers. That is, at certain times during the circadian rest phase sleep can be easily prematurely terminated. The fragmentation of the sleep period into shorter modular units, each ending with a preparatory "gate" which ensures a smooth and efficient transition from sleep to wakefulness, is much more adaptable to the ever-changing geophysical conditions than a prolonged and uninterrupted sleep period with homogenously depressed cortical activity.

What is the evidence that REM periods are "gates" of waking? First, it was shown that when subjects are relying on internal cues to wake up from sleep, they generally terminate their sleep during a REM period. Such evidence was provided in experiments conducted in environments free of time cues. Under the normal sleep-wake schedule about one-third of all morning awakenings occur during REM sleep, which is equal to the percentage of REM sleep in the final hour of sleep. When isolated from time cues, subjects end their sleep period during a REM period about 60%-70% of the time (SCHULZ and ZULLEY, 1980; WEITZMAN et al. 1980). In infants' sleep, the probability of a transition from an active sleep (corresponding to REM sleep) to waking is significantly greater than from quiet sleep to waking, particularly during the first 2 months of life (BOWE and ANDERS 1979). Furthermore, subjects who claimed to be able to wake from sleep at predetermined times without the aid of environmental cues were shown to wake up significantly more often from REM than from NREM sleep (LAVIE et al. 1979).

Second, there is evidence that awakenings from REM sleep as compared with NREM sleep are associated with higher levels of arousal and with less sleep inertia (GASTAUT and BROUGHTON 1965; FELTIN and BROUGHTON 1968; LAVIE and ZOMER 1983). Even more important, awakenings from REM sleep were shown to be associated with superior performance of the right hemisphere-related tasks of space orientation (GORDON et al. 1982; LAVIE et al. 1984; BERTINI et al. 1983). The findings that performances of tasks measuring left hemisphere-related tasks were performed better after awakening from NREM sleep than REM sleep indicate that the superior REM-related performance of the orientation tasks is a specialized characteristic of awakening from REM sleep rather than a general difference in performance capability.

Thus, it can be suggested that REM periods were selected through an evolutionary process as natural endpoints to the basic sleep units. Whenever sleep is regulated by internal cues, as opposed to alarm clocks and/or social demands, awakenings tend to cluster around REM periods to ensure a better preparedness to meet the challenges of the waking state.

SPECULATIONS AND FUTURE RESEARCH

At this stage, the functional significance of the ultradian "gating" mechanism can be only a matter of speculation. The most immediate consequence of the fact that sleep and waking are fragmented by periodic "gates" which facilitate the transition from one state to the other, is an added flexibility and consequently adaptability of the organization of the sleep-wake cycle. Sleep can be entered, with different degrees of ease, at different times during the day. Likewise, sleep can be terminated at different times during the night, without grave behavioral consequences. Obviously, from a structural point of view, a two-state system with multiple transitional points is by far more advantageous than a two-state system with only two rigidly defined transitional points. The inherent frequency characteristics of the endogenous sleep-wake cycle, which is not a replica of the 24-h geophysical cycles, demand a flexible or, better stated, a modular sleep-wake system. Is it possible that the daily phase adjustment of the endogenous free-running sleep-wake cycle is accomplished through the finer ultradian variations in sleepiness and arousal? To answer this question, we certainly need more information about the interrelationship between the ultradian and circadian rhythms.

ACKNOWLEDGMENTS. O. Tzishinsky, J. Zomer and S. Medan provided superb technical help. These studies were supported by grant No. DAJA 37-81-C-0237 from the U.S. Army Research Institute for the Behavioral and Social Sciences through its European Liaison office at the European Research Office of the U.S. Army, London, England. The opinions expressed are those of the author and do not necessarily represent those of the U.S. Army.

REFERENCES

ASERINSKY E, KLEITMAN N (1953) Regularly occurring periods of eye motility and concomitant phenomena during sleep. Science 118:273-274

BERTINI M, VIOLANI C, ZOCOLOTTI P, ANTONELLI A, DI STEFANO L (1983) Performance on a unilateral tactile test during waking and upon awakenings from REM and NREM. In: KOELLA WP (ed) Sleep 1982. Karger, Basel, pp 383-385

BOWE RT, ANDERS TF (1979) The use of the semi-markov model in the study of the development of sleep-wake states in infants. Psychophysiology 10:45-48

BROUGHTON RJ (1975) Biorhythmic variations in consciousness and psychological functions. Can Psychol Rev 16:217-235

CAMPBELL SS (1984) Duration and placement of sleep in a "disentrained" environment. Psychophysiology 21:106-113

FELTIN M, BROUGHTON R (1968) Differential effects of arousal from slow wave sleep versus REM sleep. Psychophysiology 5:231

GASTAUT H, BROUGHTON R (1965) A clinical and polygraphic study of episodic phenomena during sleep. Rec Adv Biol Psychiatr 7:197-221

GERTZ J, LAVIE P (1983) Biological rhythms in arousal indices: potential confounding effect in EEG biofeedback. Psychophysiology 20:690-695

GORDON HW, FROOMAN B, LAVIE P (1982) Shift in cognitive asymmetries between waking from REM and NREM sleep. Neuropsychology 20:99-103

HORNE JA, WHITEHEAD M (1976) Ultradian and other rhythms in human respiration rate. Experientia 32:1165-1167

KLEITMAN N (1961) The nature of dreaming. In: WOLSTENHOLME GEW, O'CONNOR M (eds) The nature of sleep. Churchill, London, p 349

KLEITMAN N (1963) Sleep and wakefulness. University of Chicago Press, Chicago

KLEITMAN N (1983) Basic rest-activity cycle - 22 years later. Sleep 4:311-317

KRIPKE DF (1972) An ultradian biological rhythm associated with perceptual deprivation and REM sleep. Psychosom Med 3:221-234

KRIPKE DF, HALBERG F, CROWLEY TT, PEGRAM U (1976) Ultradian spectra in monkeys. Int J Chronobiol 3:193-204

KRIPKE DF, SONNENSCHEIN D (1978) A biologic rhythm in waking fantasy. In: POPE KS, SINGER JL (eds) The stream of consciousness. Plenum, New York, p 321

LAVIE P (1976) Ultradian rhythms in the perception of two apparent motions. Chronobiology 3:214-218

LAVIE P (1979) Ultradian rhythms in alertness - a pupillometric study. Biol Psychol 9:49-62

LAVIE P, KRIPKE DF (1981) Ultradian circa 1 1/2 hour rhythms: a multioscillatory system. Life Sci 29:2445-2454

LAVIE P, SCHERSON A (1981) Ultrashort sleep-waking schedule. I. Evidence of ultradian rhythmicity in "sleepability". Electroencephal Clin Neurophysiol 52:163-174

LAVIE P, ZOMER J (1983) Ultrashort sleep-waking schedule. II. Relationship between ultradian rhythms in sleepability and the REM-NONREM cycles and effects of the circadian phase. Electroencephal Clin Neurophysiol 57:35-42

LAVIE P, OKSENBERG A, ZOMER J (1979) "It's time you must wake now." Percept Mot Skills 49:447-450

LAVIE P, MATANYA Y, YEHUDA S (1984) Cognitive asymmetries after wakings from REM and NREM sleep in right-hand females. Int J Neurosci 23:111-116

LUCAS EF, HARPER RM (1976) Periodicities in the rates of on-demand electrical stimulation of the mesencephalic reticular formation to maintain wakefulness. Exp Neurol 51:446-456

MANSEAU C, BROUGHTON RJ (1984) Bilaterally synchronous ultradian EEG rhythms in awake adult humans. Psychophysiology 21:265-273

NAKAGAWA Y (1980) Continuous observation of EEG patterns at night and in daytime of normal subjects under restrained conditions. I. Quiescent state when lying down. Electroencephal Clin Neurophysiol 49:524-537

OKAWA M, MATOUSEK M, PETERSEN I (1984) Spontaneous vigilance fluctuations in the daytime. Psychophysiology 21:207-211

OTHMER E, HEYDEN MP, SEGELBAUM R (1969) Encephalic cycles during sleep and wakefulness in humans: a 24-hour pattern. Science 164:447-449

PASSOUANT P, HALBERG F, GENICOT R, POPOVICIU L, BALDY-MOULINIER M (1969) La périodicité des accès narcoleptiques et le rhythme ultradien du sommeil rapide. Rev Neurol (Paris) 121:155-169

PRESSMAN MR, SPIELMAN AJ, KORCZYN AD, RUBENSTEIN AE, POLLAK CP, WEITZMAN ED (1984) Pattern of daytime sleepiness in narcoleptics and normals: a pupillometric study. Electroencephal Clin Neurophysiol 57: 129-133

SCHULZ H, ZULLEY J (1980) The position of the final awakening within the ultradian REM-NONREM sleep cycle. Sleep Res 9:124

SNYDER F (1966) Toward an evolutionary theory of dreaming. Am J Psychiatry 123:121-136

VOLK S, SIMON O, SCHULZ H, HANSERT E, WILDE-FRENZ J (1984) The structure of wakefulness and its relationship to daytime sleep in narcoleptic patients. Electroencephal Clin Neurophysiol 57:119-120

WEITZMAN ED, CZEISLER CA, ZIMMERMANN JC, RONDA J (1980) The timing of REM sleep and its relation to spontaneous awakening during temporal isolation in man. Sleep Res 9:280

Ultradian Rhythms in the Nychthemeron of Narcoleptic Patients and Normal Subjects

H. Schulz
Max-Planck-Institut für Psychiatrie, Kraepelinstrasse 10, 8000 München 40, FRG

INTRODUCTION

Sleep is a nonuniform behavioral state organized in the form of cycles which are composed of a sequence of no rapid eye movement (NREM) sleep stages with a subsequent REM sleep episode. This cyclic infrastructure of sleep is typical for all mammals, whereby cycle duration is species-specific and is associated with such factors as body size and metabolism (HARTMANN 1967; ZEPELIN and RECHTSCHAFFEN 1974).

There has been long-standing controversy regarding the question of whether this ultradian sleep cycle is strictly sleep-dependent or whether it has counterparts in waking. This latter idea has been proposed by KLEITMAN (1961, 1963) who hypothesized that a basic rest-activity cycle (BRAC) is present throughout the nychthemeron and represents the driving force for the REM-NREM cycles during sleep. Many studies have been conducted to test the BRAC hypothesis by measuring the time course of physiological and neurophysiological functions, or the behavior and performance of subjects during the waking part of the nychthemeron (cf. LAVIE 1980; KLEITMAN 1982). Although many of these studies could demonstrate ultradian components in physiological and behavioral measures during waking, these are generally less pronounced and less consistent than those during sleep. The attenuation of ultradian components during waking may be either a basic biological fact or result from the manifold interactions between the awake organism and its environment, a situation which sharply contrasts with sleep, where the organism is effectively shielded against influences from outside. In addition, it should be noted that some studies have been unable to detect any significant ultradian rhythmicity in performance (cf. KRIPKE et al., this volume, Chap. 12).

Experimental Brain Research, Suppl. 12

Yet, even the demonstration of ultradian rhythms during waking, with a frequency similar to that of the REM-NREM cycle, may not be sufficient to convincingly support the assumptions of the BRAC hypothesis, since the validity of this hypothesis has been questioned on the basis of sleep interruption and nap-sleep studies. It has been shown that sleep interruptions lengthen the REM-NREM cycle (BREZINOVA 1973; GAILLARD and TUGLULAR 1976; SCHULZ and ZULLEY 1981) while subtraction of wakefulness from interrupted cycles increases the precision of the ultradian rhythm (MOSES et al. 1978; URSIN et al. 1983). Such data challenge the concept of an ongoing driving mechanism (BRAC) and suggest that the REM-NREM cycle is sleep-dependent and is either ended or temporarily halted by an awakening.

The aim of the present study is to clarify some apparent inconsistencies in the analysis and interpretation of experimental results on the REM-NREM cycle, and to present additional data on ultradian rhythmicity in narcoleptic patients. Sleep-wake patterns of narcoleptic patients were studied for the following reason. Earlier, we reported that there might be a rather constant across-nights drift in the timing of REM sleep, detectable over many consecutive sleep recordings of a single subject (SCHULZ et al. 1975). Figure 1 represents a recent example of such an across-nights drift of REM sleep relative to clock time. Such data are in agreement with the assumption of an ongoing driving force (BRAC) which is active during sleep and wakefulness. A shortcoming of this interpretation is that there is normally no sleep during daytime and, therefore, to infer the existence of an ongoing ultradian rhythm from night-sleep data alone is hard to justify. To address this problem further, the alternation of sleep and wakefulness was studied in a group of narcoleptic patients who are known to have multiple sleep episodes during the nychthemeron. The hypothesis to be tested was that spontaneously occurring sleep, or REM sleep, during daytime is linked to the putative BRAC and hence should be (a) in the same frequency range as the REM-NREM cycle and (b) in phase with the nightly REM-NREM cycle. In other words, the tendency to fall asleep

Fig. 1. Distribution of REM sleep episodes in 14 consecutive nights of ▷ a normal subject (KUL). (a) Arrangement of the sleep episodes in clock time. (b) All sleep episodes are referenced to sleep onset (first epoch S2 sleep). (c) All sleep episodes were shifted by 6 min/24 h against clock time. In the upper part of each insert REM episodes are displayed as open bars. Lights out is represented by a dot, and the first epoch of S2 sleep by a square. The lower part of each insert represents the REM sleep frequency distribution

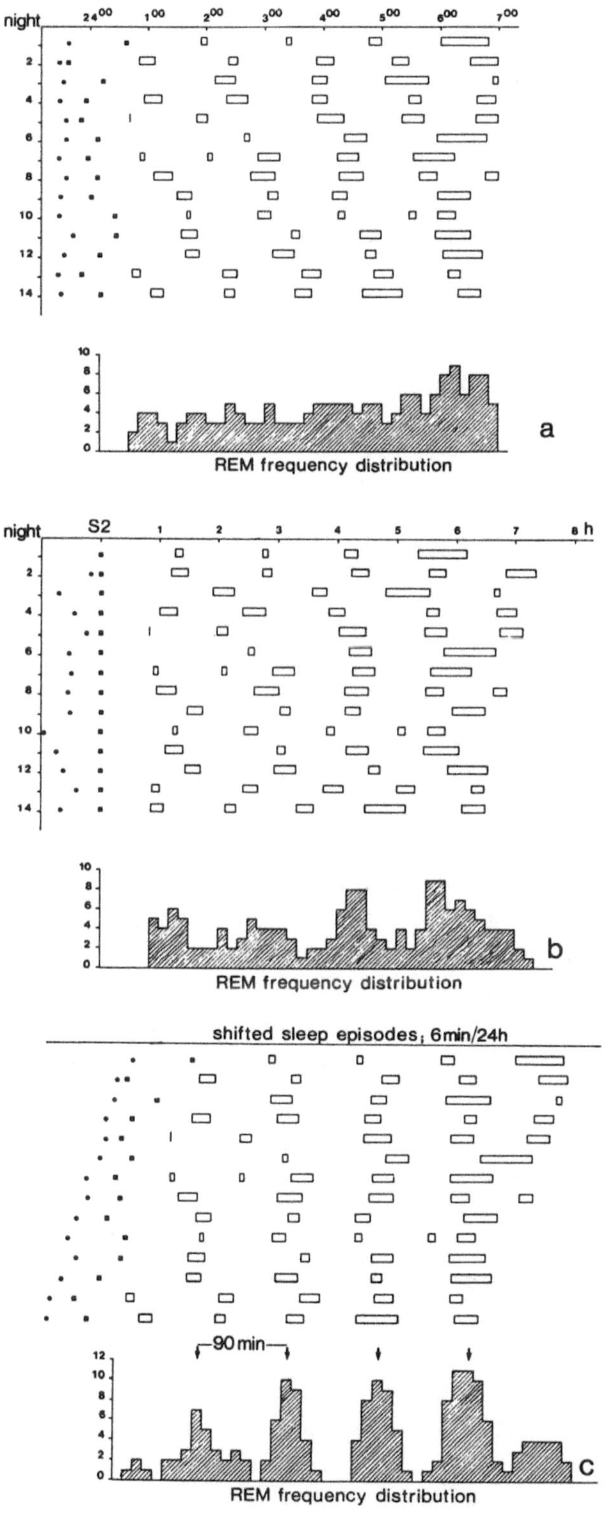
night
24⁰⁰ 1⁰⁰ 2⁰⁰ 3⁰⁰ 4⁰⁰ 5⁰⁰ 6⁰⁰ 7⁰⁰
2 4 6 8 10 12 14
10 8 6 4 2 0
a
REM frequency distribution
night
S2
1 2 3 4 5 6 7 8 h
2 4 6 8 10 12 14
10 8 6 4 2 0
b
REM frequency distribution
shifted sleep episodes; 6min/24h
90 min
12 10 8 6 4 2 0
c
REM frequency distribution

during daytime should not be randomly distributed but recur at equidistant intervals. Indeed, regularity in the temporal occurrence of spontaneous day sleep episodes in some narcoleptic patients has been reported (PASSOUANT et al. 1969; BALDY-MOULINIER et al. 1976). As will be explained later (cf. Methods section), the type of sleep (REM or NREM) during daytime in narcoleptic patients depends to a large degree on the body position during the sleep episode.

REM SLEEP RECURRENCE: SLEEP-DEPENDENT OR TIME-DEPENDENT?

To make a decision as to whether the recurrence of REM sleep is sleep-dependent or time-dependent, some assumptions on which the two hypotheses are based must be clarified. The sleep dependency hypothesis asserts that the REM-NREM cycle is triggered by sleep onset and is halted by, or ends, with the end of sleep (cf. cases II and III, respectively, in Fig. 2). This assumption is in agreement with experimental data which show (a) that sleep cycles which are interrupted by wakefulness are longer than uninterrupted ones (BREZINOVA 1973), and (b) that the timing of the first REM episode after a longer sleep interruption cannot be predicted at better than chance (BREZINOVA et al. 1975).

In addition, it has been shown repeatedly that the periodicity of REM sleep recurrence is more pronounced for compressed than for real-time sleep data (MOSES et al. 1977; MOSES et al. 1978; JOHNSON 1980). This finding in humans was supported by URSIN et al. (1983) who analyzed REM sleep recurrence in cats. Compressing sleep is a technique used to treat sleep-wake data before extracting periodic components, for example, by autocorrelation analysis (NAITOH et al. 1973). While in real-time presentation of the data, all states (REM sleep, NREM sleep, and waking) are preserved, waking is removed in the compressed sleep presentation. However, a serious problem in comparing real-time vs compressed sleep data is that the proportion of REM sleep, which is always higher in compressed sleep, influences measures of the strength of a periodic process. Thus, the index of rhythmicity of an autocorrelation function, which is defined as the difference between the minimum and maximum percent of agreement (NAITOH, this volume, Chap. 16), is biased in favor of stronger periodicity if the proportion of REM sleep increases (FAGIOLI et al. 1980). This methodological problem hampers a straightforward interpretation regarding the sleep dependency of REM sleep.

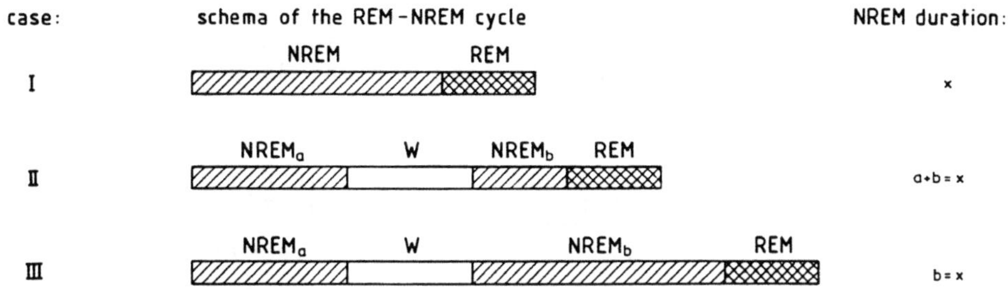

Fig. 2. Three examples reflecting uninterrupted (case I) and interrupted (cases II and III) REM-NREM cycles. Case II represents an interrupted cycle which is governed by a generator with a memory, thus there is negative correlation between the pre- and the postawakening NREM parts (NREM a and NREM b, respectively). The ongoing REM-NREM cycle is halted but not ended by the sleep interruption (W = wakefulness). In case III the cycle is interrupted by W und reset after the end of the sleep interruption

THE REM-NREM CYCLE: A PROCESS WITH A MEMORY

In addition to the normal case of the uninterrupted REM-NREM sleep cycle (case I, Fig. 2), two cases of interrupted cycles can be envisioned (Fig. 2). In one (case II), it is assumed that the REM-NREM cycle is halted during episodes of wakefulness, and resumes with the next sleep onset. In this case, information on the timing of the last REM episode before the awakening is stored and influences the triggering of the first REM episode following the sleep interruption. This model presupposes a neuronal or neurochemical memory whose content is not eliminated by intervening wakefulness. In contrast, case III represents a system without memory. Intermittent wakefulness terminates the cycle which is then reset and starts anew after the interruption. Negative correlation between the pre- and the postawakening part of the REM-NREM cycle has to be expected in case II, but not in case III.

In a previous study (SCHULZ and TETZLAFF 1982), the sleep of ten normal subjects, who slept in the laboratory for 7-30 consecutive nights (median, 14 nights), was deliberately interrupted once per night at 0230 for about 10 min. This design permitted the generation of correlations between the pre- and postawakening parts of the REM-NREM cycle. For each subject, Spearman rank correlations were computed between the NREM interval immediately preceding and that immediately succeeding the sleep interruption. The first epoch of stage 2 sleep after the awakening was defined as the starting point of the latter interval. As Table 1 shows, all individual correlations were negative,

Table 1. Individual Spearman rank correlations between the duration of preawakening and postawakening non-REM interval for ten normal subjects (age median, 46 years; range, 23-65 years)

Subject no.	No. of sleep recordings	Correlation coefficient
32	17	-0.57[a]
31	7	-0.56
35	14	-0.89[c]
37	14	-0.59[a]
38	13	-0.36
33	11	-0.71[a]
34	12	-0.52
40	14	-0.80[c]
39	14	-0.72[b]
36	30	-0.59[c]

[a] $p \leqq 0.05$

[b] $p \leqq 0.01$

[c] $p \leqq 0.001$

resulting in a highly significant mean correlation for the group of subjects ($\bar{r}_s = -0.63$, n = 146 cycles, $p \leqq 0.001$). This result is in agreement with case II and contradicts the assumptions of case III. During the interval between two consecutive REM sleep episodes a steady build-up of a REM sleep-inducing process seems to take place. This putative process may be halted, but it does not decay, during waking. This holds at least for short sleep interruptions.

As Table 2 suggests, only the model of a REM generating process without a memory (case III) would be incompatible with the BRAC hypothesis. On the other hand, cases II and III are compatible with a concept of sleep dependency in its broadest sense. However, on the basis of the compressed sleep analysis case II is clearly favored by the proponents of the sleep dependency hypothesis. For example, URSIN et al. (1983) state "... that the REM-NREM cycle is a periodically driven process but that it is run by an oscillator that is dependent on sleep" (p. 7). Since case III seems not to be supported by experimen-

Table 2. Correspondence between three cases of REM-NREM cycles and two hypotheses on the generation of REM sleep

Case description	Case compatible with	
	Sleep-dependency hypothesis	BRAC hypothesis
I. Cycle uninterrupted	Yes	Yes
II. Cycle interrupted; process with memory	Yes	Yes
III. Cycle interrupted; process without memory	Yes	No

tal data, a decision on the appropriateness of the BRAC or the sleep dependency hypothesis cannot be based on night-sleep data alone but should include observational data which are collected throughout the whole nychthemeron. Hence, data on the interaction of REM sleep, NREM sleep and waking in narcoleptic patients will be presented in what follows.

ULTRADIAN RHYTHMS IN NARCOLEPSY

Subjects and Procedure

Thirteen patients (5 males, 8 females) with narcolepsy-cataplexy were studied for 32 consecutive hours each. Their mean age was 48 years (range, 33-58 years). All patients were without medication during the study. Seven patients had never been treated with psychotropic drugs while the medication was stopped in five patients several days before the study. In one patient the medication could not be fully stopped but was reduced to a very low level.

During the study interval, sleep and wakefulness were continuously recorded using polygraphic methods (EEG, EMG, EOG, ECG, respiration, body movements, actogram, and rectal temperature). All patients displayed sleep onset REM episodes during night sleep.

Seven patients were out of bed during daytime, sitting at a large table. They were allowed to do whatever they wanted to (read, write, listen to music, engage in a hobby, etc.). Snacks and beverages, excluding alcohol and coffee, were continuously available. This group of patients was described in some detail earlier (VOLK et al. 1984).

Six additional patients were confined to bed during the whole investigation, with the exception of trips to the lavatory. Again, these patients were allowed to do whatever they wanted to, as long as they stayed in bed. For both groups, no instructions were given regarding when to sleep during daytime. Time cues were reduced but not eliminated.

Recording pauses of variable duration were caused by (a) shifts of the condition (e.g., bed rest to sitting at the table), (b) trips to the lavatory, and (c) longer recording pauses in the evening hours for those patients who felt uncomfortable due to the long recording session during daytime.

The lights-out period began between 2200 and 2300 and ended between 0700 and 0800.

Data Analysis

Polygraphic recordings were visually analyzed for sleep stages (RECHTSCHAFFEN and KALES 1968) and waking stages (SIMON et al. 1977; VOLK et al. 1984). Data from the two experimental conditions (bed rest vs sitting at a table) were pretreated differently before time series techniques were applied, since the amount and composition of sleep depends on the experimental condition. Patients sitting at a table had less total sleep and less REM sleep, if any, compared with patients who remained in bed. This observation is in accordance with a study by HISHIKAWA et al. (1978) who found a clear reduction of REM sleep in an upright compared with a supine position. Thus, when a narcoleptic patient is unable to stay awake, the type of sleep he enters depends to a large extent on external conditions such as the degree of muscle relaxation in a certain body position. For each patient a binary data set was constructed with the following weighting of behavioral states.

Sitting at a table condition. For the nighttime periods, when patients were in bed, a weight of 0 was assigned to NREM sleep and wakefulness, while a weight of 1 was assigned to REM sleep. For the daytime period, which was spent in an upright position, a weight of 0 was assigned to all epochs of wakefulness and pauses, while a weight of 1 was assigned to all epochs of sleep (NREM or REM). The sitting condition caused either a sharp reduction or absence of REM sleep (see Fig. 3), presumably due to an inability of muscular relaxation in the upright position. This impression was supported by video observations of the subjects.

Finally, the binary data sets were treated by a 15-min filter in such a way that clusters of 30-s epochs of a state (REM or sleep) were transformed into coherent episodes. Thus, a REM episode represents a block of episodes with the weight 1 from the beginning to its end.

Continuous bed rest condition. Two binary data sets per subject were constructed for this condition.

(a) All epochs of wakefulness, NREM sleep and pauses received the weight 0, all epochs of REM sleep received the weight 1. This treatment of the data was applied to the whole measurement period of 32 h. Thereafter, data sets were temporally filtered as described above.

(b) Treatment of the data by the same procedure as for the sitting at a table condition.

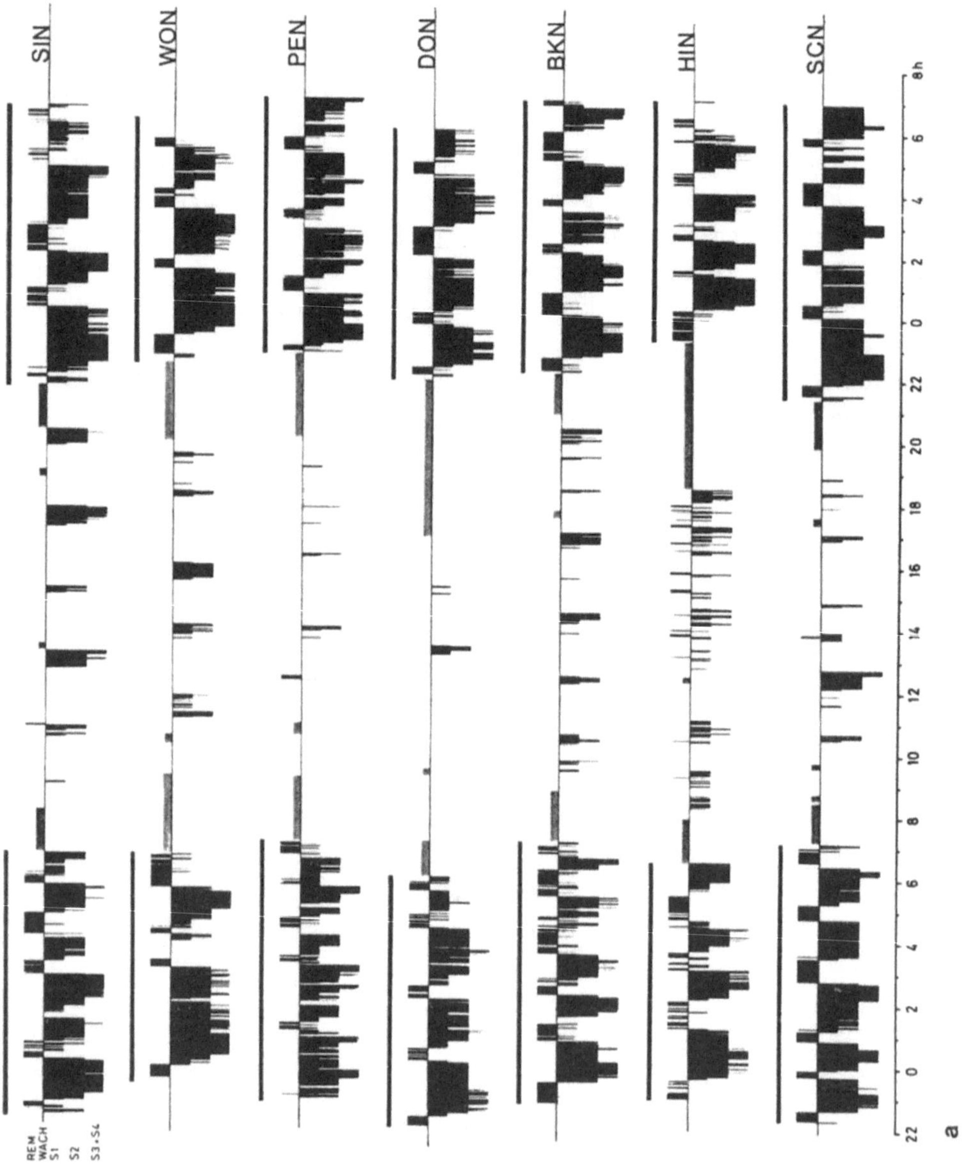

Fig. 3. Temporal distribution of wakefulness (W) and sleep stages ▷ during recording time for 13 narcoleptic patients. (a) Data from seven patients who were out of bed during daytime. The thin middle line in each record represents W. Above this line REM sleep is displayed while the other sleep stages S1, S2 and S3+4 are arranged below the reference line. Lights-out time is indicated by a black bar above the nighttime sleep profile. Recording pauses are indicated by dashed areas above the reference line. (b) Data from six patients who were in bed during the whole recording session. Symbols as above

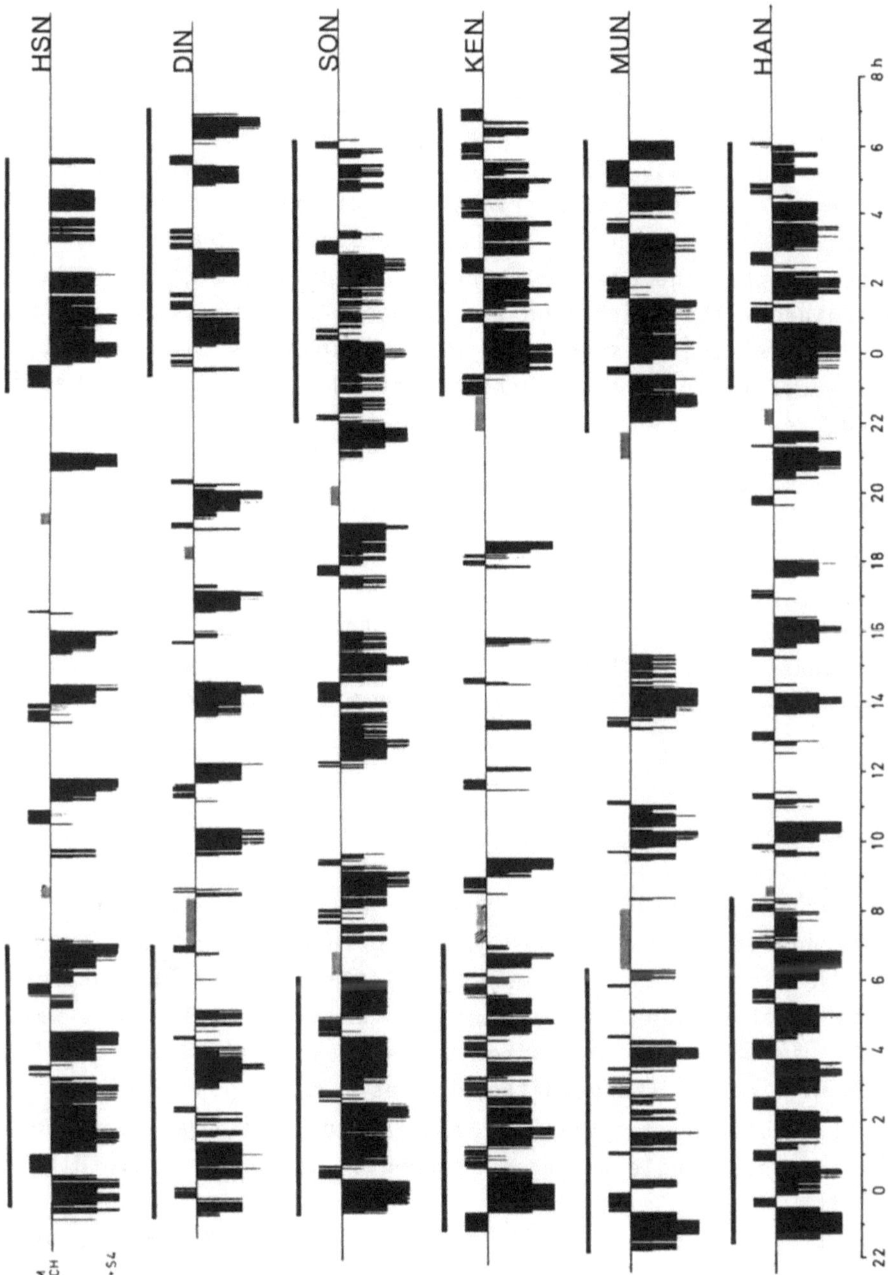

b

The binary data strings were analyzed using the following methods to estimate inherent rhythmic components in the ultradian frequency range. (a) The data were analyzed using ordinary autocorrelation analysis, and spectral analysis after a detrending of slow components. (b) The binary data strings were matched with a strictly periodic process which consisted of a grid with equidistant intervals. By changing period and phase of the periodic process systematically within adequate limits, the grid which best agreed with the binary data set was determined. The criterion for goodness-of-fit was the difference between hits and misses when a given grid was compared with observed data. Details of this concurrence method have been described elsewhere (SCHULZ et al. 1980).

RESULTS

According to the rationale of the BRAC hypothesis, one should expect (a) that the REM-NREM cycle and its putative waking counterpart are similar in duration and (b) that they are phase coupled.

Cycle Duration

A total of 95 completed REM-NREM cycles from nighttime recordings were compared with 54 sleep-wake cycles which were recorded during daytime. The mean cycle duration amounted to 108.4 $\pm$ 30.7 min for the REM-NREM sleep cycles and 119.9 $\pm$ 56.9 min for the daytime sleep-wake cycles. Cycle duration did not differ between these nighttime and daytime ultradian cycles (Mann-Whitney U test, $p \leq 0.05$). Since nighttime and daytime ultradian cycles are similar in their mean duration they were treated as a single data set, encompassing 32 h of recording time.

The temporal distribution of wake, REM, and NREM sleep episodes for each subject are shown in Fig. 3. As can be seen, sleep or REM sleep episodes during daytime occur rather regularly in some subjects (e.g., SIN or HAN), while they are infrequent (DON) or placed at irregular intervals in others (e.g., HIN or KEN). The degree of regularity with which sleep or REM sleep recurred was estimated by the autocorrelation function, the power spectrum, and the concurrence method. The results of these methods are represented in Table 3. There is good agreement

Table 3. Estimation of ultradian periods (min) using three different methods

Subject		Concurrence method	Auto-correlation	Difference Max.-Min.	Power spectrum (Main peak)
SIN		141	144	0.5	143
WON		143	144	0.3	143
PEN		130	132	0.3	133 (69)[b]
DON		130	132	0.4	133
BKN		95	96	0.4	100
HIN		88	-[c]	-	190 (87)[b]
SCN		113	108	0.4	108
SON	A[a]	110	132	0.3	-[c]
	B[a]	252	-[c]	-	250 (200)[b]
HSN	A	152	168	0.3	154
	B	152	144	0.3	143
DIN	A	116	108	0.3	143
	B	116	-[c]	-	118 (154)[b]
MUN	A	100	108	0.4	105
	B	100	216	0.4	222 (105)[b]
KEN	A	101	96	0.5	95
	B	96	96	0.3	95
HAN	A	87	96	0.6	87
	B	87	96	0.4	87

[a]A, binary data pattern; REM = 1, NREM + W = 0
B, binary data pattern; REM + NREM = 1, W = 0

[b]Two main components in the spectrum

[c]No distinct peak in the autocorrelation function or power spectrum

between the period estimates by the different methods. The duration of the ultradian cycles varies between 87 min (HAN, concurrence method) and 152 min (HSN, concurrence method).

Separate analyses for the daytime and nighttime parts of the data sets were not computed, for two reasons. (a) Day sleep episodes were too infrequent (DON, cf. Fig. 3) or too scattered (HIN, cf. Fig. 3) in some subjects to allow meaningful analysis for all subjects. (b) Since nighttime sleep of narcoleptic patients is disturbed in many cases, this would result in irregular ultradian rhythms in part of the nights (e.g., WON first night, cf. Fig. 3). Such disturbances, against a background of only 4-6 REM-NREM cycles per night, makes any period analysis questionable. In other cases, sleep episodes during daytime and REM episodes during nighttime recur at similar intervals (e.g., SIN and HAN, cf. Figs. 3 and 4).

Phase Relationship

The phase relationship between REM episodes during nighttime and REM episodes or sleep episodes during daytime can be inferred from Fig. 4. To construct this figure, total recording time, using the concurrence method, was divided into segments of equal length which corresponded to the best-fitting period. The segments were then superimposed and summed up in the lower part of each panel. To stress ultradian rhythmicity, if present, all events are double-plotted. The difference between the maximum and the minimum of each frequency distribution is a measure of the regularity of the best-fitting ultradian cycle.

Figure 4 suggests that at least in some of the subjects sleep or REM sleep episodes during daytime do not occur at random intervals but are in phase with the nighttime REM sleep episodes. This tendency is evident, e.g., for subjects SIN and WON in the sitting condition and for DIN and HAN in the bed rest condition. In other subjects this conclusion is far less convincing since either the REM episodes from both nights are not in phase (e.g., KEN) or sleep episodes during daytime are dispersed so irregularly that no definite phase relationship can be established (e.g., HIN).

DISCUSSION

One of the intriguing questions of sleep research concerns the mechanism which regulates the cyclic occurrence of REM sleep. Starting from a developmental conceptualization of sleep and wakefulness, KLEITMAN (1963) hypothesized that the alternation of REM and NREM sleep is only the sleep-dependent expression of an ongoing basic rest-activity cycle (BRAC) which operates during sleep and waking. During the activity phase of the putative BRAC, drive-oriented behavior is activated, the aim of which is self-preservation of the individual and of the species. In a similar vein of reasoning, ENGELSTED (1977; ENGELSTED unpublished manuscript) speculated that the REM state is the sleep-dependent manifestation of an ongoing drive cycle. According to this hypothesis, the special features of REM sleep allow endothermic mammals to reduce their muscular activity during sleep while preserving the drive cycle ongoing during sleep and wakefulness. Congruent with this hypothesis it has been shown that disinhibition of the motor system during REM sleep, by specific lesions of brain-stem inhibitory centers, results in a broad spectrum of overt behaviors during REM sleep, including drive-oriented actions (JOUVET and DELORME 1965; MORRISON and REINER 1985). The next question is whether there are waking counterparts of the postulated drive cycle. Most relevant in this respect are reports on ultradian components in food intake (OSWALD et al. 1970) and general oral behaviors (FRIEDMAN and FISHER 1967; FRIEDMAN 1968, 1972) in humans, and performance cycles under conditions of food or electrical self-stimulation reward in animals (STERMAN et al. 1972; LUCAS and STERMAN 1974). In humans, as in cats, the duration of these ultradian cycles in waking behavior was indistinguishable from the duration of the REM-NREM cycles during sleep which last for about 100 min in humans and for about 20 min in the cat. LAVIE (1980) and KLEITMAN (1982) reviewed additional experiments demonstrating ultradian fluctuations in performance or psychophysiological measures. Although such experimental results would be in line with the predictions of the BRAC hypothesis, the concept of BRAC has been questioned on the basis of the great variability of REM recurrence, especially in altered sleep-wake schedules. JOHNSON (1980) stated: "If the BRAC concept is to be useful, it has to be capable of predicting the approximate timing of certain events. One interpretation of the BRAC is that REM should occur at approximately equidistant time intervals throughout the altered sleep-wake schedule. Our data clearly do not support this interpretation" (p. 305). If precision in

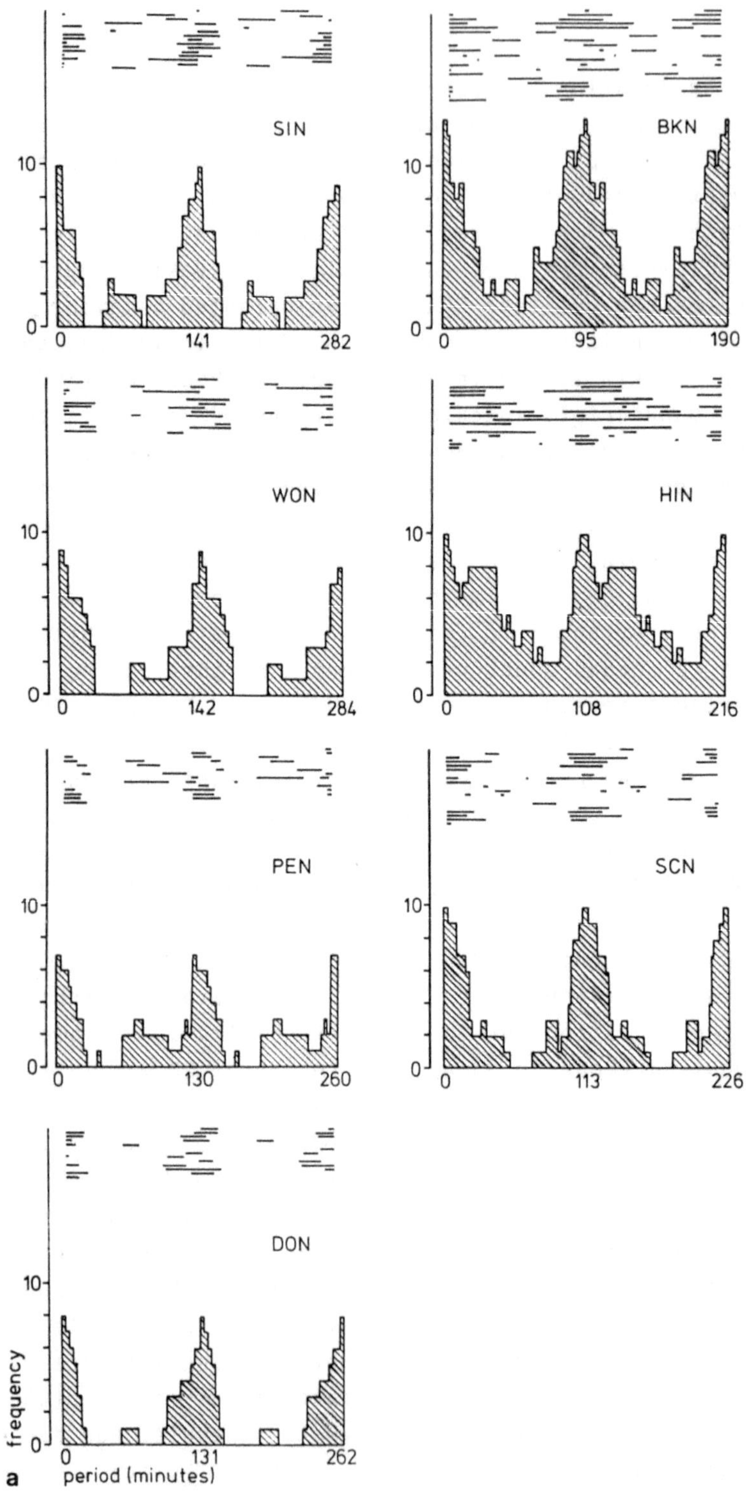
SIN
BKN
WON
HIN
PEN
SCN
DON
10
0
141
282
95
190
142
284
108
216
130
260
113
226
131
262
frequency
period (minutes)
a

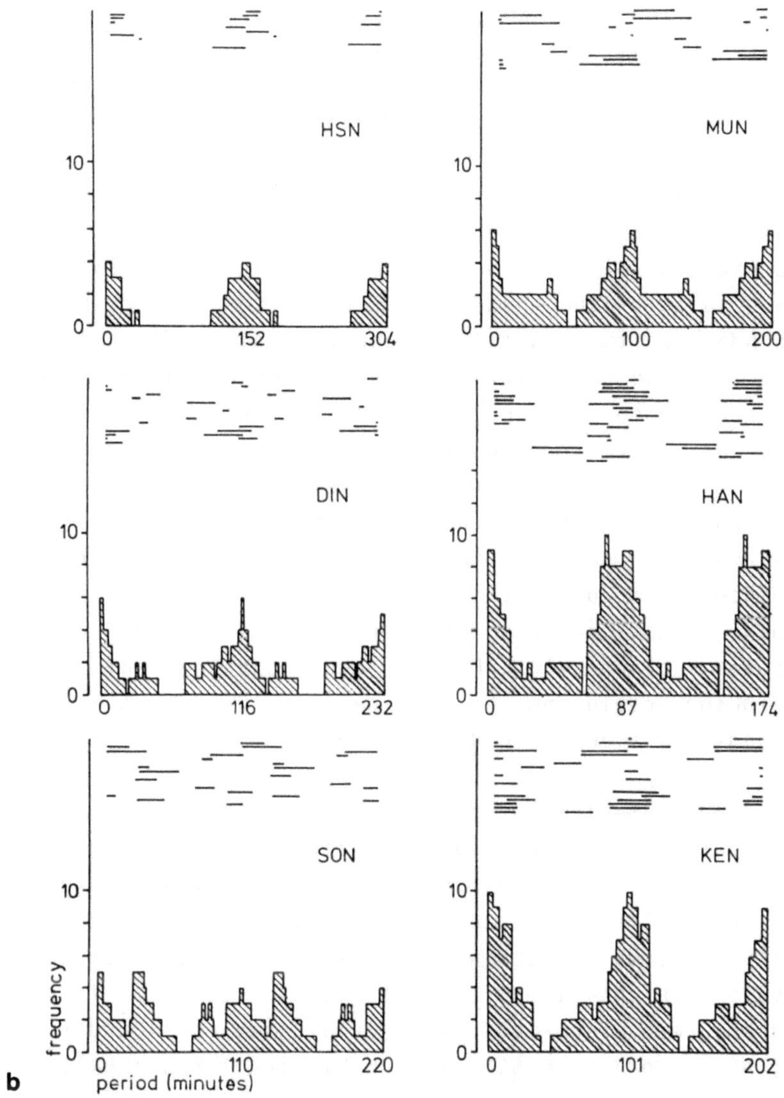

Fig. 4. (a) Representation of nighttime REM sleep and daytime sleep episodes of seven patients who were out of bed during daytime. The whole recording session was divided into segments whose duration was determined by the concurrence method (see text). The segments were then superimposed. The lower part of each insert gives the frequency distribution of the sleep episodes. (b) Representation of nighttime and daytime REM sleep episodes of six patients who stayed in bed during the whole recording session. All other explanations as above

the timing of REM sleep is viewed as a critical factor in deciding between the BRAC and the sleep dependency hypotheses, we are confronted with a dilemma since, on the one hand, precision is rather low for uninterrupted (LEWIS 1974) as well as for interrupted cycles while, on the other hand, the negative correlation between pre- and postawakening NREM parts of a REM-NREM cycle shows that the REM generator has a memory and therefore is not totally imprecise. This interpretation fits with the observation of MIYASHITA and INUGAMI (1982) who also found that REM latency after a short sleep interruption depended on the position of the awakening within the interrupted cycle. It may be that ultradian rhythms do not have to be very precise in a narrow temporal range, because, in contrast to circadian rhythms, they do not have to adapt to any external reference system but only to putative internal reference systems which may enable coupling of different functions which display ultradian fluctuations. While many observations support the notion of relative imprecision of REM recurrence from one cycle to the next, longitudinal observations across many nights suggest long-term stability of REM recurrence in at least some subjects (cf. Fig. 1 and SCHULZ et al. 1975), although this finding could not be replicated by CARMAN et al. (1984) who analyzed long-term sleep recordings of two subjects.

To overcome some of the problems which occur when the BRAC hypothesis is tested by night-sleep data alone, we investigated the sleep and REM sleep occurrence during the whole nychthemeron in narcoleptic patients. For the group of patients, sleep or REM sleep episodes during daytime occurred at intervals which were similar to the REM-REM intervals of night sleep. This ultradian modulation in sleep propensity has also been reported for normal subjects under adequate conditions (LAVIE and SCHERSON 1981; LAVIE and ZOMER 1984). The degree of regularity of sleep-wake and REM-NREM (including waking) cycles during daytime clearly differs between the patients, with only a few displaying regular recurrence patterns. This is in agreement with earlier reports (BALDY-MOULINIER et al. 1976).

Besides the question of regularity of ultradian cycles, it was also possible to test the phase relationship between REM sleep episodes during nighttime and sleep or REM sleep episodes during daytime. This was done by (a) determining the best-fitting ultradian period and (b) aligning all sleep or REM sleep episodes to this ultradian reference system. In many cases the critical episodes had a tendency to be in phase over the whole nychthemeron. This phase relationship between REM during night sleep and day sleep episodes is comparable with the results of LAVIE and ZOMER (1984) who found, in normal subjects, indications for phase continuity between REM sleep in the night and sleep episodes in the morning hours.

ACKNOWLEDGEMENTS. The investigation of narcoleptic patients was conducted together with S. Volk and J. Wilde-Frenz. We thank Mrs. J. Gunst and Mrs. L. Meyrahn for skillful technical assistance.

REFERENCES

BALDY-MOULINIER M, ARGUNER A, BESSET A (1976) Ultradian and circadian rhythms in sleep and wakefulness. In: GUILLEMINAULT C, DEMENT WC, PASSOUANT P (eds) Narcolepsy. Spectrum, New York, pp 485-497

BREZINOVA V (1973) Effect of intervening wakefulness on the NREM-REM sleep cycle. In: KOELLA WP, LEVIN P (eds) Sleep. Physiology, biochemistry, psychology, clinical implications. Karger, Basel, pp 435-438

BREZINOVA V, BECK U, OSWALD J (1975) Sleep cycle duration and timing of REM periods in interrupted night sleep. Int J Chronobiol 3:81-87

CARMAN GJ, MEALEY L, THOMPSON ST, THOMPSON MA (1984) Patterns in the distribution of REM sleep in normal human sleep. Sleep 7:347-355

ENGELSTED N (1977) Evolution søvn og depression. Psychologisk Skriftserie Nr 12. Akademisk, Kobenhavn

FAGIOLI I, RICOUR C, SALOMON F, SALZARULO P (1980) Paradoxical sleep periodicity in early human malnutrition. In: TROJAN S, STASTNY F (eds) Ontogenesis of the brain, vol 3. Universitas Carolina Pragensis, Prague, pp 239-243

FRIEDMAN S (1968) Oral activity cycles in mild chronic schizophrenia. Am J Psychiatry 125:743-751

FRIEDMAN S (1972) On the presence of a variant form of instinctual regression: oral drive cycles in obesity-bulimia. Psychoanal Q 41:364-383

FRIEDMAN S, FISHER C (1967) On the presence of a rhythmic, diurnal, oral instinctual drive cycle in man: a preliminary report. J Am Psychoanalytic Assoc 15:317-343

GAILLARD JM, TUGLULAR I (1976) The orthodox-paradoxical sleep cycle in the rat. Experientia 32:718-719

HARTMANN E (1967) The biology of dreaming. Thomas, Springfield

HISHIKAWA Y, KOIDA H, YOSHINO K, WAKAMATSU H, SUGITA H, IIJIMA S, NAN'NO H (1978) Characteristics of REM sleep accompanied by sleep paralysis and hypnagogic hallucinations in narcoleptic patients. Waking Sleeping 2:113-123

JOHNSON LC (1980) The REM cycle is a sleep-dependent rhythm. Sleep 2: 299-307

JOUVET M, DELORME R (1965) Locus coeruleus et sommeil paradoxale. C R Soc Biol (Paris) 159:895-899

KLEITMAN N (1961) The nature of dreaming. In: WOLSTENHOLME GEW, O'CONNOR M (eds) The nature of sleep. Churchill, London, pp 349-364

KLEITMAN N (1963) Sleep and wakefulness (Rev Ed). University of Chicago Press, Chicago

KLEITMAN (1982) Basic rest-activity cycle - 22 years later. Sleep 5: 311-317

LAVIE P (1980) Ultradian rhythmicity (ca. 90 min) in physiologic processes - evidence for multioscillatory phenomena. J Interdiscipl Cycle Res 11:171-177

LAVIE P, SCHERSON A (1981) Ultrashort sleep-waking schedule. I. Evidence of ultradian rhythmicity in "sleepability". Electroencephalogr Clin Neurophysiol 52:163-174

LAVIE P, ZOMER J (1984) Ultrashort sleep-waking schedule. II. Relationship between ultradian rhythms in sleepability and the REM-Non-REM cycles and effects of the circadian phase. Electroencephalogr Clin Neurophysiol 57:35-42

LEWIS A (1974) The paradoxical sleep cycle revisited. In: SCHEVING LE, HALBERG F, PAULY E (eds) Chronobiology. Igaku Shoin, Tokyo, pp 487-490

LUCAS EA, STERMAN MB (1974) The polycyclic sleep-wake cycle in the cat: effects produced by sensorimotor rhythm conditioning. Exp Neurol 42:347-368

MIYASHITA A, INUGAMI M (1982) A study of nocturnal sleep process interrupted by a short period of awakening. Clin Electroencephalogr (Osaka) 24:178-182

MORRISON AR, REINER PB (1985) A dissection of paradoxical sleep. In: McGINTY DJ, DRUCKER-COLIN R, MORRISON AR, PARMEGGIANI PL (eds) Brain mechanisms of sleep. Raven, New York

MOSES J, LUBIN A, JOHNSON LC, NAITOH P (1977) Rapid eye movement cycle is a sleep-dependent rhythm. Nature 265:360-361

MOSES J, NAITOH P, JOHNSON LC (1978) The REM cycle in altered sleep/ wake schedules. Psychophysiology 15:569-575

NAITOH P, JOHNSON LC, LUBIN A, NUTE C (1973) Computer extraction of an ultradian cycle in sleep from manually scored sleep stages. Int J Chronobiol 1:223-234

OSWALD I, MERRINGTON J, LEWIS H (1970) Cyclical "on demand" oral intake by adults. Nature 225:959-960

PASSOUANT P, HALBERG F, GENICOT R, POPOVICIU L, BALDY-MOULINIER M (1969) La périodicité des accès narcoleptiques et le rhythme ultradien du sommeil rapide. Rev Neurol 121:155-164

RECHTSCHAFFEN A, KALES A (eds), BERGER RJ, DEMENT WC, JACOBSON A, JOHNSON LC, JOUVET M, MONROE LJ, OSWALD I, ROFFWARG HP, ROTH B, WALTER RD (1968) A manual of standardized terminology, techniques and scoring system for sleep stages of human subjects. Public Health Service, U.S. Government Printing Office, Washington

SCHULZ H, TETZLAFF W (1982) Distribution of REM latencies after sleep interruption in depressive patients and control subjects. Biol Psychiatry 17:1267-1376

SCHULZ H, ZULLEY J (1981) The consistency and persistency of the REM-NREM sleep rhythm. In: KARACAN I (ed) Psychophysiological aspects of sleep. Proceedings of the third international congress of sleep research. Noyes Medical, Park Ridge, pp 40-47

SCHULZ H, DIRLICH G, BALTESKONIS S, ZULLEY J (1980) The REM-NREM cycle: renewal process or periodically driven process? Sleep 2:319-328

SCHULZ H, DIRLICH G, ZULLEY J (1975) Phase shift in the REM sleep rhythm. Pflügers Arch 358:203-212

SIMON O, SCHULZ H, RASSMANN W (1977) The definition of waking stages on the basis of continuous polygraphic recordings in normal subjects. Electroencephalogr Clin Neurophysiol 42:48-56

STERMAN MB, LUCAS EA, MACDONALD LR (1972) Periodicity within sleep and operant performance in the cat. Brain Res 38:327-341

URSIN R, MOSES J, NAITOH P, JOHNSON LC (1983) REM-NREM cycle in the cat may be sleep-dependent. Sleep 6:1-9

VOLK S, SIMON O, SCHULZ H, HANSERT E, WILDE-FRENZ J (1984) The structure of wakefulness and its relationship to daytime sleep in narcoleptic patients. Electroencephalogr Clin Neurophysiol 57:119-128

ZEPELIN H, RECHTSCHAFFEN A (1974) Mammalian sleep, longevity, and energy metabolism. Brain Behav Evol 19:425-470

The Basic Rest-Activity Cycle Revisited: Some New Perspectives

M. B. Sterman
V. A. Medical Center, Spulveda, California and School of Medicine, University of California, Los Angeles, CA, USA

In his pioneering attempt to organize the young but rapidly expanding field of sleep research in the early 1960s, KLEITMAN (1963) first elaborated the concept of a "basic rest-activity cycle" (BRAC), proposing that the periodic recurrence of the rapid eye movement or REM sleep state reflected a fundamental physiologic periodicity which must also modulate brain functions in wakefulness. Kleitman considered REM sleep to be a state of "primitive wakefulness" and its periodicity was thought to reflect a phylogenetically older neural integration concerned with the regulation of activity and rest. From the perspective of chronobiologists interested in circadian mechanisms, it is easy to see how this hypothesis might have been interpreted as suggesting a circadian-like ultradian periodicity with precise temporal characteristics. However, from the point of view of sleep researchers, the suggestion of a functionally older sleep-wakefulness system still invoked an appreciation for the many factors which could modulate such regulation. Accordingly, from the outset we never envisioned a precise timing mechanism in this regard.

To test the concept of a phylogenetically older, intrinsic modulating influence on behavioral states we decided first to look at behavior and state-defining physiologic patterns in a lower animal. For this purpose prolonged polygraphic records were obtained from the adult salamander and evaluated in terms of state organization and related EEG characteristics (LUCAS et al. 1969). We found, as have others, that the EEG pattern associated with quiescence in this animal was one of relatively low voltage, high frequency activity, whereas that observed during active periods showed higher voltage activity with mixed and slower frequencies but dominated by 4-7 Hz theta frequency activity. This reversal of state-related EEG characteristics in comparison with mammalian neocortical patterns was consistent with the fact that

Experimental Brain Research, Suppl. 12

they were recorded from paleocortical structures which in mammals show a similar trend. These polygraphic data also showed a state organization consisting of alternating periods of quiescence and activity, indeed defining a rather stable rest-activity cycle with a period of approximately 3 h. Thus, despite the absence of a neocortex and associated polygraphic indices of mammalian sleep states, this animal still demonstrated an ultradian cycle corresponding to a "primitive" sleep-wakefulness rhythm as suggested by Kleitman. Moreover, while motor activity was increased during active periods, corresponding phasic activity and EEG patterns resembled the characteristic components of mammalian REM sleep.

As a follow-up to this study, we investigated the development of somatic activity patterns in the human fetus during gestation (STERMAN 1967; STERMAN and HOPPENBROUWERS 1971). We were interested in determining when and with what characteristics a rest-activity cycle might appear ontogenetically. Periodic epochs of somatic activity began to appear at about 20 weeks postconception and became increasingly clear during the third trimester of pregnancy. Power spectral analysis of activity data recorded from the maternal abdomen during sleep disclosed two peaks of periodicity, one ranging from 80-120 min and a second from 20-57 min (Table 1). The slower frequency was clearly attributed to the maternal REM cycle while the faster was thought to reflect the fetal REM cycle. Subsequent intrauterine studies in animals and investigations employing sonographic imaging in the human fetus abundantly confirmed this conclusion.

Using these data and those reported by other investigators in human infants, children, and adults, an overview of REM cycle maturation was approximated (STERMAN 1972). These findings suggested that a 40-60 min periodicity characterized the REM cycle during the first year of life but that this period lengthened to the familiar 90-min modal value by adulthood. The exact course of this lengthening process was unclear, but available data suggested a marked change after puberty.

Our final examination of the BRAC concept during this period of investigation focused on a comparison of operant performance and sleep-state periodicities in the cat (STERMAN et al. 1972). The object of this study was to evaluate operant response characteristics in the cat using several different performance situations in an effort to disclose any consistent underlying periodicity. This periodicity would

Table 1. Summary of fetal recordings and mean periodicities in fetal motility obtained using power-spectral analysis (from STERMAN and HOPPENBROUWERS 1971)

Subject	No. of recording nights	Gestation period (weeks)	1st Cycle (min)	2nd Cycle (min)
EL[a]	2	21-23	20.0	96.0
PS	5	21-37	38.7	89.5
SW	5	21-39	57.0	105.0
JS	5	21-39	34.3	87.8
KK	3	25-35	50.3	121.3
HL	4	29-40	39.3	85.3
EM	3	31-37	47.0	81.2
SP	3	32-36	30.0	104.2
N=8	N=30	Range = 21-40 weeks	Mean = 39.6 SD = 11.8	Mean = 96.4 SD = 13.2

[a]Premature delivery at 32 weeks gestation

then be compared with simultaneously obtained REM sleep cycle data. Four different operant conditioning paradigms were studied, involving various trained responses for food or positive (hypothalamic) brain stimulation. Both continuous and intermittent performance was evoked, with the resulting rate or interval of responses evaluated.

Performance in all of these test paradigms was indeed periodic (Fig. 1). The mean period of both REM and performance cycles was found to be identical. We concluded that this cycle reflects a more basic regulating influence than the sleep-wakefulness cycle itself, one which most likely is mediated by some fundamental metabolic pacemaker that seeks to regulate function in accordance with tissue needs. Appropriately, therefore, during wakefulness this pacemaker also exerts an influence on brain function that may be of significant importance to an understanding of the regulation of conscious behavior.

In our more recent work, we have attempted to examine this conclusion from a different and more practical perspective. In the real world of human behavior, goal-directed activities are conditioned to a variety of complex factors, the most systematic of which are established by

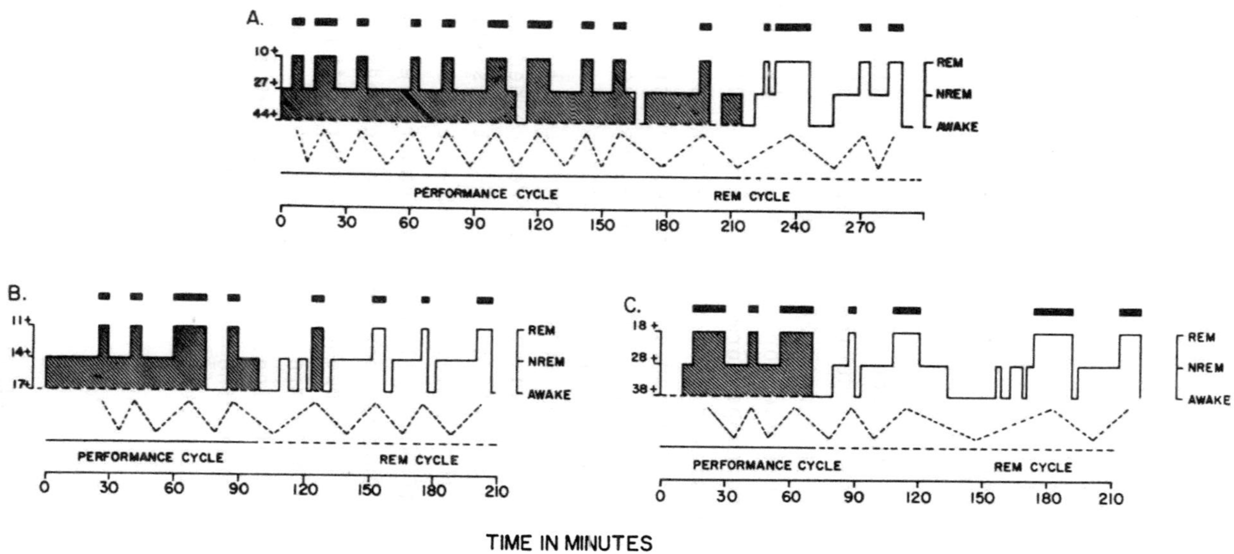

Fig. 1. Graphic display of performance and REM cycles in three food deprived cats. Operant performance was essentially continuous until satiety, at which time sleep ensued. Performance (crosshatched plot) is depicted in categories of shortest, intermediate, and longest response latencies, in seconds, and sleep (open plot) in categories of REM, NREM, and awake patterns. Solid bars mark peak performance and REM episodes. Dotted line traces cycle measurement. (A) Prolonged, cyclic performance with normal transition to sleep. (B) Equal amounts of cyclic performance and sleep with interrupted transition. Unstable initial NREM sleep terminated by period of peak performance instead of REM. (C) Brief period of cyclic performance followed by prolonged sleep. Measurement of periodicity was temporarily broken in sleep by extended period of waking without performance. (from STERMAN et al. 1972)

the workplace. Thus, while man functions according to differing social structures, he is, nevertheless, most often directed by a work schedule which requires performance within a specified time frame. These schedules themselves impose important ultradian influences on human physiology which, in turn, must interact with intrinsic biologic modulation. Ultimately, any effort to examine or utilize the concept of a BRAC within the realm of normal, complex human behavior must consider this interaction.

We have begun to examine this approach as part of an ongoing study of physiologic response characteristics during aircraft operations. Seven adult male subjects, ranging in age from 30 to 48 years, were employed in this study. Each subject was monitored polygraphically during 6 h of alternating performance and nonperformance in a mock-up of an airplane cockpit. Performance consisted of achieving a series of assigned flight paths through manipulation of orientation and velocity controls in a simulated video flight scenario. Subjects were instructed to complete the assigned course change as quickly as possible and to cease "flying" activity until the next adjustment was ordered. Instructions were delivered through a radio headset from a prerecorded tape, and both timing of performance and compliance with instructions were carefully monitored. The resulting sequence consisted of 18 periods of task execution, one every 15 min, followed by comparable periods of verified nontask activity. During nonperformance periods, activity varied from quiet resting with eyes open or closed to reading of magazine or textbook materials. A 15 min "break" was provided at 45-min intervals to provide for stretching, eating, or toilet activities. During these periods the monitoring cable was disconnected from a master input plug to allow egress from the recording situation without removal of electrodes.

Fig. 2. Mean integrated (15 s) bandpass response (Z-score transform) ▷ in two frequency bands from EEG data recorded over two different cortical sites across 18 trials of alternating performance and nonperformance epochs in a visual-motor "flying" task. Data from parietal-occipital cortex (P_3-O_1) at A show decreased density in 4-7 (+) and 8-11 (□) Hz activity during performance epochs (indicated by trial numbers) and increased density during nonperformance epochs (indicated as points between trial numbers). Data from central cortex (C_1-C_5) at β show the opposite pattern but primarily in the 8-11 Hz band (+). At this cortical site, density was increased during performance relative to nonperformance. Periodicity here was imposed by the experimental work schedule

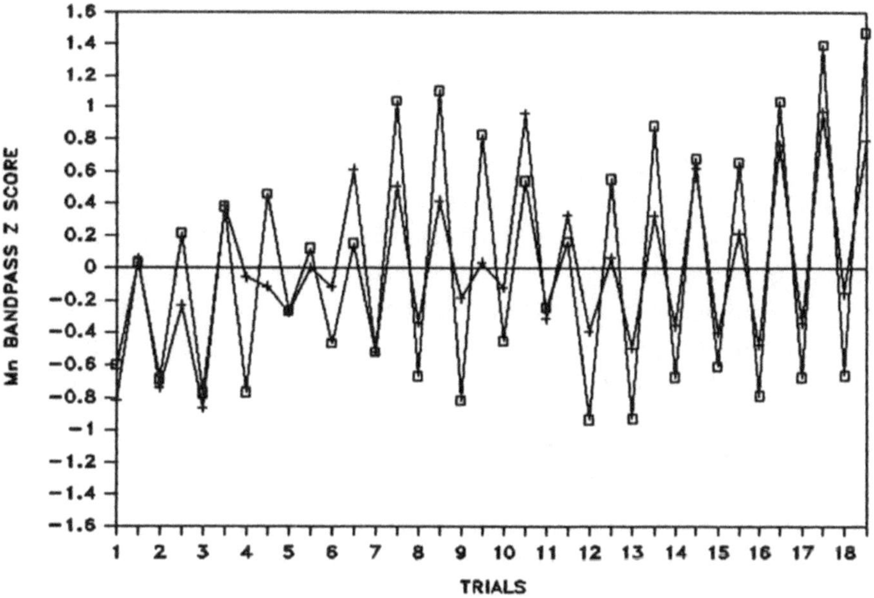

B. CENTRAL 4–7 AND 8–11 Hz.

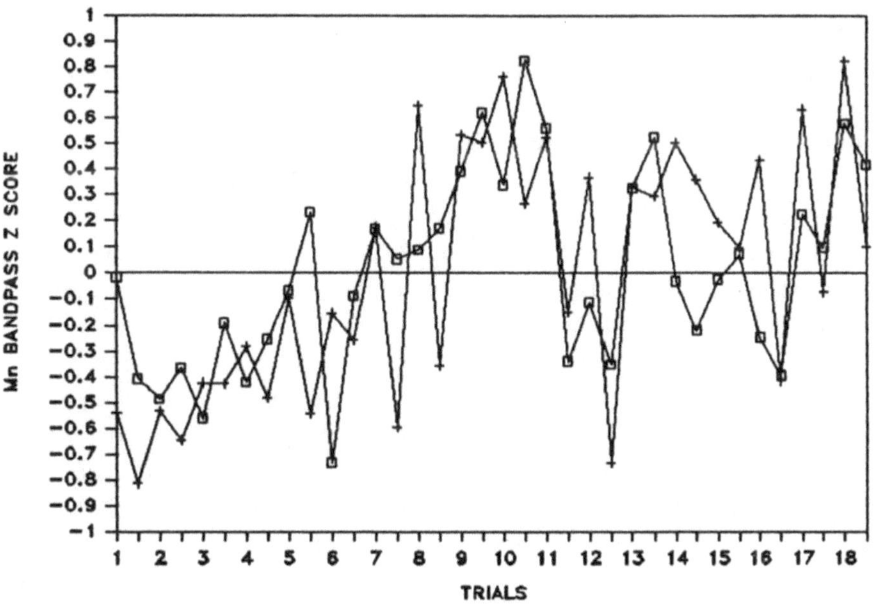

Monitored parameters included selected EEG leads, a chin (digastric muscle) EMG, and end-expired CO_2 percentage in respiratory gas. The present discussion will be limited to a review of EEG findings only. Two bipolar EEG placements were standard, and included central cortex (C_1 - C_5) and parietal-occipital cortex (P_3 - 0_1) leads, according to the International 10-20 System. These leads, together with an ear-clip ground electrode, were fed to a Grass polygraph and a bandpass frequency analyzer which automatically registered activity in two selected frequency bands (4-7 Hz and 8-11 Hz) from each cortical recording site every 15 s. From these data, mean values were determined for each performance and nonperformance epoch of the entire 18-trial sequence. In order to combine data from subjects with differing EEG baseline voltages, bandpass values from each subject were converted to standard Z-scores, with reference to the grand mean of all performance and nonperformance epochs.

Analysis of these data showed clearly that the sequence of performance and nonperformance across the 18 trials of this design imposed an ultradian periodicity on all of the frequency bands studied (Fig. 2). In the parietal-occipital cortex recording, this periodicity was clearest in the 4-7 Hz band with performance-suppressing activity and nonperformance associated with enhancement. Conversely, in the central cortical recording, periodicity was clearest in the 8-11 Hz band and this activity was generally enhanced during performance and suppressed in nonperformance epochs. The consistency of this cycling was particularly surprising in view of the variable activity characteristic of nonperformance periods, and suggests a clear dichotomy between scheduled and nonscheduled activities. Additionally, a clear reciprocity was observed between parietal-occipital and central cortical EEG characteristics in this regard (Fig. 3), and indicates that quite different functional activities are occurring in these two cortical areas in relation to this dichotomy.

If one examines this periodicity closely it is apparent that the reciprocity of frequency modulation between performance and nonperformance epochs at both cortical sites is not constant. Accordingly, difference scores were calculated for the central 8-11 Hz and parietal-occipital 4-7 Hz bands by subtracting the nonperformance epoch bandpass values from the preceding performance epoch value for each band across trials (Fig. 4). This analysis disclosed a marked ultradian periodicity in both bands, with an estimated cycle length ranging

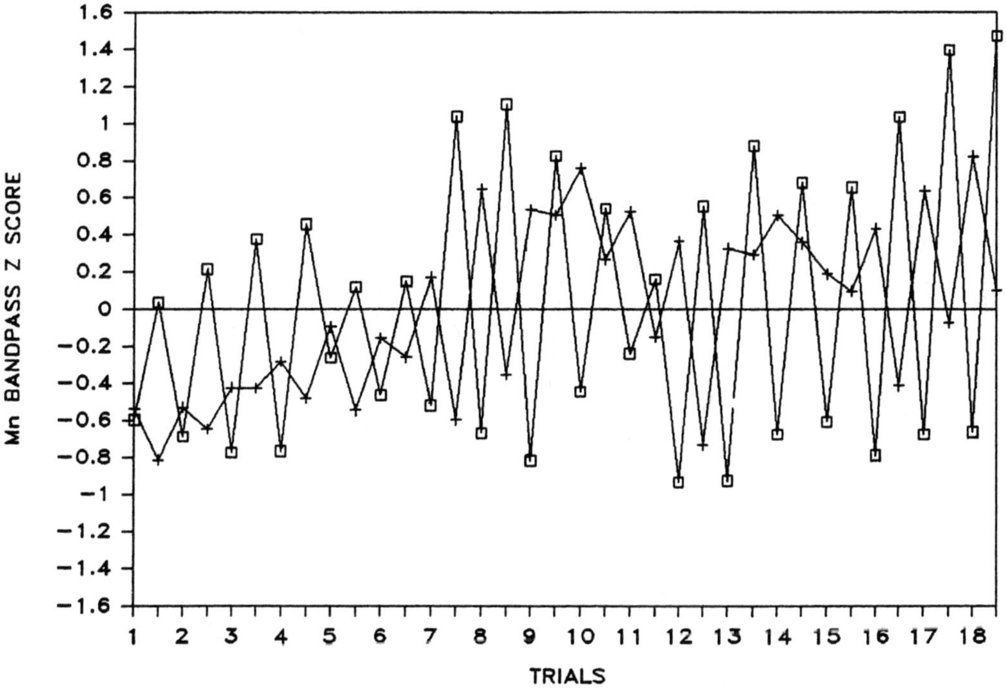

Fig. 3. Mean integrated bandpass data as in Fig. 2 but showing parietal-occipital 4-7 Hz response (□) combined with central 8-11 Hz response (+). Note relatively consistent reciprocity in these two EEG measures in relation to the sequence of performance and nonperformance epochs across trials

from 75 to 90 min. This periodicity was to some extent opposite in the two bands, suggesting the same reciprocity as was seen in the imposed periodicity described above.

In an effort to determine the origins of this slower, intrinsic cycle, a different analysis was explored. In this case sequential bandpass values from performance and nonperformance epochs were plotted separately across trials (Fig. 5). Considering performance epochs only (Fig. 5B), a general trend was observed in all data such that the activity in central cortical EEG frequency bands showed a gradual increment across trials, which leveled off by the middle of the 6-h session, while the parietal-occipital frequency bands showed a more modest and reciprocal decline. In marked contrast to this, the pattern of nonperformance epochs was periodic, with higher power apparent in the parietal-occipital data (Fig. 5A). This periodicity also showed an estimated cycle range of 75-90 min. From these data we may conclude

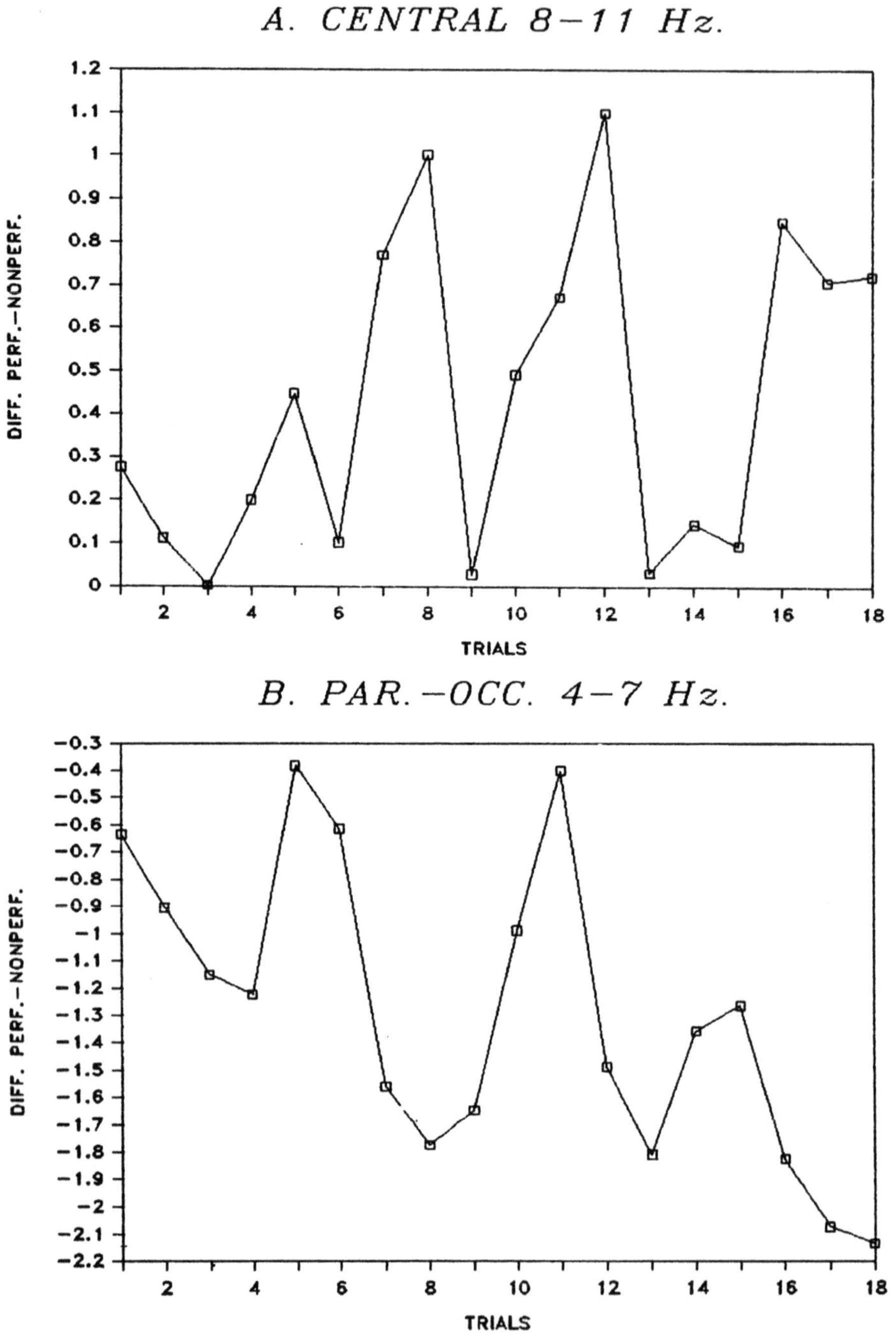

Fig. 4. Plot of successive "difference scores" derived by subtracting bandpass density values obtained during performance epochs from those registered during subsequent nonperformance epochs. Scores from central 8-11 Hz data show a marked periodicity ranging from 75 to 90 min, with peak values trending upwards across trials, while those obtained from parietal-occipital 4-7 Hz data suggest a reciprocal periodicity, with troughs trending downwards across trials

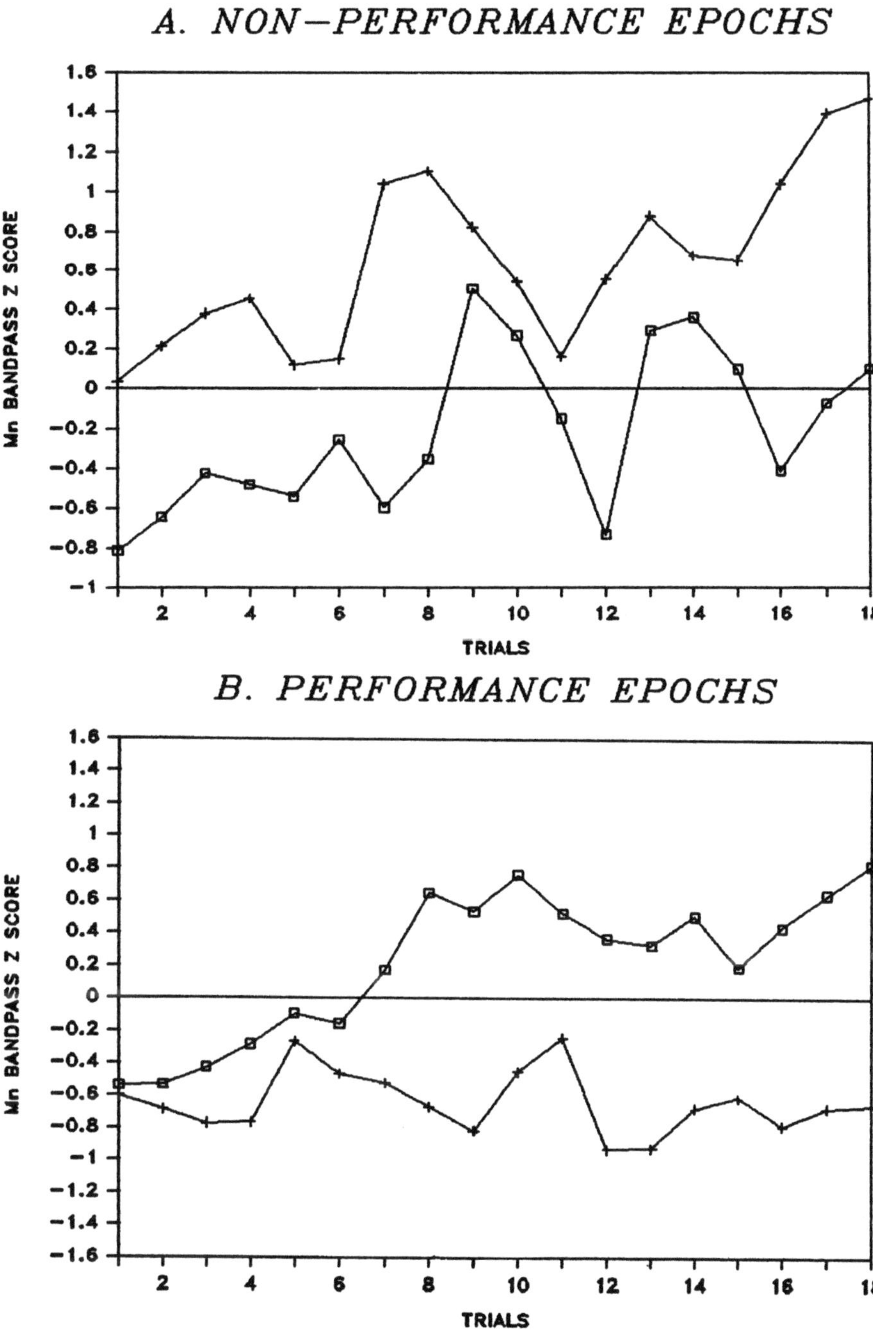

Fig. 5. Sequential bandpass analysis values, as in Figs. 2 and 3, separating nonperformance (A) and performance (B) data points for parietal-occipital 4-7 Hz activity (+) and central 8-11 Hz activity (□). Note that performance values show reciprocal and more or less linear trends across trials, with central 8-11 Hz density increasing, while nonperformance values describe a parallel periodicity of 75-90 min. Moreover, nonperformance data indicate a reversal in relative densities since parietal-occipital values are highest

that the linear trends noted across trials in the previous analysis were due to changes associated with performance epochs while the periodicity in difference scores resulted from modulation within nonperformance epochs.

It should be pointed out that learning was occurring over the course of performance in this study. While all subjects had been given the opportunity to practise this task, none had yet achieved stable proficiency. Indeed, a preliminary measure of performance speed and accuracy in this task indicated a progressive improvement across trials which appeared to be closely correlated with the trend noted in central 8-11 Hz EEG activity across performance epochs. It is reasonable to assume, therefore, that the demand character of the task dominated neural functioning during performance periods. The resulting cerebral engagement would be expected to mask or perhaps override underlying functional periodicity. During nonperformance epochs, however, attentional demand was greatly reduced. Under this circumstance, intrinsic periodicity was evidenced.

A number of conclusions can be drawn from these findings. First, as might be expected, state changes during wakefulness also systematically influence physiology. In this case, the state changes were dictated by an experimental task schedule. While this is not a surprising finding, it is indicative of the many influences which can confound the study of biologic periodicity outside of the laboratory. Additionally, when more complex physiologic variables are of interest, such as the EEG in this case, experimentally imposed periodicity can define the limits of response within a given context, aid in selecting appropriate quantitative measures, and provide important clues concerning the functional substrates of the response observed. With regard to the latter, it is apparent from these data that rhythmic EEG activity in central cortex, resembling the mu (GASTAUT 1952) and sensorimotor (STERMAN et al. 1974) patterns related to movement suppression, was specifically enhanced during a finely controlled visual-motor task. Conversely, rhythmic activity over visual cortical areas was suppressed. This reciprocity was also observed to increase with practice at this task and may correlate with improved performance.

These data again suggested an intrinsic ultradian periodicity of approximately 90 min during a prolonged period of sustained wakefulness. This periodicity was disclosed in the modulation of EEG characteristics during recurrent periods of unscheduled activity. It is reasonable to propose that attention was reduced during these periods and that this condition provided the most appropriate circumstances for detection of a rest-activity cycle in this situation. One of our primary objectives in these studies is to identify physiologic measures which disclose this modulation. Tracking such measures may provide a basis for anticipating performance decrement and/or sustaining optimal response characteristics.

The discovery of yet another "qualified" manifestation of a 90-min modulation in waking physiology again underlines the oversimplification of earlier conceptions. In this case it appears that the vigilance required for learning or coping with a difficult task tends to mask or possibly suppress this modulation. A similar interpretation could be applied to the findings reported in studies of sustained performance by KRIPKE et al. (this volume, Chap. 12). However, in addition to the question of factors which obscure the expression of this basic cycle is the issue of its instability. As stated previously, the facts associated with the study of the sleep-wakefulness cycle make it clear that this process is readily modifiable by both physiologically imperative and cognitive priorities. If indeed the REM cycle represents the manifestation of an even more basic behavioral state-organizing mechanism, reflected as well in waking physiology under appropriate conditions, then this mechanism might also be expected to respond to such influences.

New evidence in support of this conclusion was obtained from an evaluation of data derived from brain transection studies in the cat (Fig. 6). Previous work in the intact animal has documented a broad distribution of cycle intervals with a dominant period ranging from 20 to 30 min (LUCAS and STERMAN 1974; STERMAN et al. 1965, 1972; URSIN 1968). In animals sustained after complete transection of the neuraxis at the mesodiencephalic level (STERMAN et al. 1974; STERMAN 1974), this distribution was altered in two ways. First, the average period lengthened to a peak of approximately 40 min; second, observed periodicity was clearly more stable than that seen in the intact cat. In contrast to this, animals sustained after similar transections at the pontomedullary level (SIEGEL et al. 1984) showed an abbreviated cycle with an average period of approximately 10 min. Moreover, cycle variability was again clearly reduced.

Collectively, these findings suggest a number of new considerations. There may, indeed, exist at the rostral pontine level of the mammalian

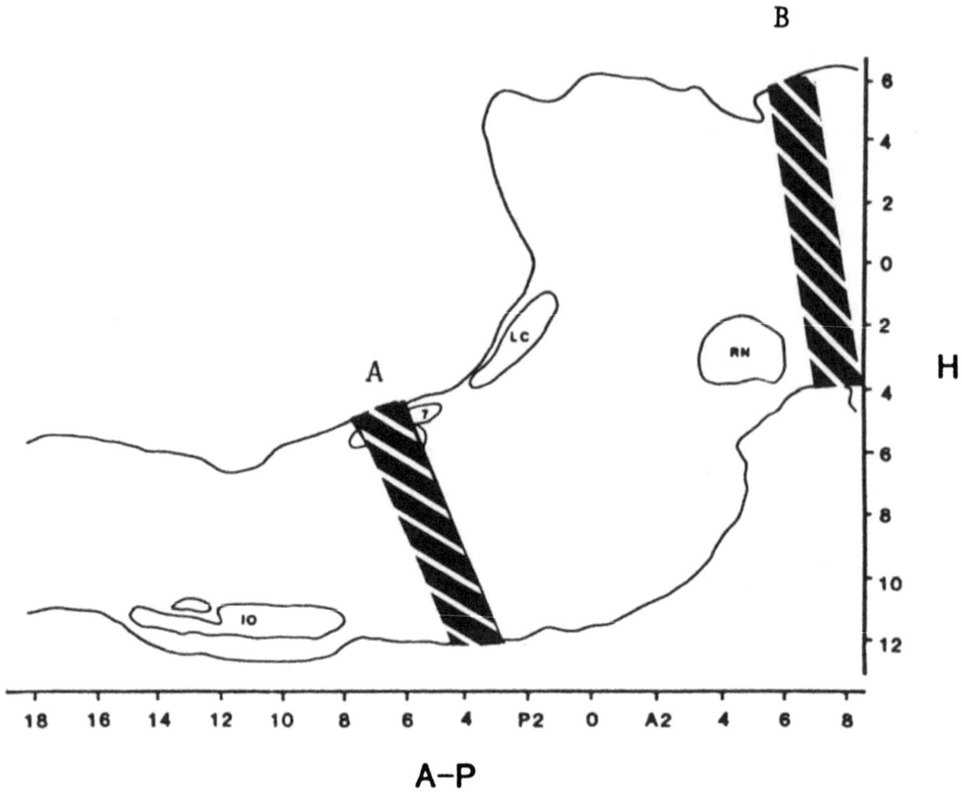

Fig. 6. Diagrammatic representation of saggital plane 1.2 mm lateral to midline in the brain-stem of the cat. The crosshatched area at A shows the range of transections made by SIEGEL et al. (1984) resulting in a stabilized REM cycle with a mean period of approximately 10 min. The crosshatched area at B shows the range of transections made by STERMAN et al. (1974) which also produced a stabilized REM cycle but with a mean period of approximately 40 min. The mean REM cycle in the intact cat ranges from 20 to 30 min. Theoretical implications of these findings are discussed in the text

brain-stem a rather stable ultradian modulating mechanism. Whether or not this influence derives from a precise timing mechanism or a physiologic feedback loop cannot be determined at this time. It does seem clear, however, that the input from more caudal structures, and probably primarily the medulla, lengthens this intrinsic cycle, while that from more rostral structures has a period-shortening influence, since separation of these connections resulted in the opposite effects. In the intact organism the combined effect of these contrasting modifiers could account for the empirical variability of this cycle in a given species. For example, sleep deprivation achieved by forced activity in the cat clearly reduced the REM cycle period (LYDIC et al. 1984). Since the deleterious effects of stress and prolonged wakefulness are most likely mediated by forebrain changes, it could be argued that the observed shortening of the REM cycle reflected a predominant forebrain role in recovery. Conversely, imposed chronic hypoxia in the cat has been observed to prolong the REM cycle period (BAKER and McGINTY 1979). Since blood gas regulation is primarily the province of the medulla, it could likewise be argued that the observed slowing of the cycle resulted from a facilitation of medullary influences on the pontine generator.

This model of a pontine ultradian cycle generator regulated by medullary and forebrain modifiers may also be grossly oversimplified. It does, however, derive from empirical findings and could contribute to an explanation for the intrinsic variability and elusiveness of the hypothetical basic-rest activity cycle. Certainly, it raises questions that are accessible to further laboratory investigation.

REFERENCES

BAKER TL, McGINTY DJ (1979) Sleep-waking patterns in hypoxic kittens. Dev Psychobiol 12(6):561-575

GASTAUT H (1952) Etude électrocorticographique de la réactivité des rythmes rolandiques. Rev Neurol 87:176-182

KLEITMAN N (1963) Sleep and wakefulness, 2nd ed. University of Chicago Press, Chicago

LUCAS EA, STERMAN MB, McGINTY DJ (1969) The salamander EEG: a model of primitive sleep and wakefulness. Psychophysiology 6(2):230

LUCAS EA, STERMAN MB (1974) The polycyclic sleep-wake cycle in the cat: effects produced by sensorimotor rhythm conditioning. Exp Neurol 42:347-368

LYDIC R, McCARLEY RW, HOBSON JA (1984) Forced activity alters sleep cycle periodicity and dorsal raphe discharge rhythm. Am J Physiol 247:R135-R145

SIEGEL JM, NIENHUIS R, TOMASZEWSKI KS (1984) REM sleep signs rostral to chronic transections at the pontomedullary junction. Neurosci Lett 45:241-246

STERMAN MB (1967) Relationship of intrauterine fetal activity to maternal sleep stage. Exp Neurol (Suppl) 4:98-106

STERMAN MB (1972) The basic rest-activity cycle and sleep: developmental considerations in man and cats. In: CLEMENTE CD, PURPURA DP, MAYER FE (eds) Sleep and the maturing nervous system. Academic, New York, pp 175-197

STERMAN MB (1974) Sleep. In: DICARA LV (ed) Advances in limbic and autonomic nervous system research. Plenum, New York, pp 395-417

STERMANM MB, HOPPENBROUWERS T (1971) The development of sleep-waking and rest-activity patterns from fetus to adult in man. In: STERMAN MB, McGINTY DJ, ADINOLFI AM (eds) Brain development and behavior. Academic, New York, pp 203-227

STERMAN MB, KNAUSS T, LEHMANN D, CLEMENTE CD (1965) Circadian sleep and waking patterns in the laboratory cat. Electroencephalogr Clin Neurophysiol 19:509-517

STERMAN MB, LUCAS EA, MacDONALD LR (1972) Periodicity within sleep and operant performance in the cat. Brain Res 38:327-341

STERMAN MB, MacDONALD LR, STONE RK (1974) Biofeedback training of the sensorimotor EEG rhythm in man: effects on epilepsy. Epilepsia 15: 395-417

STERMAN MB, McGINTY DJ, IWAMURA Y (1974) Modulation of trigeminal reflexes during the REM state in brain transected cats. Arch Ital Biol 112:278-297

URSIN R (1968) The two stages of slow wave sleep in the cat and their relation to REM sleep. Brain Res 11:347-356

Ultradian Rhythms During Sustained Performance

D. F. Kripke, D. J. Mullaney, and P. A. Fleck
Department of Psychiatry, University of California, La Jolla, San Diego Veterans Administration Medical Center, San Diego, CA 92093, USA

INTRODUCTION

Since KLEITMAN (1963) first proposed the concept of the "basic rest-activity cycle" (BRAC), numerous attempts to observe this hypothesized rhythmic process have met with mixed success (LAVIE and KRIPKE 1981; KRIPKE 1982). On the one hand, several intriguing descriptions have suggested that the waking human experiences cyclic physiologic variations in roughly the same frequency range as the rapid eye movement (REM) sleep cycle during sleep, that is, one cycle per 90-100 min. On the other hand, some attempts to identify a BRAC in humans have been unrewarding (OKUDAIRA et al. 1984). Studies of monkeys have shown that the prominent frequencies of monkey waking behavior are slower than the monkey's cycle of REM periods within sleep (KRIPKE et al. 1976).

Several positive reports describing BRAC-like phenomena may have resulted from lack of conservatism in statistical methods. Our laboratory's experience in this regard presents a useful example. Some years ago, we described an apparent approximately 90-min rhythm in human fantasy activity both in a series of isolated laboratory subjects and in a separate group of normal subjects studied in their everyday habitats (KRIPKE and SONNENSCHEIN 1978). Each of these studies included about 100 h of data collection. Even 100 h provided a limited amount of data to obtain adequate spectral resolution, and therefore the variance spectra used to confirm the presence of the predicted rhythms had very low resolution after hanning (spectral smoothing). Use of detrending may have interacted with hanning to bias the results toward a false-positive conclusion.

Experimental Brain Research, Suppl. 12

We have now combined these old data with three separate additional experiments performed in our laboratory by Drs Peretz Lavie, John Hiatt, and Mark Bernstein. With the combined results for 39 subjects, it was possible to compute an averaged variance spectrum without detrending and at higher spectral resolution. The averaged spectrum shown in Fig. 1 demonstrates a spectral peak near 7.2 c/d, suggesting an approximately 200-min rhythm, not an approximately 100-min rhythm. This analysis suggesting a 200-min rhythm is inconclusive, nevertheless, not only because the frequency of the spectral peak was not predicted prospectively, but also because even the improved spectral resolution does not permit adequate separation of a 7.2 c/d peak from the linear, curvilinear, and circadian trends in the data. The results presented in Fig. 1 comprise approximately 400 h of data collection.

KLEIN and ARMITAGE (1979) presented evidence for an approximately 96-min rhythm in performances thought to reflect the alternating dominances of the right and left cerebral hemispheres. Our attempt to replicate this study, a precursor of the current series of experiments, was unsuccessful (KRIPKE et al. 1983). We have discussed methodologic problems in the KLEIN and ARMITAGE (1979) study.

To describe rhythms in waking human performances, cerebral dominance, and subjective sensations with greater confidence, we completed an extensive series of experiments, collecting about 2400 h of waking data. Sixty subjects participated in 48-h experiments including up to 42 h of continuous recorded performance from each volunteer. By combining a series of computerized performance tests with automated interrogations, we sought to examine subjective sensations and several performance rhythms simultaneously, including tests intended to reflect functions of both the right and the left cerebral hemispheres. In addition, we systematically examined certain behaviors which could 1ontaminate our performance measures, including eating, drinking, trips to the lavatory, and sleep attacks while the subjects were assigned to perform continuously. The influence of a social factor, human isolation, was also systematically explored. Finally, the duration of our experiments allowed us to examine the influences of fatigue, sleepiness, and circadian rhythms upon the ultradian rhythmicity of the variables measured.

Demonstration of an ultradian rhythm in waking behavioral or physiologic functions is not sufficient to confirm the BRAC hypothesis. It is necessary to demonstrate that observed waking rhythms arise from the

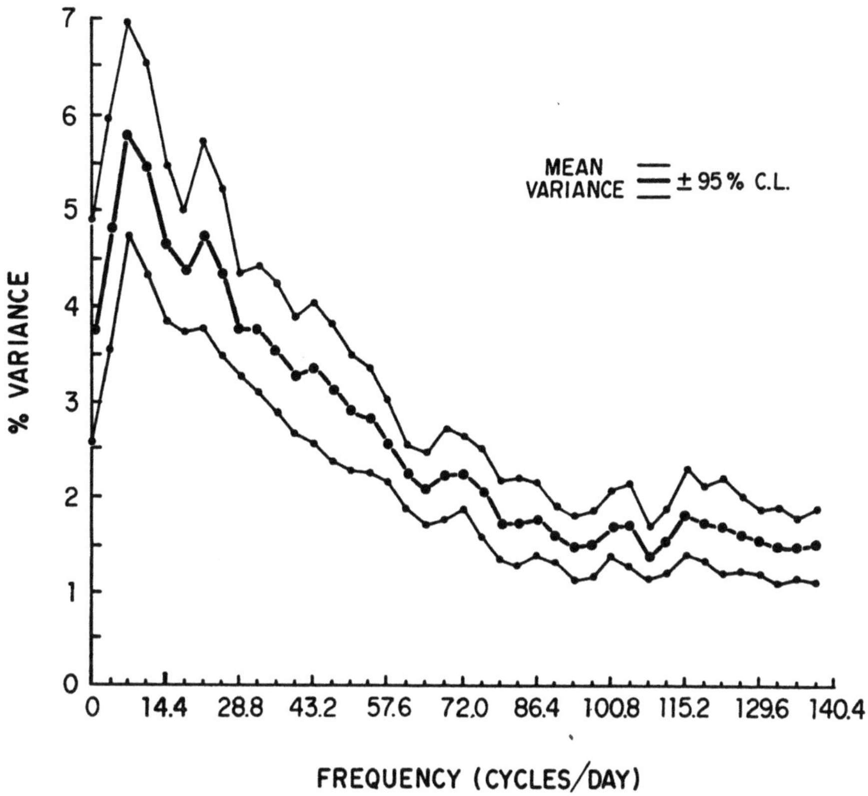

Fig. 1. For the intensity of fantasy, an averaged spectrum is shown for 39 subjects combined from five previous experiments

same oscillatory process which controls the alternation of REM and NREM EEG patterns during sleep. If waking behavioral rhythms are a continuation of the REM-NREM oscillation, then it should be possible to demonstrate a consistent phase relationship between the REM-NREM rhythm during sleep and subsequent behavioral rhythms after awakening. Such phase consistency would support the concept of a continuing oscillatory process. In these 48-h experiments, sleep was recorded for 6 h prior to performance testing in order to compare the phase of the REM-NREM rhythm to the phase of subsequent performance rhythms. In addition, 10 subjects were permitted one 6-h nap in the middle of the performance testing to allow replicated testing of sleep-wake phase continuities, and 10 subjects were allowed six 1-h naps in order to examine the relationship of REM sleep during naps to subsequent performance rhythms.

METHODS

Two 48-h experiments were conducted in which subjects were requested to perform behavioral tasks continuously for up to 42 h. We have described elsewhere the detailed methodology of these experiments and the general effects of this continuous performance paradigm upon sustained performance (MULLANEY et al. 1983a,b).

Subjects

Subjects were 60 male volunteers aged 18-35 years (mean, 21.3 years), mostly university students, recruited by advertisements in a campus newspaper. Potential subjects were screened for suitable personality and motivation. The majority completed short-form MMPIs in advance of the experiment. Subjects with highly disturbed psychological test scores or impaired motivation were not accepted.

Procedure

Each study began with a 6-h period for EEG sleep recording, commencing at the subject's usual bedtime. The 6-h sleep period duration was selected so that virtually all subjects would sleep until aroused for the commencement of performance testing, which began immediately after subjects were awakened. In the first experiment, 10 subjects were randomly assigned to perform continuously for 42 h (42-straight group), 10 subjects were permitted one 6-h nap, from 18 to 24 h after the beginning of performance testing, followed by 18 final hours of testing (18-and-6 group), and 10 subjects were permitted six 1-h naps after each 6 h of performance (6-and-1 group). All of these subjects worked alone in a well-lighted 3 x 3.5 m room. They were occasionally interrupted by staff visits, which were necessary to provide food and drink, repair equipment, and assure the health and safety of the subjects. In the second experiment, all 30 subjects were asked to perform continuously for 42 h. In this second experiment, 10 subjects were asked to work alone largely isolated from human contact, as in the first experiment. The remaining 20 subjects worked in a similar room, together in pairs. Thus, these 20 subjects always had social contact available.

Twelve of the 40 subjects asked to perform continuously for 42 h dropped out toward the end of the protocol, complaining of sleepiness, momentary physical discomfort, or loss of motivation to continue. Thirty of these 40 subjects reported that mental impairments such as hallucinations, delusions, and loss of concentration affected their performance. All 20 of the subjects allowed 6 h of napping within the 42-h performance period were able to complete the study, and only six reported such psychological impairments.

Performance Tasks and Behavioral Measurements

In both experiments, subjects were asked to perform a repetitive 10-min sequence of tasks. The sequence of tasks was presented continuously every 10 min throughout the experiment. Three performance tasks were common to both experiments. First, subjects performed a two-dimensional tracking game which resembled a pilot guiding a randomly moving target into his gun sights. Subjects were rewarded for "hits" on target. In the second task, a 5 x 5 grid of squares was presented followed by two randomly modified patterns. The subject's response as to whether the patterns were altered tested visual pattern memory. This was intended as a test of nondominant hemisphere function. Third, subjects were requested to add two randomly generated three-digit numbers plus the number 7 in their heads and then enter the sum on a keyboard. This addition task was intended as a test of dominant hemisphere function. For each of these three tasks, subjects received immediate feedback as to whether performance was correct or incorrect. A complex monetary reward system provided rewards for continuing performance. Bonuses were provided for successful performances and penalties were incurred, not for errors but for failures to respond to the tasks. The reward system maintained a high level of continuous performance in most subjects throughout the 42 h of testing.

In the first experiment, each of these three tasks lasted 3.0 min, and each was accompanied by a simultaneously presented Wilkinson Auditory Vigilance Task. In the second experiment, auditory vigilance was not tested, and each of the three performance tasks lasted 2.5 min. At the end of the sequence of three tasks, in both experiments, subjects were asked to provide input on visual analog scales of self-rated sleepiness (Stanford Sleepiness Scale) and an attention-fantasy scale, similar to a fantasy scale previously used to measure daydreams (KRIPKE and SONNENSCHEIN 1978). Numeric keyboard inputs were also requested to identify episodes of food ingestion, drinking, and lavatory visits. Finally, in the second experiment, time was allowed for the subject to write out on a card a verbal summary of "what has been going on in your mind" during the past 10 min. These cards were subsequently rank-ordered by both the subjects and the experimenters along a dimension from perceptual-motor attentiveness to dream-like reverie (KRIPKE and SONNENSCHEIN 1978). The mean of the rankings of the subject and the experimenter were then used as a daydream score for each 10 min.

EEG Measures

In the first experiment, EEG, eye movements, and muscle tone (EMG) were recorded continuously during performance. From these recordings, the seconds of EEG alpha per 10-min interval and the seconds of any sleep per 10-min interval were scored.

Spectral Methods

To detect the presence of any stable ultradian rhythms in these data, we employed the variance spectrum method. Spectral smoothing (hanning) was employed to reduce filter side-lobes. Several criteria were established to avoid false interpretations of spectral results:

1. A hypothesized rhythm should produce a peak in the variance spectrum at a prospectively predicted frequency. The prospectively predicted frequency was 1 c/100 min or 14.4 c/d. For a discrete rhythm to be demonstrated, the variance in this prospectively predicted frequency would have to be greater than the variance in adjacent frequencies.

2. A peak in the predicted frequency should be presented without removal of linear trends, circadian rhythms, or other slow frequencies. Detrending is essentially a high-pass filtering procedure which can produce spurious peaks in data consisting only of "pink" noise. Detrending should not be necessary if the spectra have adequate resolution after spectral smoothing.

3. To weight subjects equally, the individual spectra had to be normalized, e.g., converted to variance per frequency as a percentage of total variance. The significance of peaks at the prospectively predicted frequency was examined with tests for the group of subjects, where each subject counted as one degree of freedom. In one-tailed comparisons, a reliable spectral peak should be greater than the mean variance for all frequencies, that is, the expectation for white noise, and it should be significantly greater than the adjacent frequencies.

To examine these data for rhythms, variance spectra were computed for each subject, and then normalized for each variable taken separately. Mean variance spectra were computed for each group of 30 subjects, with the two experiments treated as replications. In addition, in the first experiment, data were examined separately for the subjects studied for 42-straight, for the 18-and-6 schedule, and for the 6-and-1 schedule. Within the 42 h of testing, separate spectral analyses were also performed for consecutive 6-h subintervals to determine whether spectral patterns changed during the course of the study, either due to fatigue or due to circadian effects. All data were examined with a spectral frequency resolution of 1 c/300 min, that is, the spectral frequencies examined were 0 c/d, 4.8, 9.6, 14.4, 19.2 ... c/d. Thus, we prospectively predicted a peak in the 14.4 c/d band, which had half-amplitude limits (after spectral smoothing) of 9.6 to 19.2 c/d (75-150 min/c). In addition, for the second experiment, spectra of double this resolution were also computed, that is, 0 c/d, 2.4, 4.8, 7.2, 9.6, 12.0, 14.4, 16.8 ... c/d. For these higher resolution spectra, the peak was thus prospectively predicted in a 14.4 c/d spectral band which had half-amplitude limits of 12.0-16.8 c/d (86-120 min/c).

Phase Relationship Tests

To examine whether ultradian rhythms in waking performances might be related to the same oscillator which controls REM sleep, we examined whether there was a consistent phase relationship between the REM-NREM rhythm during sleep and subsequent rhythms in performance after the subject had awakened. If the two rhythms are expressions of the same oscillator, there should be a lawful relationship between their phases, assuming that the phase of the oscillator would not be totally reset at the moment of awakening.

To examine such phase relationships, the phase of the REM-NREM rhythm was determined using least-squares cosine fitting during the early morning hours before the subjects ended their sleep interval, e.g., midnight to 0500. The phase of the performance rhythm was then determined from the time of awakening forward. Periods of 50, 100, and 150 min were fit independently. It was prospectively predicted that the 14.4 c/d (1 c/100 min) frequency would show more phase consistency than the other frequencies. In addition, in the first experiment, it was possible to compute phase-angle relationships after the 6-h breaks for the 18-and-6 subjects. Likewise, it was possible to reference the phase of stage REM from each nap to the subsequent performance for the ten 6-and-1 subjects who were allowed 1-h naps. Since no REM sleep occurred in many of the naps, it was arbitrarily assumed that in naps without any REM sleep, the nadir of the REM-NREM cycle corresponded to the midpoint of the nap. As we have noted elsewhere (MULLANEY et al. 1983a,b), consistent trends in several variables occurred in the first few 10-min performance cycles after awakenings. These trends seemed to reflect wake-up effects and, therefore, it was thought that they might obscure any underlying cyclicity. Therefore, the phases of performance cycles were also computed starting with the performance data commencing 10, 20, 30, 40, and 50 min after awakenings. For each subject and variable, the phase-angle difference was computed at each specified frequency between the REM-NREM rhythm and the subsequent performance rhythm. Then the Rayleigh test (BATSCHELET 1965) was used to examine whether the phase-angle differences for the group of subjects were significantly consistent, and thus, nonrandom. A significant group phase-angle concentration would indicate a consistent phase-angle relationship between the REM-NREM cycle during sleep and the subsequent performance cycle during wakefulness.

As a control, the phase consistencies for both the REM-NREM rhythms and the performance rhythms were determined as referenced to the wake-up times themselves.

RESULTS

Spectral Peaks

Because so many spectra were computed for so many intercorrelated variables for several different groups of subjects and several different time intervals, all spectral results cannot be presented in complete detail. In general, the long-term trends and circadian components in the data were the major components in the variance, as described in our previous presentations (MULLANEY et al. 1983a,b). All variables had variance components in the 14.4 c/d frequency band; however, these components were never prominent, regular, and distinct either in the plots of raw data for single subjects or in the averaged plots for groups of subjects. The spectral analyses showed no variance peak in the 14.4 c/d frequency band in any variable for any grouping of subjects in either experiment, with the sole exception of drinking in the 18-and-6 group (first experiment only). This was not a significant peak. At least one such peak was expected by random chance. Even examining the intervals at night or after the subjects had become fatigued, the variance did not peak in the 14.4 c/d frequency band. No notable difference was found between subjects studied alone or in pairs in this regard. Although most of the variance was concentrated in the very slowest frequencies, the 9.6 c/d variance was substantial.

The measures with the most substantial variance in the predicted 14.4 c/d frequency band were drinking episodes and then eating episodes; however, in neither variable was a reliable spectral peak at the predicted frequency found. The mean daydream ranking in the second experiment had a nonsignificant secondary peak at 19.2 c/d, not 14.4 c/d (Fig. 2). Lavatory trips had rather flat spectra throughout all frequencies, whereas sleepiness, sleep attacks, EEG alpha counts, tracking, pattern memory, and addition task performances were entirely dominated by the slowest frequencies. Examination of the ratio of addition performance to pattern memory performance to test hemispheric dominance also produced no spectral evidence for a rhythm at the predicted frequency.

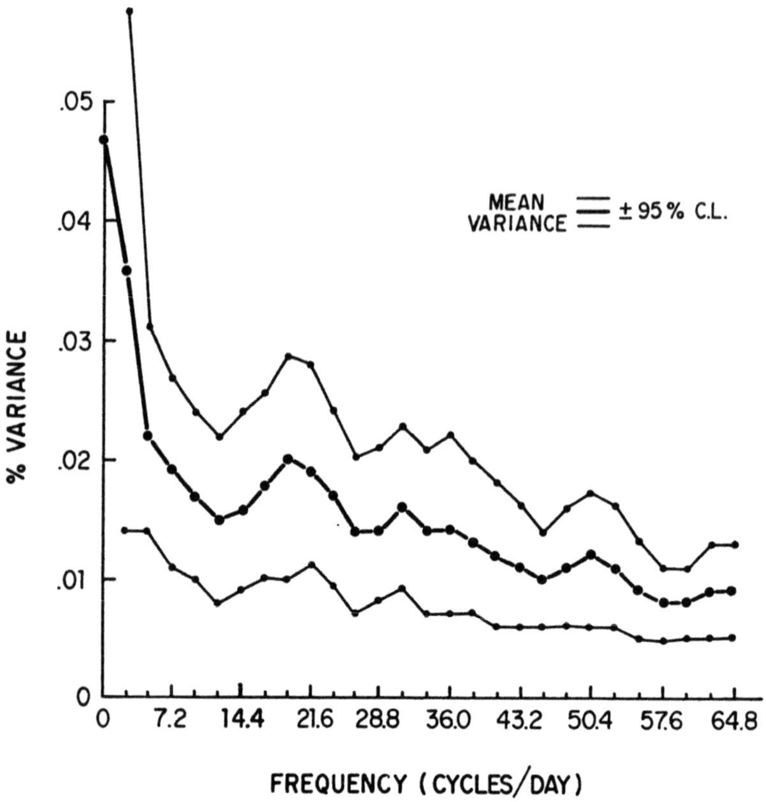

Fig. 2. For the daydream rank scores, an averaged spectrum is shown for the 30 subjects participating in experiment 2

Phase Consistency after Awakening

In the first experiment, the phase consistencies of performance variables referenced to awakening times could be measured in 13 variables for eight different intervals, i.e., after the first awakenings (for 30 subjects), after the 6-h naps (for ten 18-and-6 subjects), and after the six 1-h naps (for ten 6-and-1 subjects). This gave a total of 13 x 8, or 104 tests for significant phase consistency. In general, higher phase consistency was noted when examining the performance which began immediately after awakenings than when considering the performance commencing 10-50 min after awakenings. Thus, to minimize effects of transient awakenings, the phase analyses were applied to performance data from 50 min after awakening to 350 min after awakening.

Of the 104 comparisons computed for the 14.4 c/d frequency, 11 were significant with $p \leq 0.05$. In several cases, the phase consistency was significant with $p \leq 0.001$. Nevertheless, of the 11 variables which were examined at the 14.4 c/d frequency in both experiment 1 and experiment 2, only correct addition responses had significant phase consistency in both experiments, considering data starting 50 min after the first awakening. In experiment 1, significant ($p \leq 0.05$) phase consistency was also noted in this variable at 14.4 c/d after two of the seven nap awakenings.

Of the 104 comparisons computed for a 9.6 c/d frequency in the first experiment, 18 were significant with $p \leq 0.05$. Nevertheless, of 11 variables considered, only correct addition responses had significant phase consistency starting 50 min after the first awakening in both experiments. This variable also showed significant ($p \leq 0.05$) phase consistency at 9.6 c/d after three of seven nap awakenings. The phase consistency was also numerically somewhat greater at 9.6 than at 14.4 c/d. Rather similar phases were computed in both experiments at both frequencies (218 degrees, 207 degrees, 217 degrees, 222 degrees, respectively), suggesting the phases could have been determined by initial curvilinear trends.

Phase Consistency of REM Periods before Awakening

Phase consistency of the REM-NREM cycle during sleep as referenced to the subsequent awakening times was examined in the first experiment for the eight possible comparisons. Significant phase consistency was found for seven of eight awakenings for the 9.6 c/d frequency band, for six of eight awakenings for the 14.4 c/d frequency band, and for only one nap awakening for the 28.8 c/d frequency band. Similar results were obtained in the second experiment.

Phase Consistency of Performance Referenced to REM-NREM

In the first experiment, it was possible to make 104 comparisons of the phase of performance variables starting 50 min after awakenings to the phase of the REM-NREM cycle preceding these awakenings. There was significant ($p \leq 0.05$) phase consistency in 14 of 104 comparisons with the 9.6 c/d frequency, in 9 of 104 comparisons with the 14.4 c/d frequency, and in 14 of 104 comparisons with the 28.8 c/d frequency.

Nevertheless, no performance variable showed significant phase consistency to the prior REM-NREM cycle in both experiments at either the 14.4 or 28.8 c/d frequency. At 9.6 c/d, only drinking behaviors had significant phase consistency to the prior REM-NREM cycle in both experiments (phi = 98 degrees, z = 4.9, N = 30, $p \leq 0.01$ and phi = 14 degrees, z = 3.14, N = 30, $p \leq 0.05$, respectively). Such significant phase consistency in drinking was also noted after one of seven nap awakenings in the first experiment. Drinking was thus significantly phase-locked to the prior REM-NREM cycles, but drinking was not phase-locked to the time of awakening at 9.6 c/d in any comparison in either experiment.

Considering the performance beginning immediately after the first awakenings, at 9.6 c/d, tracking and pattern memory were also significantly phase-consistently referenced to the prior REM-NREM cycle in both experiments. However, the phase consistency for these variables referenced to the awakening time was greater than the phase consistency referenced to the REM-NREM cycle. No such phase consistency was noted at 14.4 or 28.8 c/d.

DISCUSSION

To summarize these results, the variance spectrum analyses showed no evidence for a distinct rhythmic process in the predicted 14.4 c/d frequency band. With respect to drinking, food intake, and daydreaming, the results might be regarded as equivocal; however, they certainly cannot be regarded as positive. It is possible that the periodic phenomena in these data averaged closer to 1 c/150 min (9.6 c/d) than 1 c/100 min (14.4 c/d), but even retrospectively examining this slower frequency band, there was hardly evidence for a definite rhythm.

These observations do not imply that ultradian rhythms in waking humans do not exist, for much evidence for such rhythms has already been described. Rather, these results suggest that such rhythms are not favorably measured in the automated performance tests which were the focus of this experiment. It also appears that ultradian rhythms are not facilitated in their expression by our sustained continuous performance model. In the experience of our laboratory, ultradian rhythms seem stronger in isolated subjects who are given almost nothing to do. The extreme performance pressure of these protocols seems to have reduced the relative variance in the 14.4 c/d frequency band compared with that observed in similar subjects in other protocols. For example, rhythms have been noted in brief sleep attacks and in the alpha rhythm in previous protocols (KRIPKE 1972), but such rhythms were not noted in equivalent measurements in this protocol.

As previously stated, performance measurements do not seem to be particularly favorable aspects of human function in which to perceive ultradian rhythmicity. The alternation of right and left hemisphere cerebral dominance as measured in performance is not, in our hands, a favorable reflection of ultradian rhythms. Nevertheless, in the experiment of KLEIN and ARMITAGE, the subjects were pressured to perform rapidly for only 3 min of each 15 min; therefore, their protocol may have more closely resembled protocols which facilitate the expression of ultradian rhythms.

The mental content reports of these volunteers contained, on average, less rich fantasy material than we have observed in previous isolation experiments where the occurrence of fantasy was specifically encouraged. Even though the protocols produced delusions, visual illusions, or hallucinations among most subjects (MULLANEY et al. 1983a,b), the general quantity of fantasy per unit time was not high. Also, the correlation between the subjects' and the experimenters' rankings of mental content reports ($r = 0.67$) was substantially below the reliabilities we have obtained between raters in previous experiments. This likewise suggests that the richness and range of fantasy reports was reduced by the current protocol. It was striking that the $r = 0.28$ correlation between the retrospective mental content rankings of free verbal reports and the visual analog scale for fantasy was quite low. Perhaps this visual analog scale does not effectively tap the same mental dimension.

Even if waking ultradian rhythms were not of sufficient amplitude to create distinct peaks in variance spectra, it is possible that significant phase consistency between the REM-NREM cycle and subsequent 14.4 c/d performance components could demonstrate a common underlying oscillator. The phase analyses to test this yielded complex and confusing results.

It is clear that much of the phase-angle consistency noted was due to synchronizing effects of the experimental protocol, namely, the effects of awakenings. Awakenings produced transient curvilinear trends in performance, especially in the first 10-30 min after awakening. Since curvilinear trends in the first 10-40 min would influence the phases of cosine rhythms fitted to such data, this was certainly one reason why rhythms in performance appeared synchronized after awakenings. Also, although we had not expected the REM-NREM cycle to display consistent phases across a group of subjects in the hours shortly before awakening from a 6-h sleep period, we did observe synchronization of the REM-NREM rhythm. Perhaps the synchronization of bedtimes produced synchronization of the REM-NREM cycles within sleep which persisted for as long as 6 h.

Because most performance variables were more tightly and significantly synchronized to wake-up times than to the prior REM-NREM cycle phases, and because the amount of estimated synchronization progressively decreased as 10-50 min of the initial performance data were eliminated from the consideration of the performance rhythm phases, any apparent synchronization between the REM-NREM cycles and subsequent performance in tracking and pattern memory was probably due to synchronizing effects of the experimental protocol on both rhythms. The data do not suggest that the tracking or pattern memory rhythms were mutually synchronized with the REM-NREM cycle independent of wake-up effects. Also, the phase consistency was notably greater at 9.6 c/d than at 14.4 c/d, which had been the prospectively predicted frequency of the hypothesized rhythms.

Although the number of significant comparisons at 14.4 c/d exceeded random expectation for the REM-NREM vs performance comparisons, it did not exceed the number of such comparisons at 28.8 c/d, which was examined as a control frequency thought not to contain valid rhythms. This would suggest that there was no valid rhythm phase relationship between the REM-NREM cycle and subsequent performance rhythms. Nevertheless, the drinking rhythm did display a significantly consistent phase-angle relationship to the prior REM-NREM rhythm at 9.6 c/d in both experiments. Because this drinking rhythm was not consistently phase-locked to wake-up times at this frequency, this observed phase-angle locking could not be an artifact of experimental synchronizers. Further, both in this sustained performance model and in prior studies, we have found drinking to reflect ultradian rhythms more than various other measures (KRIPKE 1972; KRIPKE et al. 1983). Since 11 variables were considered in both experiments, and only one seemed positive, this could be a random false-positive result. The computed phase-angle relationships between drinking and the REM-NREM cycles seemed quite different in the two experiments, which seems illogical. We had not prospectively predicted either that drinking would be the most favorable behavioral measure in which to detect such phase consistency nor that the most favorable frequency would be 9.6 c/d. Thus, this result must be regarded as inconclusive unless it can be further replicated.

In summary, the evidence for phase consistency between the REM-NREM cycle and subsequent performances seemed somewhat greater than chance, but was not best demonstrated at the frequency and in the variables we had prospectively predicted. The critical contrasts were contaminated by an unexpected degree of synchronization with wake-up times for both the REM-NREM cycle and the performance variables. Thus, we suspend judgment as to whether or not the REM-NREM cycle predicts subsequent performance rhythms until further experiments can be completed.

Clearly, the predictive relationship between the REM-NREM cycle and subsequent performance is neither strong nor simple. It can be contaminated by influences of experimenter-induced awakenings. It is conceivable that the process of awakening partially resets the phase and frequency of an underlying ultradian oscillator or basic rest-activity cycle. Previous evidence for such phase resetting was observed in monkey data collected some years ago (KRIPKE et al. 1976). One might then expect not an exact phase correspondence between the preceding REM-NREM cycles and subsequent performance cycles, but rather, some sort of lawful phase response curve determined by the moment of awakening. Should the phase response curve for ultradian rhythms be at all analogous to circadian phase response curves, the present experimental design would be far too naive to elucidate its dynamics.

ACKNOWLEDGEMENTS. Laverne C. Johnson, Ph.D. and Paul Naitoh, Ph.D. contributed to the planning of these experiments. Dr. John Webster, Dr. Nobuyuki Okudaira, Mr. Sam Messin, Mr. William Mason, and Ms Martha Martell provided technical help. These studies were supported by ONR N00014-79-C-0317, by NIMH RSDA MH00117 (to DFK), and by the Veterans Administration.

REFERENCES

BATSCHELET E (1965) Statistical methods for the analysis of problems in animal orientation and certain biological rhythms. American Institute of Biological Sciences, Washington

KLEIN R, ARMITAGE R (1979) Rhythms in human performance: 1 1/2-hour oscillations in cognitive style. Science 204:1326-1328

KLEITMAN N (1963) Sleep and wakefulness, 2nd edition. University of Chicago Press, Chicago

KRIPKE DF (1972) An ultradian biological rhythm associated with perceptual deprivation and REM sleep. Psychosom Med 34:221-234

KRIPKE DF (1982) Ultradian rhythms in behavior and physiology. In: BROWN FM, GRAEBER RC (eds) Rhythmic aspects of behavior. Erlbaum, Hillsdale, p 313

KRIPKE DF, FLECK PA, MULLANEY DJ, LEVY ML (1983) Behavioral analogs of the REM-NREM cycle. In: MENDELWICZ J (ed) Advances in biological psychiatry, vol 11. Karger, Basel, p 72

KRIPKE DF, HALBERG F, CROWLEY TJ, PEGRAM V (1976) Ultradian spectra in monkeys. Int J Chronobiol 3:193-204

KRIPKE DF, SONNENSCHEIN D (1978) A biologic rhythm in waking fantasy. In: POPE K, SINGER J (eds) The stream of consciousness. Plenum, New York, p 321

LAVIE P, KRIPKE DF (1981) Ultradian circa 1 1/2 hour rhythms: a multioscillatory system. Life Sciences Mini Review 29:2445-2450

MULLANEY DJ, KRIPKE DF, FLECK PA (1983a) Sleep loss and nap effects on sustained continuous performance. Psychophysiol 20:643-651

MULLANEY DF, KRIPKE DF, FLECK PA (1983b) Effects of sustained continuous performance on subjects working alone and in pairs. Percept Mot Skills 57:819-832

OKUDAIRA N, KRIPKE DF, WEBSTER JB (1984) No basic rest-activity cycle in head, wrist, or ankle. Physiol Behav 32:843-845

Three Central Issues Concerning Ultradian Rhythms

R. J. Broughton

Division of Neurology, University of Ottawa and Ottawa General Hospital, 501 Smyth Road, Ottawa, CDN K1H 8L6

INTRODUCTION

In this chapter three somewhat interrelated issues concerning ultradian rhythms are considered. (a) Which factors influence the amplitude of the elusive human daytime ultradian rhythm in performance at the KLEITMAN (1961, 1963) basic rest-activity cycle (BRAC) rate? Two sections describe experiments which together support the proposal that such performance rhythms reflect ultradian waking fluctuations in alertness, and that the latter become accentuated in conditions of increased sleep need, lowered motivation, and lowered performance demand, i.e., in decreased "arousal". (b) What is the evidence for an approximately 12-h (circasemidian) rhythm in sleep tendency, as proposed earlier (BROUGHTON 1975)? Current evidence for this rhythm, which is of considerable theoretical, biological, and clinical interest, as will be seen, is very strong. The integer ratio of 2:1 between the 24-h circadian sleep rhythm and the 12-h rhythm is emphasized. (c) A briefer final section discusses the issue of the overall temporal relationships between the various ultradian biological rhythms. It proposes that, in fact, there is a general tendency for biological rhythms to exhibit such integer ratios and that this tendency may have survival value for species evolution.

EVIDENCE FOR HUMAN ULTRADIAN RHYTHMS IN DAYTIME PERFORMANCE

The existence of waking ultradian rhythms in performance would, if marked, have obvious relevance for educational and work activities. To date evidence for such rhythms has been equivocal. GLOBUS et al. (1971) were first to describe ultradian performance rhythms using fairly prolonged periods of repeated testing. ORR et al. (1974) had

Experimental Brain Research, Suppl. 12

subjects perform a visual vigilance task over many hours. A 90- to 100-min ultradian rhythm was just visible in their raw data but was made evident by treating the time series with complex demodulation. The ultradian rhythms revealed using this technique increased in amplitude with time on task. This increase was attributed to cumulative stress or fatigue. On the other hand, the elusive nature of these rhythms is underlined by the negative results of KRIPKE et al. (this volume, Chap. 12). He found no evidence for ultradian rhythms using an extensive protocol which involved continuous testing of multiple performance measures for over 42 h, with and without prior sleep deprivation, and during sustained wakefulness throughout the usual period of nocturnal REM sleep.

My own experience with Keith Busby in attempting to document ultradian rhythms in performance (and movement) was instructive. The study (BUSBY and BROUGHTON 1983) involved hyperkinetic and control male children, aged 8-12 years (N = 11, each group). We postulated that hyperkinetic children might have ultradian variations in performance and/or movement which were either greater in magnitude or shorter in period than controls. This would help explain several features of the hyperkinetic syndrome including the greater variability of performance, the shortened attention span, and the increased motility itself.

The performance test chosen was the Rosvold-Mirsky visual continuous performance test (a vigilance-type task). Subjects were told to respond by button press to a preselected sequence of two letters (e.g., A-M) appearing in a series of letter presentations reflecting all possible combinations of 12 letters. The letters flashed on a screen for 0.2 s every 1.5 s. During testing the subjects sat on a stabilometric chair which recorded global body movements and wore an accelerometer/accumulator watch on the dominant leg which measured segmental limb movement. After an appropriate training session, the children were tested on two consecutive days every 15 min for 5 min across a 6-h period from 0800-1400.

Each of the four dependent measures (detections and false positives in performance, number of global body movements, and number of segmental body movements) of each test day of each child was treated as an independent time series. The raw data were individually plotted. Some individual time series contained evident ultradian rhythms in performance and, at times, in movement at the predicted period of approximately 75 min. There was also regular occurrence of an 180^0 out of phase relationship between the number of detections and the number of false positives on the visual vigilance task. This phenomenon can be clearly seen in Fig. 1 on which overnight sleep and subsequent morning performance in a hyperkinetic child are plotted.

To then clarify whether an ultradian rhythm was, in fact, present and statistically significant, each time series was individually standardized (converted to Z scores), detrended, and subjected to power spectral analysis. A relatively short time period of testing had been rendered necessary by the age of the children and the inherent nature of the disability. Six centre frequency bands were therefore chosen: 0, 9.6, 19.2, 28.9, 38.5 and 48.1 c/d, each ± 2 c/d, thereby sacrificing frequency resolution for spectral stability. The 19.2 c/d band represented the predicted period range of 64-84 min for BRAC in that age range.

Statistical comparisons of the magnitude of the variance in the 19.2 c/d range with the means of the neighbouring ranges (9.6 and 28.9 c/d) and the means of all other ranges (0, 9.6, 28.9, 38.5 and 48.1 c/d) confirmed that individual time series could have significant peaks at the expected rate. A similar number of subjects in both groups showed a significant peak in the predicted range. A subject could show a significant peak for one variable on one day and for another the next. The appearance of significant peaks in the predicted range in fact appeared random across subjects and across days. Significant peaks at other frequency ranges were also encountered. There were no significant differences between days 1 and 2 for any dependent measure in either subject group. Each group's data was therefore pooled. The pooled group data for both hyperkinetic and normal children indicated that there was no significant peak for either at the predicted range for any variable, so that the experimental hypotheses could not be confirmed. There were, however, between group differences in overall levels of performance and movement.

It therefore appeared that ultradian rhythms in performance (and movement) could be significantly present in individual time series, but that some factor or factors were related to their intermittent appearance and disappearance around noise levels (unmasking and masking). As there were only 24 sample points, which for an anticipated 75-min biorhythm sampled every 15 min would fall over only 24 : 5 = 4.8 cycles, we were concerned about artifactually significant results from power spectral analysis. Another concern was whether the

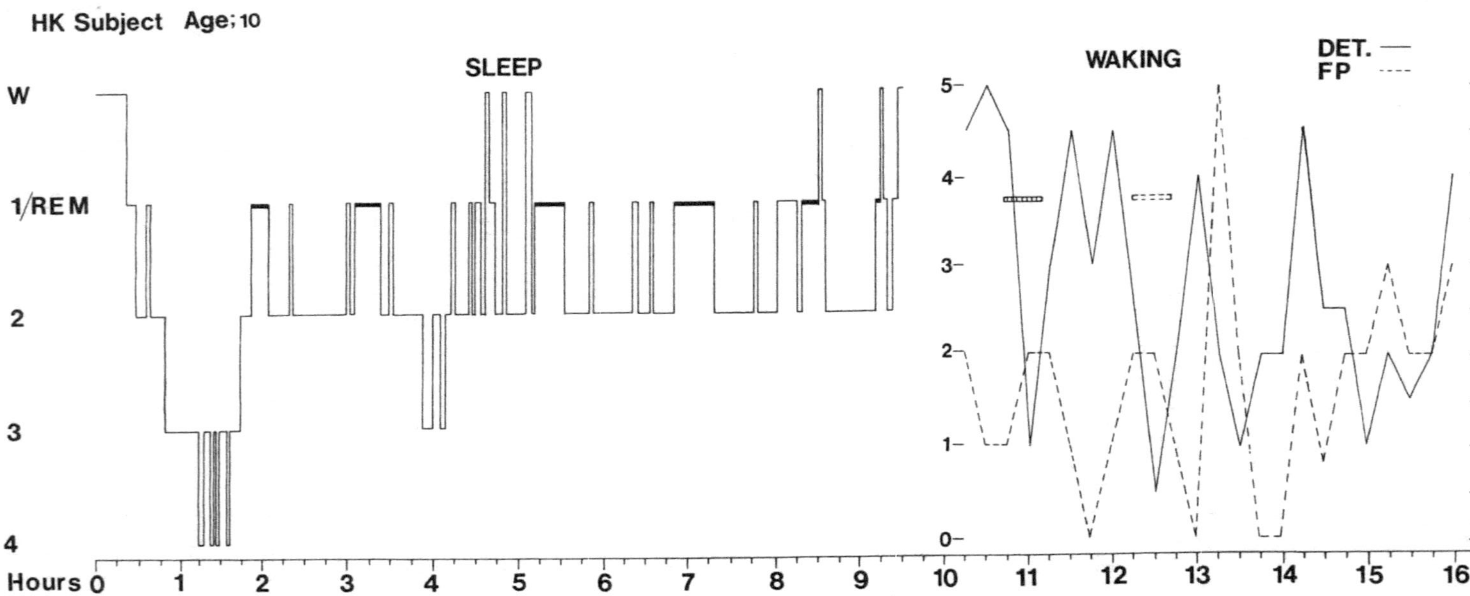

Fig. 1. Nocturnal sleep and subsequent performance on a visual vigilance task done by a 10-year-old hyperkinetic boy. The 45-min break after the overnight recording permitted breakfast, minimized "sleep inertia" effects on performance, and gave time to move to the performance test area. Horizontal bars in performance data represent roughly predicted times of further REM periods, were sleep to continue. The raw performance data shows a clear 75-90 min (approx.) periodicity in both signal detections and false positives. The latter were essentially out of phase, a pattern typical of fluctuations in alertness. Results in control subjects were similar. (BUSBY and BROUGHTON, unpublished observations)

repeated application of a performance test at a submultiple of the anticipated period (i.e., every 15 min for an approximately 75-min rhythm) might not either actively induce an artificial ultradian rhythm or obscure (or suppress) a true endogenous one.

One robust result strengthened our belief that the intermittent appearance of an ultradian performance rhythm in individual time series was a real phenomenon. When such a rhythm was evident in the raw data, it was generally found to be present for both detections and false positives; moreover, these two performance variables were regularly out of phase. This consistent pattern could not be expected by chance. The association in vigilance tasks of fewer detections with more false positives has been widely reported to characterize sleepiness. This is true both for sleepiness in normal subjects (WILLIAMS et al. 1959; WILKINSON 1970; GLENVILLE et al. 1978) and for the pathological sleepiness of narcoleptics (compared with controls), even at times of EEG patterns of wakefulness (VALLEY and BROUGHTON 1983).

It was concluded therefore that the appearance of significant ultradian fluctuations in sleepiness/alertness probably underlay the presence of ultradian performance rhythms. We believed that motivation was the main factor relating to increased and more even alertness, and so to suppression of the performance rhythm: "Subjects who displayed high motivation in the CPT (continuous performance task) may have suppressed any cyclic modulation of these diurnal processes. In contrast subjects with less motivation and little interest in outcome might have allowed more variability in response to be manifested, and conversely, these rhythms may have been more clearly expressed" (BUSBY and BROUGHTON 1983).

ULTRADIAN EEG RHYTHMS IN UNMOTIVATED NON-PERFORMING HUMAN SUBJECTS

In order to help clarify the apparent unmasking/masking phenomenon of the ultradian performance rhythm described in the literature and also present in the previous study, it was decided to attempt to document ultradian rhythms of alertness in nonperforming, unmotivated, awake subjects. Absence of a performance task was used to avoid the possible induction or suppression of an underlying ultradian rhythm in alertness, and the lack of any known motivating factors in the design was hoped to help unmask such a rhythm. Analysis of spontaneous EEG changes over time was chosen as a reasonably sensitive approach to detect variations in alertness; a protocol was designed which also would test the hypothesis (BROUGHTON 1975) of an interhemispheric shift inherent in BRAC.

The project, performed with Claude Manseau, involved the study of seven normal adult subjects seated in a partial isolation chamber (sound proofed, air conditioned, no time cues) across 8 h from 0930-1730 (MANSEAU and BROUGHTON 1984). The EEG was recorded in unipolar fashion from left and right frontal and parietal (international 10/20) electrode sites referred to linked mastoids. Eye movements were also recorded and appropriate precautions were taken to exclude EMG, head movement or ocular artifacts. The EEG was sampled every 15 min for 100 s. Before each sample, subjects were told by intercom to relax and fixate visually on a dot in front of them.

Each 100-s sample of EEG was taped and later analysed by a Nihon-Kohden frequency analyser into five frequency bands: delta (2-4 Hz), theta (4-8 Hz), alpha (8-13 Hz), beta-1 (13-20 Hz), and beta-2 (20-30 Hz). The output of each EEG band and of overall EEG (4-20 Hz, thereby further minimizing any possible artifacts of low frequency ocular or high frequency EMG origin) was rectified, A-D converted and stored on disk. Each of the six time series (five narrow EEG bands and overall EEG) for each of the four electrodes (i.e., 24 time series per subject) was standardized and submitted to power spectral analysis. The centre frequencies were 0, 8, 16, 24, 32, 40 and 48 c/d, the 16 c/d (range 12-20) band corresponding to a period of 72-120 min and therefore including the adult BRAC range. The 8-h sampling period covered 5.3 cycles for a 90-min rhythm or 4.0 cycles for a 120-min rhythm. As in the previous behavioral study, the magnitude of the variance in the predicted range was compared statistically with the means of the two neighbouring peaks and the means of all other peaks.

The results confirmed that there was a significant ultradian EEG rhythm in group data of nonperforming unmotivated subjects. Significant 16 c/d peaks were present at both frontal electrodes for overall 4-20 Hz EEG (Fig. 2). In the individual EEG frequency bands, only right frontal theta activity approached the stringent level ($p < .0008$) of significance. It was of interest that there were also frequent marked peaks at 8 c/d (3-h periodicity).

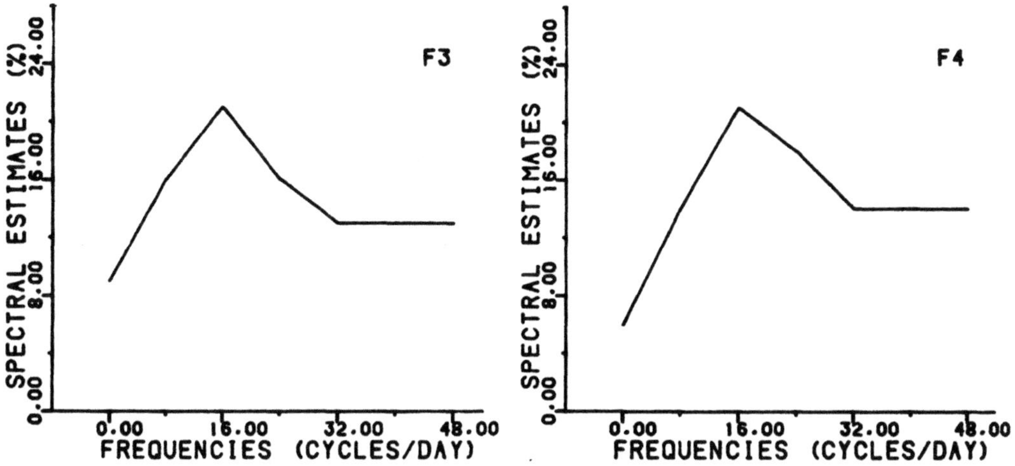

Fig. 2. Group mean autospectra of overall EEG (4-20 Hz) in the left and right frontal regions of human adult subjects. Statistically significant ultradian peaks were present at 16 c/d (range 12-20 c/d). This represents a period range of 72-120 min, essentially that of the NREM/REM sleep cycle and BRAC rate in adulthood. (Results from MANSEAU and BROUGHTON 1984)

EEG synchronization and especially an increase in theta activity, mainly frontal, characterize sleepiness (LINDSLEY and WICKE 1974) and represent scoring criteria for stage 1 drowsiness (RECHTSCHAFFEN and KALES 1968). The ultradian EEG variations were therefore interpreted as reflecting 16 c/d fluctuations in alertness or sleep tendency. This interpretation was further supported by visual analysis of the monitored polygraphic recordings. Others have provided evidence for recurrent sleepiness at the BRAC rate for normal subjects in either such sleep-facilitating conditions as bed rest (OTHMER et al. 1969) and perceptual deprivation (KRIPKE 1972), or with instructions to attempt to fall asleep every 20 min for 5 min (LAVIE and SCHERSON 1981).

A number of behavioral and psychological parameters were correlated with a physiological measure of the magnitude of the ultradian EEG rhythm. The latter was defined as the number of times the 16 c/d peak ranked first or second for all 24 autospectra of each subject. The behavioral-psychological parameters included items from a daily habits questionnaire, the Eysenck Personality Inventory, the Cattell Anxiety Scale, and the Shostrom Personal Orientation Inventory. Of the 35 possible correlations, only two were significant. The ultradian EEG rhythm was greater for subjects who assessed themselves as poor vs good sleepers and for subjects who scored high on neuroticism in the Eysenck Personality Inventory. It would therefore appear that night sleep of good quality leads to the ability to maintain more even daytime alertness, and that the latter may be associated with the personality trait of greater emotional stability.

The results of these experiments with graduate students indicate that significant ultradian rhythms in alertness definitely exist at least intermittently, and that it is these alertness/sleepiness rhythms which (mainly) determine the presence and the nature of the ultradian rhythms in performance (at least for those of vigilance type). Any factors which lead to a sustained even level of daytime alertness, and perhaps especially those causing tonic high arousal levels, would tend to override the relatively weak endogenous rhythm in alertness and would be associated with difficulty or inability to document ultradian performance rhythms (i.e., with masking). These factors include stress, strong motivation and continuous high performance demands.

Such factors were present in the KRIPKE et al. (this volume, Chap. 12) study which could not find significant ultradian performance rhythms. In the ORR et al. (1974) study, the increased magnitude of such rhythms over the long testing period could have been equally related to the appearance of drowsiness as to accumulated stress or physical fatigue. Indeed the authors mention a diminishing level of alertness in their subjects. Finally, as SCHULZ (this volume, Chap. 10) has stated, the remarkable tendency of patients with narcolepsy-cataplexy to show ultradian rhythmicity in diurnal sleep episodes seems due to their overall chronic impaired arousal which unmasks BRAC during the daytime. This interpretation is greatly strengthened by the presence of similar ultradian periodicity in sleep attacks of isolated (NREM) narcolepsy (BALDY-MOULINIER et al. 1976) for which a role for causal cyclic REM pressure can be excluded.

It appears increasingly probable from recent research that the only daytime rhythms which represent direct continuation of the NREM-REM sleep cycle into wakefulness, and so reflect the KLEITMAN BRAC, may be those of sleepiness/alertness and variables directly or mainly dependent on the latter (e.g., vigilance-type performance tasks). Activity cycles in other body functions (e.g., renal, gastro-intestinal), on the other hand, do not appear precisely coupled to variations in the ultradian sleep/wake systems (LAVIE 1982) and thus do not express the KLEITMAN BRAC.

CIRCASEMIDIAN (ABOUT 12-H) RHYTHM IN HUMAN SLEEP TENDENCY

Some years ago, I proposed that adult human sleep has an inherent biorhythmic tendency towards a bimodal expression about every 12 h per day which, in normally entrained subjects, usually occurs as a major nocturnal sleep period and a briefer period of afternoon sleepiness or overt sleep (BROUGHTON 1975). Moreover, it was further hypothesized that this reflected an approximately 12-h rhythm of pressure for slow wave sleep (SWS; deep NREM sleep) with two usual maxima (acrophases) about 180^0 out of phase, one in the first third of the nocturnal sleep period and the second in the early to mid-afternoon. This appears to have been the first proposal of a human 12-h circasemidian sleep tendency.

Evidence of the inherent propensity for a second sleep period occurring (in normally entrained individuals) in the afternoon has subsequently become quite compelling. It may be summarized as follows:

1. Ontogenic nap studies indicate that the last nap to be given up in growth and development (as the neonatal polyphasic sleep pattern is replaced by a monophasic one) is always the afternoon nap (KLEITMAN 1963). This typically occurs at about 5-6 years of age, when the child is required to go to school for the full day. In college students naps are unexpectedly frequent and are typically taken at that time (EVANS et al. 1977). And in later life, after retirement, when demands to sustain daytime wakefulness are lessened, nighttime sleep fragmentation increases, and sleep can be scheduled quite freely, napping becomes almost universal and once again is most frequent in the mid-afternoon (WEBB 1983).

2. The siesta typical of some cultures is regularly observed in the afternoon rather than at other times. It characterizes societies (usually in hot climates) that restrict nocturnal sleep, often eating supper, and retiring, late, while maintaining a normal hour of rising. A nocturnal sleep deficit is therefore made up in the day (BROUGHTON 1983).

3. Performance abilities during the day studied by repeated testing show transient deterioration in the afternoon. The phenomenon was first clearly documented by BLAKE (1967) who called it the "post-lunch dip". Subsequent unpublished studies by Blake done prior to his untimely death showed, however, that it occurred whether or not a mid-day meal was taken; although Blake placed the performance dip at 1230-1300, his own data actually indicated a somewhat later peak at around 1400-1500 (cf. COLQUHOUN 1972).

4. Objective sleepiness in normal subjects recurs at this time. This is expressed as shorter sleep latencies in multiple measures of sleep tendency across the day. In the Multiple Sleep Latency Test (MSLT; MITLER et al. 1982b) subjects are put to bed usually every 2 h from 1000-1800 and told to try to sleep. Shortest sleep latencies around 1400-1600 typify pubertal adolescents (CARSKADON et al. 1980), young and middle-aged adults (RICHARDSON et al. 1978) and the elderly (RICHARDSON et al. 1982). The same timing for maximum daytime sleep tendency is encountered in an inverse type of sleepiness test in which subjects are told to try to remain awake (MITLER et al. 1982a; HARTSE et al. 1982).

5. Pathological daytime sleepiness in medical sleep disorders peaks with the same mid-afternoon timing. This has been shown by shorter MSLT sleep latencies in narcoleptics (RICHARDSON et al. 1978; REYNOLDS et al. 1982) and in such hypersomnias as the sleep apnea syndrome (REYNOLDS et al. 1982).

6. Temporal isolation studies, which represent the best approach to define endogenous biorhythms, have recently shown a strong tendency for human adult sleep to become distributed in a bimodal fashion with a major sleep period occurring at the time of the minimum circadian body temperature and a shorter sleep period (nap) occurring about 180° out of phase, i.e., about 12 h later (CAMPBELL and ZULLEY, this volume, Chap. 14). In entrained subjects these poles normally occur at night and in the afternoon, respectively.

The evidence supporting afternoon pressure for deepening NREM sleep as the basis of the bimodal sleep tendency comes from three main sources. (a) Polygraphic recordings of voluntary naps in adults taken at different times of the day have a maximum probability of containing SWS in the early and mid-afternoon (WEBB 1971). (b) "Sleepability" studies on a 15-min wake 5-min sleep schedule showed that, although stage 1 drowsiness exhibited an ultradian periodicity of approximately 90 min, stage 2 sleep (little delta was recorded) exhibited a single peak which was at 1400-1600 (LAVIE and SCHERSON 1981). (c) Extended sleep patterns in normal subjects able to sleep sufficiently long typically contain two SWS peaks, the first beginning some 45 min after sleep onset and the second just over 12 h later. This phenomenon was first described anecdotally by WEBB (1977) and noted in isolation studies by WEITZMAN et al. (1980). It was subsequently the object of direct study by GAGNON and DE KONINCK (1981, 1984) who found the second delta peak in each of the six of eight subjects who were able to extend sleep duration to 15 h.

Several mechanisms for the bimodal SWS distribution have been suggested. These include the effects of prior wakefulness, presence of a 12-h biorhythm of SWS pressure, and a release phenomenon related to minimal pressure for REM sleep.

Although plausible for daytime naps after a morning spent awake, a causal role for prior wakefulness certainly can be excluded for the extended sleep data in which an insufficient amount of wakefulness (sometimes almost none) is present before the reappearance of SWS. It also cannot explain Lavie and Scherson's single stage 2 peak on an ultrashort sleep schedule. The critical test of the SWS biorhythm hypothesis, however, involves the effects of abrupt phase shifting of sleep, and was performed with my colleagues Gagnon and De Koninck. When the onset of extended sleep was delayed by 4 h from 2400 to 0400, sleep patterns again contained the appearance of two major SWS peaks, one beginning within 1 h of falling asleep and the other about 12 h later. Initially, there was also a third peak at approximately 1300, representing a residual expression of the second peak prior to shift. After 3 days of maintaining the shifted extended sleep pattern, only the two SWS peaks separated by 12 h persisted (GAGNON and DE KONINCK 1982; GAGNON et al. 1985).

These results strongly support the existence of a 12-h biorhythm of SWS, as initially proposed (BROUGHTON 1975). They cannot be interpreted as a release phenomenon related to a nadir for REM sleep, because it would take several days for the circadian acrophase of REM sleep to realign. In both the baseline and the shifted schedules of extended sleep the mean interval between SWS peaks was about 12 h 20 min, i.e., almost exactly one-half of the period of the freerunning human circadian sleep rhythm. This suggests that, whereas the first SWS peak of the bipolar circasemidian rhythm is time-locked to sleep onset, the second peak exhibits within-day freerunning.

In normally entrained humans with a nocturnal sleep pattern there is, therefore, a recurrent tendency for afternoon sleep which reflects an endogenous biorhythm. It is arguable whether this 12-h rhythm should in fact be classed as an ultradian rhythm or whether it should be considered an inherent property of the circadian sleep rhythm to manifest itself with two peaks. It reflects pressure towards SWS.

There is little evidence for any significant drop in body temperature at the time which might relate to the second sleep period (RICHARDSON et al. 1982). Sleep and low body temperature at any rate do not consistently correlate with each other, as they are easily dissociated by phase shifting sleep and may run free at different periods in isolation studies. At present there seems little necessity to propose an independent transient enhancement of sleep-inducing mechanisms to explain the afternoon sleep tendency (LAVIE, this volume, Chap. 9), although such a mechanism would not be mutually exclusive with that of pressure for NREM (slow wave) sleep.

The existence of a circasemidian sleep tendency is not restricted to man. It is a trait that he shares with many other species (ASCHOFF 1966). The functional significance of a bimodal sleep pattern remains obscure but may relate to more efficient conservation of energy or to the importance of restricting excessive exposure to the mid-day sun and/or to predators. It is a powerful biorhythm that in man is manifestly of greater magnitude than the ultradian rhythm of sleep tendency. The amplitudes of both, however, are strongly and positively correlated with daytime sleep need. Both increase when daytime sleep need increases (as in normals because of sleep debt from restricted or fragmented night sleep, or in pathological sleepiness as part of a medical sleep disorder). Conversely, the presence of sufficient normal sleep of good quality, as well as other factors which lead to sustained even levels of alertness (e.g., motivation, continuous high demands), will reduce the expression of both the 12-h and the ultradian 90-min rhythms of sleep tendency.

BIOLOGICAL RHYTHMS AND INTEGER RATIOS

Human sleep/wake rhythms show a strong tendency to have constant integer ratios. This tendency was emphasized earlier (BROUGHTON 1975), based upon evidence for 2:1 ratios between the approximately 24-h circadian rhythm, the 12-h rhythm of sleep tendency (discussed above) and the occasional appearance in freerunning conditions of an infradian 48-h sleep rhythm (CHOUVET et al. 1974). These were interpreted as reflecting harmonic relationships conceived as a subharmonic halving of the basic circadian rhythm (1 c/d) leading to the circabidian or "bicircadian" rhythm (0.5 c/d), and a superharmonic doubling to the circasemidian rhythm (2 c/d). Evidence in man is accruing which indicates that other biological rhythms show related 2:1 integer ratios. There is, for instance, some evidence for about 4 c/d (6-h) rhythms in raw EEG data (TSUJI et al. 1981) and for 8 c/d (3-h) EEG rhythms in both raw data (TSUJI et al. 1981) and in data treated using spectral analysis (MANSEAU and BROUGHTON 1984). Of course, in man the BRAC rate itself is (coincidentally?) close to the next submultiple, and a large number of approximately 16 c/d (1.5 h) rhythms have been documented (cf. KLEITMAN 1982; LAVIE 1982).

This pattern of biological rhythms exhibiting apparently privileged periodicities at 24, 12, 6, 3 and 1.5 h with the common integer ratio of 2:1 is also not unique to man. The tendency appears in fact to be common in nature. It is not possible in this short paper to fully substantiate this claim, but the following may be briefly said. First, the evidence for circadian biological periodicities is, of course, overwhelming. Second, as already mentioned, the tendency of sleep to split into two peaks with a circasemidian distribution has been found to characterize many species (ASCHOFF 1966). And, third, a large number of rhythms documented to date mainly in mammals show periodicities of around 6, 3 and 1.5 h. In this symposium, for instance, a 3-h period is prominent for urine flow and osmolality in dogs (GORDON and LAVIE, this volume, Chap. 7), and the sleep and neuroendocrine measures in rats reported by KIMURA et al. (this volume, Chap. 4) show consistent multiple peaks with periods of about 6, 3 and 1.5 h. (The NREM-REM cycle BRAC rate, of course, varies widely across species and exhibits periodicities independent of the above seemingly privileged rates.)

It is crucial to note that these periods recur in the raw data. A number of techniques often used to study biorhythms such as power spectral analysis are prone to give artifactual harmonics of the major periodicity, which is often circadian. Other statistical approaches for defining biological rhythms (e.g., Enright's periodogram) appear to minimize or avoid this flaw.

The belief that these periodicities with integer ratio relationships represent biological harmonics of the basic circadian rhythm is probably questionable, although the recent data of CAMPBELL and ZULLEY (this volume, Chap. 14) may be said to support this interpretation for three major sleep periodicities. These authors reported that, by changing experimental instructions in the disentrained state, subjects could show a sleep pattern which becomes first circasemidian, then circadian, then circabidian.

When considering the large variety of biorhythms showing the relationships described, it seems much more likely that they reflect (at least semi-independent) multiple oscillators at these particular periods. No doubt in evolution biological oscillators have appeared with all possible periodicities. An organism with oscillator subsystems at constant integer ratios would have the ability to more easily maintain consistent phase relationships between them. This in turn would permit a more efficient temporal organization of the organism's major functions, lead to greater overall bio-energetic efficiency, and so have survival value. Those organisms whose main oscillating subsystems had constant integer ratios which also closely matched the environmental physical zeitgebers would be particularly favored. Thus, an organism with oscillating biological subsystems with periods of about 24, 12, 6, 3 and 1.5 h would have definite advantages compared with another with oscillators at, for instance, 22, 13, 7, 5 and 3 h which could not be consistently phase related both to each other and to the major environmental periodicities.

In short, it is proposed that:

1. Organisms tend to exhibit biorhythms with periods at constant integer ratios, the most common of which is 2:1.
2. A constant integer ratio of periods permits constant phase relationships between the subsystem oscillators within the shortest time duration (i.e., the period of the slowest oscillator).
3. Ability to have constant phase relationships between the various body subsystems within the shortest time duration allows a more consistent and efficient temporal organization of an organism's functions.
4. More consistent and efficient temporal organization of an organism's functions is more energy efficient and has greater survival value.
5. The optimal temporal organization for species survival includes the capabilities to have such constant phase relationships between both the endogenous subsystem oscillators and the major natural periodicities in the geophysical environment.
6. The latter combination underlies the particularly privileged nature of biorhythms at 24, 12, 6, 3 and 1.5 h periodicities (and perhaps other similar ratio patterns, e.g., the 3:1 ratio). It therefore

explains their high frequency of persistence in biological evolution.

ACKNOWLEDGEMENTS. I thank Dr. Claudio Stampi for constructive criticism of the manuscript, Wendy Buchanan and Nadia Dall'Aglio for typing assistance, and the Medical Research Council of Canada for research support.

REFERENCES

ASCHOFF J (1966) Circadian activity with two peaks. Ecology 47:657-662

BALDY-MOULINIER J, ARGUNEN A, BESSET A (1976) Ultradian and circadian rhythms in sleep and wakefulness. In: GUILLEMINAULT C, DEMENT WC, PASSOUANT P (eds) Narcolepsy. Spectrum, New York, pp 485-497

BLAKE MJF (1967) Time of day effects on performance in a range of tasks. Psychonom Sci 9:349-350

BROUGHTON RJ (1975) Biorhythmic fluctuations in consciousness and psychological functions. Can Psychol Rev 16:217-239

BROUGHTON RJ (1983) The siesta: social or biological phenomenon? Sleep Res 12:28

BUSBY KA, BROUGHTON RJ (1983) Waking ultradian rhythms of performance and motility in hyperkinetic and normal children. J Abnorm Child Psychol 11:431-442

CARSKADON MA, HARVEY K, DUKE P, ANDERS TF, LITT IF, DEMENT WC (1980) Pubertal changes in daytime sleepiness. Sleep 2:453-460

CHOUVET G, MOURET J, COINDET J, SIFFRE M, JOUVET M (1974) Périodicité bicircadienne du cycle veille-sommeil dans des conditions hors du temps. Etude polygraphique. Electroencephalogr Clin Neurophysiol 37:367-380

COLQUHOUN WP (1972) Circadian variations in mental efficiency. In: COLQUHOUN WP (ed) Biological rhythms and human performance. Academic Press, London, p 49

EVANS F, COOK M, COHEN H, ORNE E, ORNE M (1977) Appetitive and replacement naps: EEG and behavior. Science 197:687-689

GAGNON P, DE KONINCK J (1981) Reappearance of EEG slow waves in extended sleep. Sleep Res 10:137

GAGNON P, DE KONINCK J (1982) Reappearance of EEG slow waves in extended sleep with delayed bed-time. Sleep Res 11:101

GAGNON P, DE KONINCK J (1984) Reappearance of EEG slow waves in extended sleep. Electroencephalogr Clin Neurophysiol 58:155-157

GAGNON P, DE KONINCK J, BROUGHTON RJ (1985) Reappearance of EEG slow waves in extended sleep with delayed bed time. Sleep 8 (in press)

GLENVILLE M, BROUGHTON RJ, WING AM, WILKINSON RT (1978) Effects of sleep deprivation on short duration performance measures compared to the Wilkinson auditory vigilance task. Sleep 1:169-176

GLOBUS GG, DRURY RL, PHOEBUS EC, BOYD R (1971) Ultradian rhythms in human performance. Percept Mot Skills 33:1171-1174

HARTSE KM, ROTH T, ZORICK FJ (1982) Daytime sleepiness and daytime wakefulness: the effect of instruction. Sleep 5 (Suppl 2):107-118

KLEITMAN N (1961) The nature of dreaming. In: WOLSTENHOLME GEW, O'CONNOR M (eds) The nature of sleep. Churchill, London, p 349

KLEITMAN N (1963) Sleep and wakefulness. University of Chicago Press, Chicago, p 364

KLEITMAN N (1982) Basic rest-activity cycle - 22 years later. Sleep 5:311-317

KRIPKE DF (1972) An ultradian biological rhythm associated with perceptual deprivation and REM sleep. Psychosom Med 3:221-234

LAVIE P (1982) Ultradian rhythms in human sleep and wakefulness. In: WEBB WB (ed) Biological rhythms, sleep and performance. Wiley, New York, p 239

LAVIE P, SCHERSON A (1981) Ultrashort sleep-waking schedule: I. Evidence of ultradian rhythmicity in "sleepability". Electroencephalogr Clin Neurophysiol 52:163-174

LINDSLEY DB, WICKE JD (1974) The electroencephalogram: autonomous electrical activity in man and animals. In: THOMSEN RF, PATTERSON MM (eds) Bioelectric recording techniques, part B. Electroencephalography and human brain potentials. Academic, New York, p 3

MANSEAU C, BROUGHTON RJ (1984) Bilaterally synchronous ultradian EEG rhythms in awake adult humans. Psychophysiology 21:265-273

MITLER MM, GUJAVARTY KS, BROWMAN CP (1982a) Maintenance of wakefulness test: a polysomnographic technique for evaluating treatment efficacy in patients with excessive somnolence. Electroencephalogr Clin Neurophysiol 53:658-661

MITLER MM, GUJAVARTY KS, SAMPSON MG, BROWMAN CP (1982b) Multiple daytime nap approaches to evaluating the sleepy patient. Sleep 5 (Suppl 2):119-127

ORR WC, HOFFMAN HJ, HEGGE FW (1974) Ultradian rhythms in extended performance. Aerospace Med 45:995-1000

OTHMER E, HAYDEN MP, SEGELBAUM R (1969) Encephalic cycles during sleep and wakefulness: a 24-hour pattern. Science 164:447-449

RECHTSCHAFFEN A, KALES A (eds) (1968) A manual of standardized terminology, techniques and scoring system for sleep stages of human subjects. US Government Printing Office, Washington

REYNOLDS CF, COBLE PA, KUPFER DJ, HOLZER B (1982) Application of the multiple sleep latency test in disorders of excessive sleepiness. Electroencephalogr Clin Neurophysiol 53:443-452

RICHARDSON GS, CARSKADON MA, FLAGG W, VAN DEN HOED J, DEMENT WC, MITLER MM (1978) Excessive daytime sleepiness in man: Multiple sleep latency measures in narcoleptics and control subjects. Electroencephalogr Clin Neurophysiol 45:621-627

RICHARDSON GS, CARSKADON MA, ORAV EJ, DEMENT WC (1982) Circadian variations of sleep tendency in elderly and young adult subjects. Sleep 5 (Suppl 2):82-94

TSUJI Y, FUKUDA H, OKUNO H, KOBAYASHI T (1981) Diurnal rhythm of alpha wave activity. Electroencephalogr Clin Neurophysiol 52:S43

VALLEY V, BROUGHTON RJ (1983) The physiological (EEG) nature of drowsiness and its relation to performance deficits in narcoleptics. Electroencephalogr Clin Neurophysiol 55:243-253

WEBB W (1971) Sleep behavior as a biorhythm. In: COLQUHOUN WP (ed) Biological rhythms and human performance. Academic, London, p 166

WEBB W (1977) The spontaneous termination of sleep. Sleep Res 6:118

WEBB W (1983) presented at the annual meeting of the Society for Sleep Research, Bologna, Italy

WEITZMAN ED, CZEISLER CA, ZIMMERMAN JC, RONDA JM (1980) Timing of REM and stages 3 and 4 during temporal isolation in man. Sleep 2:391-408

WILKINSON RT (1970) Methods for research in sleep deprivation and sleep function. Int Psychiatr Clin 7:369-381

WILLIAMS HL, LUBIN A, GOODNOW J (1959) Impaired performance with acute sleepless. Psychol Monogr 73:484-510

ZULLEY J, CAMPBELL SS (1985) The coupling of sleep-wake patterns with the rhythm of body temperature. In: KOELLA WP, RÜTHER E, SCHULZ H (eds) Sleep 1984. Fischer, Stuttgart (in press)

Ultradian Components of Human Sleep/Wake Patterns During Disentrainment

S. S. Campbell[1] and J. Zulley[2]

[1] Department of Psychiatry, University of California, San Diego, CA 92161, USA

[2] Max-Planck-Institut für Psychiatrie, Kraepelinstrasse 10, 8000 München 40, FRG

"Napping got a bad reputation somewhere along the line and I resent it."
Andy Rooney

INTRODUCTION

The likelihood that napping behavior in adult humans is the expression of a part of the endogenous sleep/wake system has been recognized by numerous authors (BENOIT 1984; BROUGHTON 1975; CAMPBELL 1984; DINGES et al. 1983; KLEITMAN 1963; LAVIE and SCHERSON 1981; NAKAGAWA 1980; TUNE 1968, 1969; WEBB 1978; ZULLEY and CAMPBELL 1985a,b). Until relatively recently, experimental data regarding spontaneous napping have been lacking and it has been necessary, therefore, to rely largely on subjective reports and anecdotal evidence to support such a proposal. For example, both college students (WHITE 1974; WEBB and DUBE 1981) and individuals of retirement age (TUNE 1968) have reported increased frequencies of daytime sleep in response to their less-structured lifestyles. Similarly, the prevalence of siesta cultures has often been cited as an indication that there exists a natural tendency for a period of rest approximately halfway through the major waking interval.

These special cases notwithstanding, the usual view of human sleep/wake organization is that of a monophasic pattern, characterized by a major waking episode of about 16 h duration followed by a major sleep episode of about 8 h duration. The ubiquity with which this monophasic organization of sleep and wakefulness occurs may explain, in part, the general paucity of experimental data specifically addressing noncircadian components of the sleep/wake system, i.e., spontaneous napping, in humans.

Experimental Brain Research, Suppl. 12

Moreover, as has been noted by WEBB (1978), the very nature of naps (i.e., usually brief interruptions of relatively long waking intervals) makes them "uncontrollable and interfering sources of experimental variance" (p. 316) in most studies which are usually focused on major components of sleep, waking, and performance. Because of this focus, napping behavior is typically suppressed, either by specific experimental instructions or as a consequence of continuous performance, psychological or other testing.

Yet the available experimental evidence suggests that while naps may, indeed, interfere in the analyses of circadian components of sleep and wakefulness, the reverse may also be true. That is, by imposing a monophasic, circadian sleep regimen on subjects by circumscribed experimental protocols, the expression of putative noncircadian components of the sleep/wake system may be effectively masked. It has been shown, for example, that simply by removing restrictions on sleep in an otherwise typical time-free environment, the frequency of occurrence of sleep episodes, and the variability in the duration of episodes, increases substantially (WEBB and AGNEW 1974). In this sense, then, experimental instructions or behavioral demands which restrict the spontaneous initiation of sleep across the nychthemeron may be viewed as interfering sources of experimental variance in the examination of putative noncircadian components of adult human sleep.

Behavioral demands which may interfere with the expression of sleep tendency need not necessarily be imposed by experimental instructions, social/occupational pressures, or other external sources. Such controls may be self-imposed as well. It is not uncommon to delay, displace, or even omit typically scheduled sleep episodes in favor of self-imposed alternatives to sleep, such as finishing a good book, attending a social function, or completing an important project. It is, likewise, not unusual to attempt to overcome a period of drowsiness by a transient increase in physical activity, or, for example, by drinking a cup of coffee or tea.

Thus it is clear that the sleep behavior exhibited by a human adult may be influenced to a substantial degree by at least two environmental factors in addition to the biological system mediating the sleep response: external and self-imposed behavioral controls.

Because external and self-imposed behavioral controls may exert strong mediatory influences on the physiological sleep/wake system, it should be apparent that the behavioral manifestation of sleep and wakefulness observed in daily life, and in many experimental conditions, does not necessarily reflect underlying physiological sleep tendency. By way of analogy, the course of a river, after imposition of restraints and controls in the form of dams and levees, in an effort to structure the river for more efficient use, probably does not reflect accurately the natural dynamics of the river.

There is growing evidence that this is also the case with regard to human sleep. Thus, one approach to more closely examine properties of putative ultradian components of the human sleep/wake system would be to provide conditions under which the system was permitted to follow a course of least resistance. This would include not only the elimination of experimental instructions relative to when and when not to sleep, but also would require minimization of behavioral alternatives which may be incompatible with the sleep response.

In the following study, we have employed a "disentrained" environment in an effort to invoke such conditions. It was reasoned that, in response to the lack of behavioral options in such an environment, subjects would be more likely to respond to periods of drowsiness, not by attempting to overcome such a state, but rather by unambiguously expressing sleep tendency in the form of napping. In turn, it was hypothesized that the view of sleep and wakefulness which emerged from such an environment may provide a more accurate reflection of the biological mechanisms underlying the human sleep/wake system.

METHOD

Subjects

Nine paid subjects (4 females, 5 males) aged 19-37 years (mean age, 25.1 years) participated in the study. Subjects were selected who reported normal sleep patterns and the absence of both chronic health problems and acute illness. No subject reported habitual napping, though most subjects reported napping occasionally. Prior to participation, each subject was informed, in detail, of the environmental conditions to be used in the study and each was informed of the general aim of the study.

Disentrainment

The term "disentrainment" was selected to distinguish the current conditions from those typically employed in time-free environments, in which time cues are removed but other aspects of the environment remain essentially the same as in daily life, i.e., subjects are asked to continue daily activities and are usually requested to "structure" their days. "Disentrainment" was initially used in a previous paper (CAMPBELL 1984) to describe an environment which included enforced bed rest. In the present study, the same behavioral conditions were imposed, with the exception of enforced bed rest.

During the period of disentrainment subjects were confined individually in underground apartments shielded from all natural and artificial time cues. Subjects were allowed to move freely about the isolation unit, but were requested not to engage in strenuous exercise. Subjects were, in addition, prohibited from reading, writing, listening to music, and so on. They were specifically instructed not to structure their days, but rather to eat and sleep when inclined to do so. Food and noncaffeinated beverages, as well as kitchen facilities, were available in the isolation unit. Illumination was provided by fluorescent lighting in the ceiling and by several incandescent lamps on tables and over the bed. Control of light and darkness was at the discretion of each subject.

In summary, the physical environment used here was essentially the same as environments employed in standard human circadian research (cf. WEVER 1979). The difference was that there were no constraints placed on the initiation of sleep and behavioral options to sleep were extremely limited.

Procedure

Each subject slept for one night in the isolation unit immediately prior to the 72-h disentrainment period. Subjects were permitted to self-select bedtime on this occasion, as well as on all occasions during disentrainment. At 0800 on the morning following the baseline night, each subject was signalled that the experimental period had begun. They were instructed to acknowledge having heard the signal, but they were not required to get out of bed (i.e., they could return to sleep if so inclined).

During the entire period, including baseline night, the EEG, EOG, and EMG of each subject was continuously monitored via a Glonner Biomes 80 telemetry system. The signals were fed through a standard Siemens physiograph for output to paper. All records were subsequently scored in 30-s epochs following standard scoring procedures (RECHTSCHAFFEN and KALES 1968). Sleep onset was defined by the first epoch containing a sleep spindle or K-complex.

While awake, subjects wore the telemetry transmitter on a belt around the waist. During sleep episodes, the transmitter could be placed on a table beside the bed. Because of the relatively extended recording interval, the transmitter was powered by a 12-volt auto battery, connected by a cable approximately 6 m in length. Thus, it was unnecessary to change batteries during the disentrainment period.

Temperature was recorded each minute via a rectal probe, also connected by a 6 m long cable. Cables were of sufficient length to allow subjects complete freedom of movement around the isolation unit.

Communication with subjects was possible through an intercom system, but this was used only to request subjects to reattach electrodes when a loss of signals warranted such requests. While transmission and recording difficulties were, in most cases, corrected by the subjects, it was necessary on several occasions to enter the isolation unit to reattach electrodes or make mechanical adjustments. Such interruptions, as well as requests to subjects to attend to equipment failures, were made only during ongoing waking episodes.

RESULTS

While there were a limited number of sleep episodes with durations of less than 30 min (13.6% of total, N = 9), these sleep episodes (mean duration, 10.2 min) accounted for less than one percent of total sleep obtained. In the following analyses, therefore, a sleep episode is considered as such when sleep continued for at least 30 min. Similarly, there were a limited number of waking periods which continued for over 15 min but for less than 60 min (16.2% of total, N = 12). However, the large majority of waking episodes were of greater than 1-h duration. Furthermore, examination of sleep episodes, particularly major nocturnal episodes, made it clear that waking periods of less than 60 min could occur within otherwise coherent sleep periods. Therefore, in the following analyses, two successive sleep episodes are considered as such when separated by at least 60 min of wakefulness.

Overall Characteristics of Sleep and Wakefulness during Disentrainment

Figure 1 shows the alternation of sleep and wakefulness for all subjects, after applying the criteria for sleeping and waking episodes described above. A total of 57 sleep episodes were spontaneously initiated and terminated during the period of disentrainment. This number does not include sleep episodes taken on the baseline night, nor does it include the last sleep episodes in disentrainment which were intentionally terminated to end the experiment.

The median duration of sleep episodes was 2.4 h (mean, 4.1 h; SD, 3.7 h; range, 0.52-12.8 h). As can be seen in Fig. 2, the distribution of sleep episode durations was skewed, with 68.4% of all sleep episodes recorded continuing for less than 4 h. Only 10 of the 57 sleep periods recorded continued for as long as 8 h. Considering all sleep obtained during the period of disentrainment, i.e., from the awakening at 0800 following the baseline night to the termination 72 h later, for those eight subjects whose sleep was recorded for the entire period, an average of 31.0 h (43.1% of the period) was spent asleep (range: 20.5 - 40.2 h). (Due to equipment failure, the EEG of one subject (GLB) was recorded for only the first 48 h of disentrainment.)

Waking episodes recorded during disentrainment also tended to be substantially shortened relative to normal conditions, with a median duration of 4.9 h (mean, 5.2 h; SD, 3.1 h; range, 1.0-13.0 h). On only one occasion did a subject maintain wakefulness for longer than 12 h (see Fig. 2).

It is clear, then, that subjects in disentrainment typically slept for shorter durations and remained awake for shorter periods between sleep episodes than do humans under conditions of daily life. As a result, the average sleep/wake cycle length, measured from the onset of one sleep episode to the onset of the next, was 10.2 h (SD, 4.4 h; median, 10.6 h). Variability in duration of sleep/wake cycles was decreased by about 25% (to 3.4 h) without substantially altering cycle length

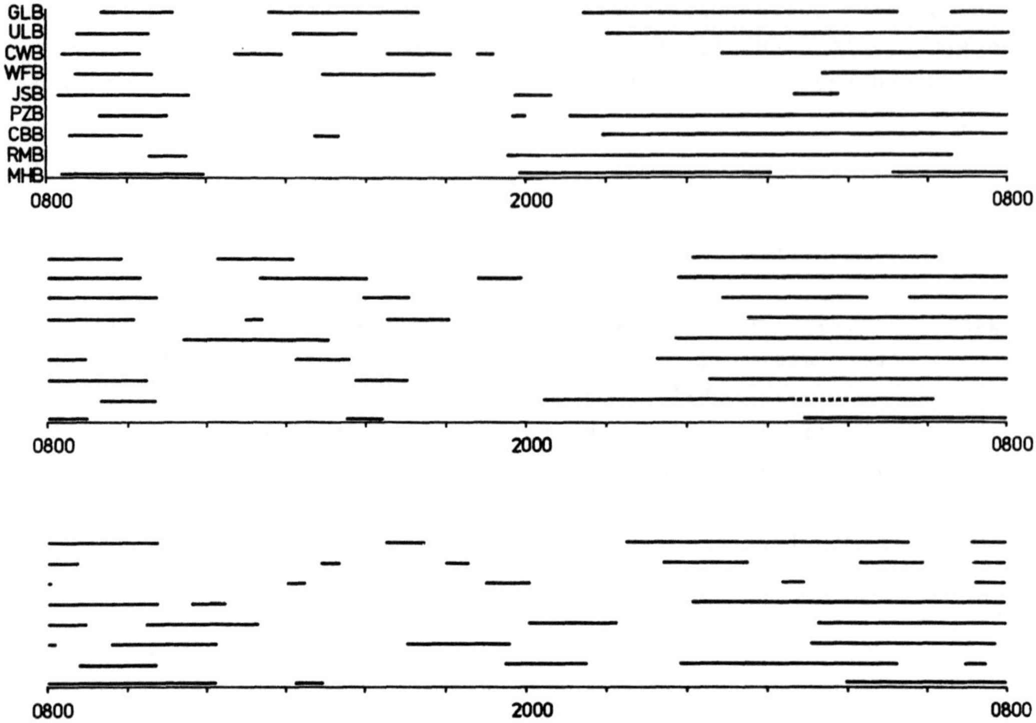

Fig. 1. Alternation of sleep (dark bars) and wakefulness for all subjects across the period of disentrainment. The EEG of one subject (GLB) was recorded for only the first 48 h of disentrainment. Broken line in fifth sleep episode of subject RMB indicates a loss of signal during that interval

(9.9 h) by calculating the cycle from the middle of one sleep episode to the middle of the next.

Temporal Characteristics

Figure 3 shows the distribution of sleep period onsets during disentrainment as a function of time of day. In order to have sufficient sample size in each cell for statistical analysis, it was necessary to group sleep onset times in 3-h blocks. The overall distribution did not differ significantly from a chance distribution (X^2 = 8.3, df = 7, ns) nor did the number of sleep onsets between 1400 and 1700, or between 2300 and 0200, differ significantly from the number of sleep onsets in adjacent 3-h blocks.

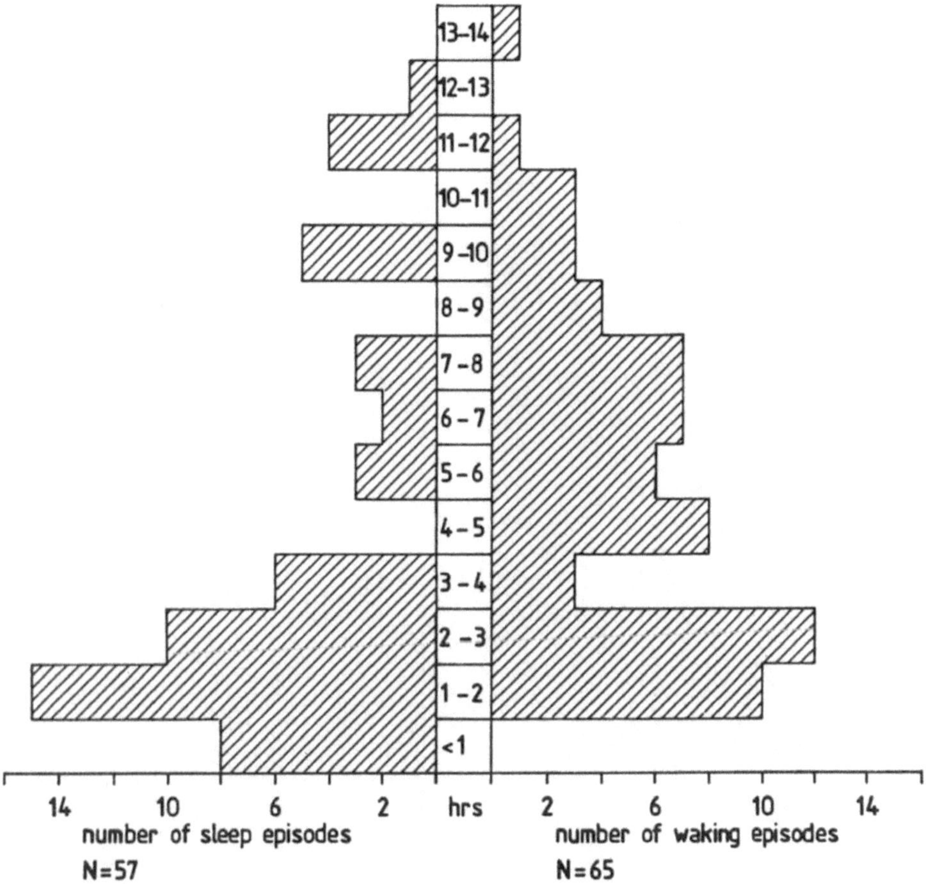

Fig. 2. Distribution of sleep and waking episodes recorded during disentrainment, as a function of their durations. Both sleep and waking episodes showed tendencies to continue for substantially shorter durations than under conditions of daily life

Since sleep onset can occur, of course, only when a subject is awake, the frequency distribution shown in Fig. 3 was influenced substantially by the circadian variation in sleep period duration. For example, the frequency of sleep onset was relatively low between 0200 and 0500 primarily because subjects had earlier initiated their major nocturnal sleep episodes. As such, sleep onset time may be a misleading index of group sleep tendency. A more meaningful representation of group sleep tendency may be derived from the examination of Fig. 4, which shows the number of sleep episodes, as a percentage of total possible sleep episodes, comprising part or all of any given hour across the nychthemeron. Placement of sleep was clearly bimodal, with

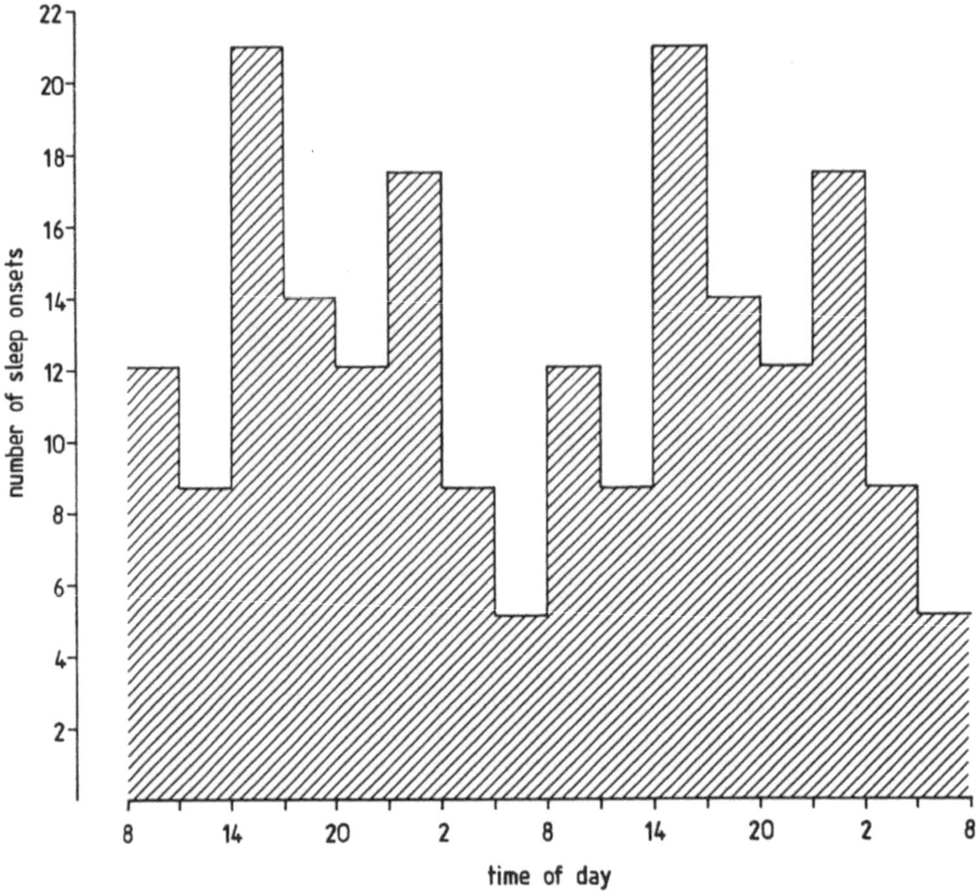

Fig. 3. Distribution of sleep period onsets recorded during disentrainment, plotted across the nychthemeron (double-plotted). For statistical analysis onsets were grouped in 3-h blocks. Distribution did not differ significantly from chance

a major peak between 0200 and 0600, and a second, less robust, peak between 1400 and 1700.

While sleep onsets occurred throughout the nychthemeron, and group tendency for sleep was clearly bimodal, the duration of sleep episodes clearly followed a circadian course, as shown in Fig. 5. Sleep episodes initiated between 0800 and 2000 continued for a median duration of 1.4 h. On only two occasions during this interval did sleep episodes continue for longer than 3 h. In contrast, sleep periods initiated between 2000 and 2400 continued for a median duration of 7.2 h, with a decreasing trend in median durations over the subsequent 8 h.

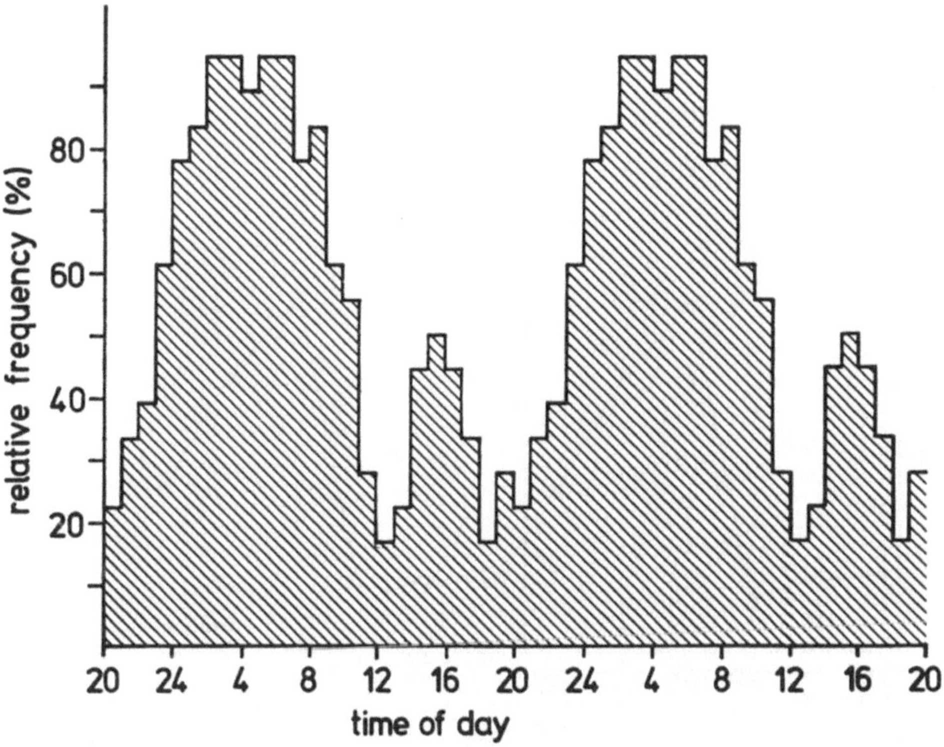

Fig. 4. Distribution of the number of sleep episodes, as a percentage of total possible sleep episodes, comprising all or part of any given hour across the nychthemeron (double-plotted)

Characteristics of Major Sleep Episodes and Naps

Results presented thus far have focused on the characteristics of sleep, in general, during disentrainment: sleep episodes have been treated as a homogeneous group. However, it is apparent from these analyses, and from inspection of Fig. 1, that two general categories of sleep episodes are distinguishable, primarily on the basis of their durations (Fig. 5) and placement within the 24-h day (Fig. 4).

Twenty-one sleep episodes (in addition to baseline sleep periods) were comparable with usual nocturnal sleep episodes in that they were of relatively long duration and were typically initiated after 2000. These sleep periods (total N = 30) have been designated major sleep episodes (MSEs) in the following analyses. Naps were operationally defined as any sleep episode occurring between two MSEs. Twenty-three sleep episodes could be designated as naps, based on this criterion.

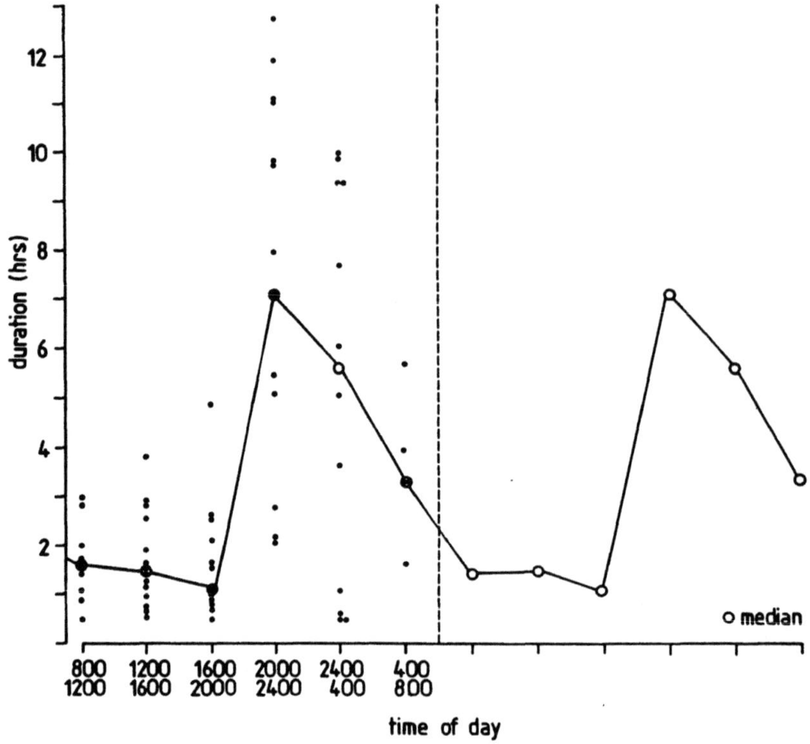

Fig. 5. Distribution of sleep episodes across the nychthemeron (in 4-h blocks, double-plotted), as a function of sleep episode duration. Each dot represents a sleep episode, baseline nights are included

Because some sleep periods resembled neither MSEs nor met the criterion for naps, they are excluded from the following analyses.

Eighteen MSEs were both spontaneously initiated and spontaneously terminated. The median duration of these episodes was 9.8 h (mean, 8.8 h; SD, 2.7 h). This is compared with a median duration for naps ($N = 23$) of 1.4 h (mean, 1.7 h; SD, 0.9 h). The difference in durations for the two groups was significant (Mann-Whitney $U = 6.5$, $p \leq 0.0001$, two-tailed). There was also a significant difference ($U = 33$, $p \leq 0.001$, two-tailed) in durations of waking episodes prior to MSEs (mean, 7.7 h; SD, 2.6 h; median, 7.5 h) vs those preceding naps (mean, 3.6 h; SD, 1.6 h; median, 3.1 h).

Temporal Relations

As a first step in the analysis of temporal relationships among MSEs and naps, the relationship between successive MSEs was examined. During the first days in a time-free environment it is not uncommon for subjects to exhibit transient states in which the onsets of successive MSEs do not necessarily occur systematically later, as is typical in the "free-running" condition. As can be seen in Table 1, this was clearly the case in the 72 h of disentrainment. Only one subject (CBB) showed systematically later sleep onset times for MSEs. However, a clear free-running trend was present in the mean group data, with the mean "day" (measured from the onset of a MSE to the onset of the next MSE) during disentrainment continuing for 25.2 h.

Also shown in Table 1 are onset times of successive naps across the 3 days of disentrainment. As was the case with MSEs, most subjects did not show tendencies for systematically later onset times for nap sleep (notably, the only subject who showed "free-running" in MSEs also exhibited "free-running" naps). However, also in correspondence with the findings of MSEs, there was a clear trend for naps to "run free" when group data were considered. Because on a few occasions a single subject exhibited two naps on a given day, it was necessary to calculate more than one mean "nap day" (measured from the mean onset of naps on one day to the mean onset of naps on the following day). However, as can be seen, values did not differ significantly, and the mean "nap day" continued for about 24.8 h.

In summary, then, slightly over 25 h separated one MSE from the next during disentrainment. Similarly, on average, slightly under 25 h separated successive nap episodes during the disentrained period.

Placement of Naps Relative to MSEs

Our next step in temporal analysis was to examine the placement of naps relative to adjacent MSEs. The average sleep/wake cycle length for all MSEs and naps (i.e., from the onset of each MSE to the onset of a subsequent nap, and from the onset of each nap to the onset of the subsequent MSE) was 12.3 h (SD, 3.8 h; median, 12.3 h). However, there was a difference in the cycle lengths measured from MSEs to naps (mean, 14.6 h; SD, 2.8 h) and those measured from nap onsets to sub-

Table 1. Onset times for successive major sleep episodes (MSEs) and for successive naps. (Blanks do not necessarily mean subject did not sleep, but rather that sleep periods could not be classified as naps or MSEs.)

Subject	Baseline	Disentrained 1 Nap	Disentrained 1 MSE	Disentrained 2 Nap	Disentrained 2 MSE	Disentrained 3 Nap	Disentrained 3 MSE
GLB	00:40	13:32	21:24	12:15	00:09	—	—
ULB	00:03	14:09	22:00	13:17/18:48	23:46	16:31	22:26
CWB	00:09	12:40/16:31	00:50	15:56	00:50	14:52/17:50	23:23
WFB	23:57	14:55	03:19	16:52	01:26	18:59	—
JSB	23:29	—	—	—	23:43	—	00:06
PZB	23:22	—	21:04	14:12	23:14	—	—
CBB	01:07	14:41	21:55	15:42	00:32	17:01	03:04
RMB	23:06	10:30	19:49	9:17	20:24	8:47	—
MHB	02:36	—	—	15:28	02:54	14:14	03:57
MSE Mean (SD)	24:17 (1.1 h)		22:37 (2.6 h)	→ 25.5 h ← (interval)	24:07 (1.8 h)	→ 24.9 h ← (interval)	24:59 (2.4 h)
Nap Mean		13:25/14:03	→ 24.7 h ← (interval)	14:07/14:49	→ 24.9 h ← (interval)	15:04/15:34	
Nap (SD)		(1.6 h) (2.0 h)	→ 24.8 h ← (interval)	(2.5 h) (2.9 h)	→ 24.8 h ← (interval)	(3.5 h) (3.7 h)	

sequent MSEs (mean, 9.9 h; SD, 3.3 h) ($U = 76$, $p \leq 0.0001$, two-tailed). In other words, the initiation of naps tended to occur about 60% through the MSE to MSE sleep/wake cycle.

Relationships to Temperature

As previously noted, group data revealed free-running trends for successive onsets of both MSEs and of naps, as well as circasemidian periodicity between MSEs and subsequent naps. We next analyzed the relationship between the circadian rhythm of body core temperature and the two groups of sleep episodes, as shown in Fig. 6. The figure shows the average temperature curve for all subjects across the 72 h of disentrainment and the baseline night. Above the temperature curve is a summation histogram showing the number of subjects asleep during all or part of any given hour during disentrainment. Shown below the temperature curve (black bars) are the average onsets and durations of MSEs and naps, as described in Table 1.

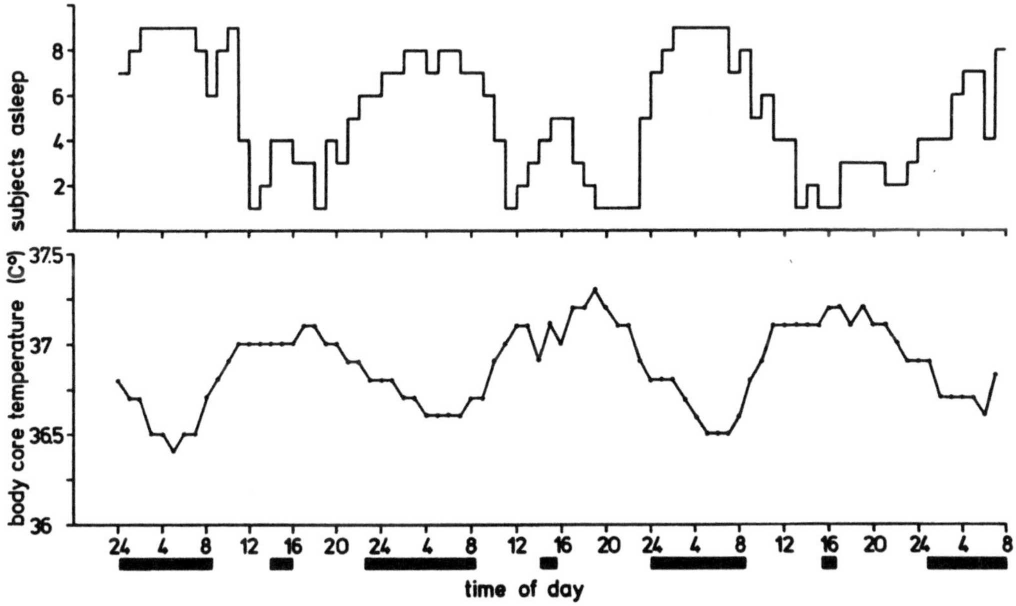

Fig. 6. Relationship between sleep episodes recorded in disentrainment and circadian rhythm of body core temperature. Above the temperature curve (averaged across all subjects) is a summation histogram of the number of subjects asleep during any given hour during disentrainment. Shown below the temperature curve (black bars) are average onsets and durations of major sleep episodes and naps

Group sleep tendency, both for MSEs and naps, exhibited a clear phase relationship with body core temperature. As in all free-running conditions, the MSEs (as reflected by average onset time as well as total number of subjects sleeping) corresponded to the troughs in temperature. Indeed, this circadian sleep tendency was essentially a mirror image of the circadian rhythm in body core temperature. In addition, there were clear tendencies for average nap times to occur in correspondence with daily maxima in temperature. This tendency is best seen in the middle of the figure (day 2 of disentrainment) when group sleep tendency was robust enough to result in a transient drop in average temperature.

Structural Characteristics

A final analysis concerned the comparison of structural parameters of MSEs vs naps. Overall structural parameters of the two groups of sleep episodes are shown in Table 2. Only REM sleep percentages differed significantly between the two groups, with naps averaging about 30% less REM sleep than MSEs. However, the variability of mean percentages associated with naps were typically about twice that of corresponding values for MSEs. Such variability is particularly evident with regard to stage 4 values, where the difference between mean and median measures is dramatic (7.8% vs 0.61%). The distribution of stage 4 values during naps was essentially bimodal with naps being characterized by either no stage 4 or over 10% of stage 4. Naps with onsets between 1300 and 1700 showed median stage 4 amounts of 5.5%. In contrast, all other naps were characterized by median stage 4 values of 0%. In general, then, naps occurring toward the middle of the nap distribution showed a greater tendency for the occurrence of stage 4.

DISCUSSION

Given conditions under which little was allowed to obstruct the unambiguous expression of a putative ultradian propensity for sleep, i.e., sleep itself, subjects exhibited sleep/wake behavior characterized by a clearly bimodal, albeit broad, distribution in the occurrence of sleep. Thus, the present examination of sleep patterns recorded under conditions of disentrainment obtained results which are consistent with, and lend considerable support to, the notion that the human

Table 2. Overall structural parameters of major sleep episodes (MSEs) and naps recorded during disentrainment

	Major sleep episodes (MSEs)			Naps		
Latency	Mean	(SD)	Median	Mean	(SD)	Median
SWS	14.3 min	(4.5)	14.5	19.1 min	(9.4)	19.0
REM	91.3 min	(44.9)	75.5	48.4 min	(14.2)	49.3
Stage (%)						
0	5.5	(3.4)	5.5	5.6	(6.7)	2.9
1	4.6	(2.2)	4.4	6.6	(6.5)	5.5
2	55.1	(6.3)	53.9	58.5	(12.2)	58.0
3	7.6	(3.9)	7.2	7.5	(8.4)	4.3
4	7.3	(3.6)	6.9	7.8	(11.3)	0.6
R	19.8	(7.1)	20.5	13.7	(10.3)	15.5
Complete (%)	100			60.9		
Usual sleep stage sequencing (%)	100			100		
Sleep duration	8.8 h	(2.7)	9.8	1.7 h	(0.89)	1.4
Prior wakefulness	7.7 h	(2.6)	7.5	3.6 h	(1.6)	3.1

sleep/wake system is characterized not only by a strong propensity to obtain a MSE once per circadian day, but also by a prominent tendency for a shorter sleep episode to occur roughly halfway between successive "night's sleep" episodes.

The finding that average nap times maintained a general phase relationship not only with MSEs, but also with the group average rhythm of body core temperature, suggests that the organization of naps within the nychthemeron was a consequence of the existence of a second preferred phase position for sleep, located around the temperature maximum. This suggestion is also supported by a recent study in which we analyzed the data of subjects who, contrary to experimental instructions, napped during isolation under typical time-free conditions

(ZULLEY and CAMPBELL 1985a,b). Sleep episodes subjectively perceived as "naps", and with durations similar to those of naps in the present study, showed a clear tendency to occur about halfway between successive minima of body core temperature.

Under entrained conditions, such a phase position for sleep tendency would correspond to the early afternoon (roughly 1400-1600), a time when the vast majority of napping occurs in populations in which napping is frequently reported (DINGES et al. 1983; OGUNREMI 1978; OKUDAIRA et al. 1983; WEBB 1978). This interval also corresponds to a time of peak occurrence of stage 2 in the EEGs of young adults attempting to sleep every 20 min across the day (LAVIE and SCHERSON 1981).

The relationship between naps and the phase of body core temperature was also reflected in predictable alterations in structural features of these sleep episodes, relative to MSEs. REM sleep amounts show clear circadian variations, with minimum percentages typically recorded during sleep episodes occurring in coincidence with maximum body core temperature (AKERSTEDT and GILLBERG 1981; ENDO et al. 1981; HUME and MILLS 1977). This circadian variation in REM sleep amount resulted in the only significant difference in sleep stage percentages between major and nap sleep episodes.

Likewise, the well-established circadian variation in REM sleep latency (CAMPBELL and ZULLEY, in press; HUME 1978; KARACAN et al. 1970; TAUB and BERGER 1973; WEBB and AGNEW 1978) resulted in significant differences between the two groups of sleep episodes in that measure as well (see Table 2).

These structural differences notwithstanding, and despite the extreme differences in average duration of naps vs MSEs (also as a result of the influence of the circadian rhythm of body core temperature; CZEISLER et al. 1980; ZULLEY 1980), the two groups of sleep periods showed remarkable similarities in overall structure and makeup. Probably as a consequence of their abbreviated durations, about one-third of the naps were characterized by the absence of slow-wave sleep (SWS), REM sleep, or both states of sleep. However, all "complete" naps (i.e., those containing both SWS and REMS) showed sleep stage sequencing characteristic of typical (nocturnal) sleep, with SWS preceding the first epoch of REM sleep at sleep onset. Furthermore, there were no significant differences in the sleep stage percentages between complete naps and MSEs. Thus, in contrast to the findings of WEITZMAN et al. (1974) regarding scheduled naps, our findings indicate that naps, taken ad libitum, quite often are miniatures of the normal 8 h sleep pattern. From these findings it can be concluded that the same general rules governing structural aspects of MSEs also hold for the composition of naps.

Both the finding that naps maintain a general phase relationship with body core temperature, and that they share structural similarities with MSEs, militate against the interpretation that the relatively brief sleep episodes observed in the disentrained environment were taken simply in response to boredom. Rather, the findings indicate that naps are regulated by the same mechanism(s) mediating nocturnal sleep, and occur as a manifestation of an underlying, periodically recurring propensity for sleep.

Additional support for this position is provided by studies of napping behavior in environments which are not characterized by such drastic minimization of behavioral alternatives to sleep. Simple instructions to "sleep when you feel sleepy", given to a subject in an otherwise normal time-free environment, resulted in a doubling of the number of sleep episodes which occurred during an interval of equal duration (10 days) during which the subject was instructed not to nap (CAMPBELL and ZULLEY in preparation). WEBB and AGNEW (1974) reported similar findings in seven subjects maintained under equivalent conditions. And, as mentioned above, instructions not to nap do not guarantee that napping will not occur. Such examples seem to confirm the ubiquity of napping behavior, provided that sufficient release from experimental or social constraints on such behavior is a feature of the environment.

Nevertheless, it is quite likely that the basal, static behavioral conditions of disentrainment contributed to an increase in the likelihood that the physiological tendency for sleep would be expressed unambiguously in the form of naps. This, of course, was the rationale behind invoking such conditions.

An additional result of the elimination of controls on sleep initiation and the minimization of behavioral options to sleep was to increase the average proportion of the nychthemeron spent asleep. Subjects averaged 10.4 h of sleep per 24 h in disentrainment compared with the 7-8 h typical of normal conditions. The increase was due both to a lengthening of MSEs as well as the addition of nap sleep. At least a portion of this increase in total sleep time is probably due to the conditions imposed.

The sleep recorded from human subjects under these conditions is comparable with the sleep recorded from other animals under standard laboratory conditions. It is generally accepted that under such conditions, in which nutritional requirements are met, no restrictions are placed on sleep, and the recording chamber is generally devoid of behavioral alternatives to sleep (despite the opportunity for free movement in the chamber), the amounts of sleep observed reflect the system working at maximum capacity rather than at a level which, in daily life, is limited by environmental demands such as predator pressure, or, in the case of humans, social and occupational pressures.

Whether such limitations on the sleep system result in chronic sleep deprivation is a matter of conjecture. WEBB and AGNEW (1975) have presented considerable data to suggest that if we are not, in fact, chronically sleep deprived, we are, minimally, on inveterate "sleep diets". (Kleitman (1963, p. 307) cites a source (Camp 1923) who "maintained that many people sleep too long and that oversleeping is just as reprehensible as overeating".)

Whether we are deprived or simply restrained by our environment, it is clear that the exclusion of naps from our usual sleep orgnization can be tolerated indefinitely (virtually an entire lifetime) without noticeably detrimental consequences. However, that we are able to function reasonably well without them should not diminish the relevance of naps in the understanding of the nature and the functions of sleep. In the introduction of this chapter, we drew a parallel between the human sleep system, as it functions under the influence of behavior controls, and a river, the course of which has been modified by the imposition of artificial barriers. As with the human system, the milieu of the river is essentially undamaged by the establishment of such controls, and its utility to society is clearly enhanced as a consequence of such structuring. As such, the imposition of behavioral controls is beneficial.

Yet, if one wishes to examine the natural history of the river, barriers such as dams and levees lose their utility and instead become impediments, probably contributing to spurious, or, at best, incomplete conclusions regarding the river's true course and characteristics. In the same way, the natural progression of sleep and wakefulness across the nychthemeron cannot be adequately evaluated in the presence of restrictions on the system which have the potential to influence its true expression.

In summary, the present findings indicate that when the obstructions are removed, what remains is a rather labile organization of sleep and wakefulness characterized by at least two intervals of elevated drowsiness per circadian day, tied to the temperature cycle, and distinguished by their differential strengths. It is the presence and relative strength of environmental restraints that determines the extent to which these noncircadian variations in sleep propensity will be expressed in the unequivocal form of naps.

REFERENCES

AKERSTEDT T, GILLBERG M (1981) The circadian variation of experimentally displaced sleep. Sleep 4(2):159-169

BENOIT O (1984) Homeostatic and adaptive roles of human sleep. Experientia 40:437-440

BROUGHTON RJ (1975) Biorhythmic variations in consciousness and psychological functions. Can Psychol Rev 16(4):217-239

CAMPBELL SS (1984) Duration and placement of sleep in a "disentrained" environment. Psychophysiology 21(1):106-113

CAMPBELL SS, ZULLEY J (in press) Rhythmic components of sleep structure during disentrainment

CZEISLER CA, WEITZMAN ED, MOORE-EDE MC, ZIMMERMAN JC, KNAUER RS (1980) Human sleep: its duration and organization depend on its circadian phase. Science 210:1264-1267

DINGES DF, ORNE EC, ORNE MT (1983) Napping in North America: a siesta rhythm? Symposium abstracts, 4th International congress of sleep research, p 86

ENDO S, KOBAYASHI T, YAMAMOTO T, FUKUDA H, SASAKI M, OHTA T (1981) Persistence of the circadian rhythms of REM sleep: a variety of experimental manipulations of the sleep-wake cycle. Sleep 4(3):319-328

HUME KI (1978) Some electrophysiological studies of sleep from subjects on varying sleep-wakefulness schedules. Thesis, University of Manchester

HUME KI, MILLS JN (1977) Rhythms of REM and slow-wave sleep in subjects living on abnormal time schedules. Waking Sleeping 1:291-296

KARACAN I, FINLEY WW, WILLIAMS RL, HURSCH CJ (1970) Changes in stage 1-REM and stage 4 sleep during naps. Biol Psychiatry 2:261-265

KLEITMAN N (1963) Sleep and wakefulness. University of Chicago Press, Chicago

LAVIE P, SCHERSON A (1981) Ultrashort sleep-waking schedule. I. Evidence of ultradian rhythmicity in "sleepability". EEG Clin Neurophysiol 52:163-174

NAKAGAWA Y (1980) Continuous observation of EEG patterns at night and in daytime of normal subjects under restrained conditions. I. Quiescent state when lying down. EEG Clin Neurophysiol 49:524-537

OGUNREMI O (1978) The subjective sleep pattern in Nigeria. Sleep Res 7:163

OKUDAIRA N, FUKUDA H, OHTANI K, NISHIHARA K, ENDO S, TORII S (1983) Napping patterns in the elderly in Vilcabamba, Ecuador. Symposium abstracts, 4th International congress of sleep research, p 88

RECHTSCHAFFEN A, KALES A (eds) (1968) A manual of standardized terminology, techniques and scoring system for sleep stages of human subjects. U.S. Department of Health, Education and Welfare, NIH, Washington

TAUB J, BERGER R (1973) Sleep stage patterns associated with acute shifts in the sleep-wakefulness cycle. EEG Clin Neurophysiol 35:613-619

TUNE G (1968) Sleep and wakefulness in normal human adults. Br Med J 2:269-271

TUNE G (1969) Sleep and wakefulness in 509 normal human adults. Br J Med Psychol 42:75-80

WEBB WB (1978) Sleep and naps. Speculations in Sci Technol 1:313-318

WEBB WB, AGNEW HW (1974) Sleep and waking in a time-free environment. Aerospace Med 45:617-622

WEBB WB, AGNEW HW (1975) Are we chronically sleep deprived? Bull Psychonomic Soc 6:47-48

WEBB WB, AGNEW HW (1978) Effects of rapidly rotating shifts on sleep patterns and sleep structure. Aviat Space Environ Med 49(2):384-389

WEBB WB, DUBE MG (1981) Temporal characteristics of sleep. In: ASCHOFF J (ed) Handbook of behavioural neurobiology, vol 4. Biological rhythms. Plenum, New York, pp 499-522

WEITZMAN ED, NOGEIRE C, PERLOW M, FUKUSHIMA D, SASSIN J, McGREGOR P, GALLAGHER TF, HELLMANN L (1974) Effects of a prolonged 3-hour sleep-wake cycle on sleep stages, plasma cortisol, growth hormone and body temperature in man. J Clin Endocrinol 38(6):1018-1029

WEVER RA (1979) The circadian system of man. Springer, Berlin Heidelberg New York

WHITE R (1974) Sleep length and variability: measurement and interrelationships. Dissertation, University of Florida

ZULLEY J (1980) Distribution of REM sleep in entrained 24-hour and free-running sleep-wake cycles. Sleep 2(4):377-389

ZULLEY J, CAMPBELL SS (1985a) The coupling of sleep-wake patterns with the rhythm of body temperature. In: KOELLA W, RÜTHER E, SCHULZ H (eds) Sleep 1984: proceedings of the 7th European congress of sleep research. Fischer, Stuttgart (in press)

ZULLEY J, CAMPBELL SS (1985b) Napping behavior during "spontaneous internal desynchronization": sleep remains in synchrony with body temperature. Hum Neurobiol (in press)

Section 3
Methods and Models

The third section of the book includes four chapters which deal with methodological and theoretical aspects of ultradian rhythms. This section opens with two chapters by Kobayashi, Tsuji and Endo and by Naitoh which discuss their approaches to the sleep REM-NREM cycle.

In Chapter 15 Kobayashi, Tsuji and Endo demonstrate an interesting relationship between the length of the REM-NREM cycle and its content. By analyzing cycles containing REM and stage 2 sleep only, or REM sleep and slow wave sleep (SWS) and stage 2 sleep separately, they conclude that SWS suppresses REM sleep whenever both sleep states appear in the same cycle. Naitoh, in Chapter 16, describes in a nonmathematical language a computational procedure for detection and defining the sleep REM-NREM cycle. He illustrates the usefulness of this method by using infants sleep data and sleep data obtained under elevated ambient temperature and noise conditions.

The last two chapters present two of the first theoretical models of the ultradian sleep cycles. McCarley and Massaquoi (Chapter 17) expand the earlier McCarley and Hobson model of REM-NREM alternation that was mathematically based on the Lotka-Volterra equations. They sketch the main features of the expanded, limit cycle model, and discuss its predictions regarding the phase-response curve to pharmacologic perturbations of the ultradian REM recurrence cycle.

Wever, in Chapter 18, shows that ultradian rhythms during sleep can be modeled by expanding his earlier model for circadian rhythms. He suggests that a system of fourth order differential equations best fits the experimental data. In addition, this final chapter offers a conceptual framework to bridge the gap between ultradian and circadian rhythms.

Sleep Cycles as a Basic Unit of Sleep

T. Kobayashi, Y. Tsuji, and S. Endo
Psychiatric Research Institute of Tokyo, 2-1-8, Kamikitazawa Setagaya-Ku, Tokyo 156, Japan

INTRODUCTION

The physiological and psychological functions of REM sleep and slow wave sleep (SWS) have been investigated using deprivation techniques (DEMENT 1960; BERGER and OSWALD 1962; AGNEW et al. 1964) and chronobiological approaches (MILLS 1964; ASCHOFF 1965; CZEISLER et al. 1980). These techniques and approaches have provided us with much useful information about sleep, but some important problems still remain to be solved. For example, not much is known about (a) whether the REM cycle is a sleep-dependent rhythm or not (KLEITMAN 1963; KRIPKE 1974; MOSES 1977), (b) the effects of wakefulness on SWS (BERGER et al. 1971; WALKER et al. 1978; SHAPIRO et al. 1981), and (c) the psychophysiological function of REM sleep (PEARLMAN 1982). One of the reasons these problems have not yet been adequately addressed is that the sleep cycle as a basic unit of sleep has not been fully understood.

The periodic appearance of REM sleep in the sleep cycle is one of the dominant characteristics of sleep, as DEMENT and KLEITMAN (1957) were the first to report. The sleep cycle is a component of the infrastructure of sleep which recurs every 90-100 min, and whose content is influenced by such factors as sleep loss and circadian effects. If we can develop an algorithm for the temporal organization and the content of the sleep cycle, we can more precisely predict changes in sleep patterns due to various conditions. For example, the distribution of REM sleep and SWS during recovery sleep after sleep deprivation could be described by means of such an algorithm.

In order to establish the algorithm, we have examined the range of the sleep cycle length, as well as its content and organization.

Experimental Brain Research, Suppl. 12

METHODS

Nocturnal sleep (140 episodes) of 35 healthy male students (mean age, 22.5 years) was polygraphically recorded from 2300 ± 10 min to 0700 ± 10 min. The EEG was derived from electrodes in frontal (F3-A1), central (C3-A1), and occipital (O1-A1) positions. Eye movements (EOG) were recorded from electrodes fixed above and below the outer canthi. Chin EMG and ECG were also recorded. The sleep recordings, which were not interrupted by more than 3 continuous min of wakefulness, were scored for sleep stages in 1-min epochs using the sleep stage manual (RECHTSCHAFFEN and KALES 1968).

Definition of a Sleep Cycle

NREM sleep cycles were analyzed in the present study because they take into account SWS with high amplitude EEG delta waves after sleep onset. Thus, a NREM sleep cycle was defined as a NREM period followed by a REM period. In order to determine what the minimal length of a NREM period should be, the interval histogram of REM episodes was analyzed for all 140 nocturnal sleep episodes. The interval histogram showed a bimodal distribution (Fig. 1). The first peak, corresponding to short REM-REM intervals, is due to typical REM sleep interruptions by stages 1 or 2. The second peak, at more than 40 min, indicated typical NREM periods with sleep stages 2, 3, and 4. Therefore, the minimum NREM length was set at the trough between the two peaks and was defined as being more than 25 min.

Thus, each cycle began with NREM sleep. The beginning of the first sleep cycle was defined by the first epoch of stage 2. The end of a sleep cycle and the beginning of the next cycle were determined by the first appearance of NREM sleep after a REM period had been completed. A REM period was considered as being complete only when the duration of the NREM episodes following the last REM episode exceeded 25 min. The duration of a REM period, except for the first cycle, was defined as being more than 5 min. Sleep stages were divided into three groups: (a) stage 2 sleep (S2), (b) slow wave sleep (SWS), and (c) REM sleep (REMS). Sleep cycles were also divided into two types; type A consisted of S2 and REMS and type B of S2, REMS, and SWS.

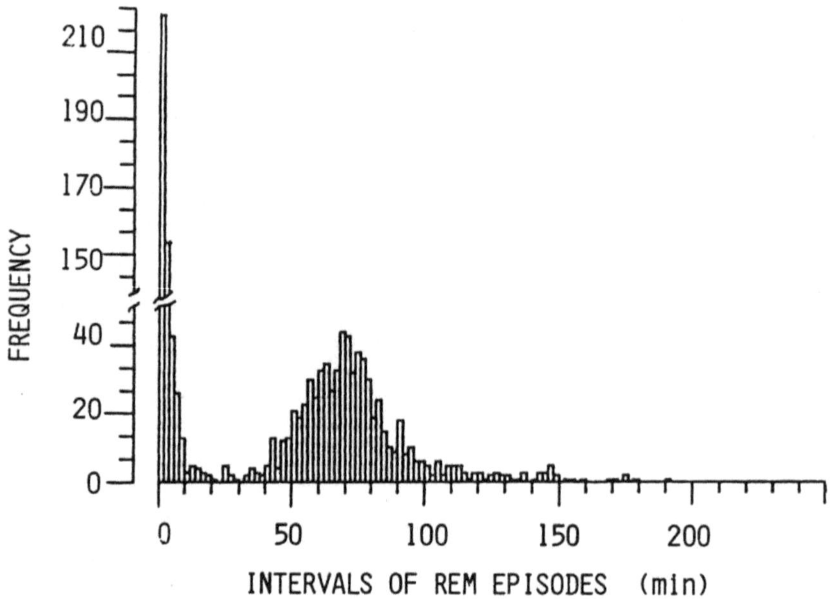

Fig. 1. Interval histogram of REM episodes

RESULTS

Cycle Content and the Range of Cycle Length

The amounts of S2 and REMS in a type A sleep cycle had a looped curve relationship between 30 and 140 min of cycle length as shown in Fig. 2. The starting point of the looped curve relationship was at 30 min, when S2 and REMS each occupied nearly 50% of the cycle. The amount of S2 in type A cycles increased from 30 to 100 min of the cycle length, but did not increase further from 100 to 140 min of the cycle. On the other hand, the amount of REMS in type A cycles did not increase from 30 to 70 min of the cycle, but increased abruptly between 70 and 140 min of the cycle. The amount of S2 and REMS again each occupied nearly 50% of a cycle length at 140 min, and the looped curve was completed there. The amount of S2 abruptly increased again from 140 to 160 min of the cycle, while that of REMS abruptly decreased in the same interval. Therefore, the composition of cycle content in type A cycles can be described by the looped curve relationship.

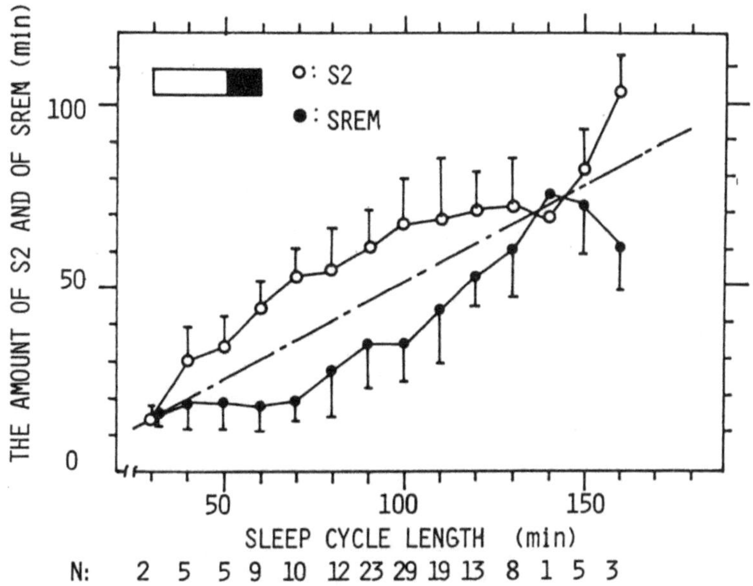

Fig. 2. Changes in content of type A cycles in relation to the cycle length. There is a looped curve relationship from 30 min to 140 min of the cycle length. The dotted line corresponds to 50% of cycle length. N, the number of cycles at each cycle length

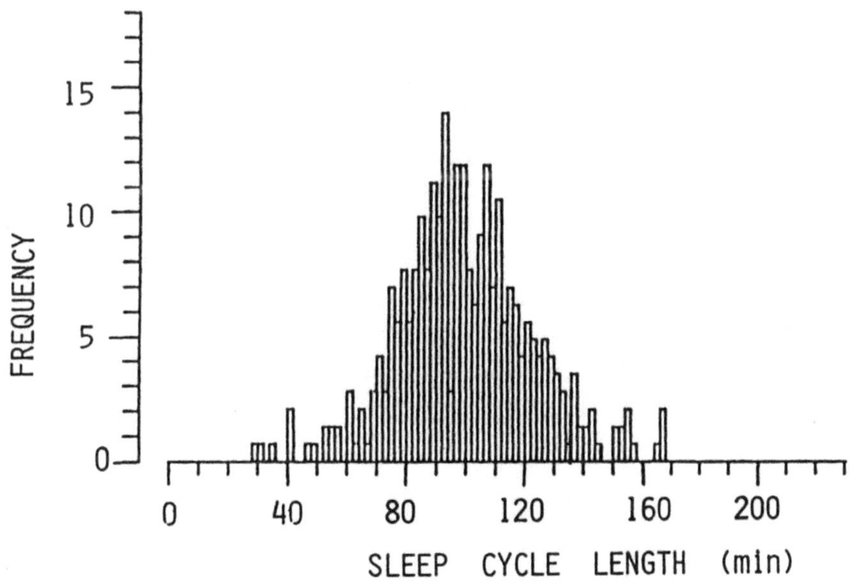

Fig. 3. Frequency distribution of the length of type A and type B sleep cycles excluding the first cycle (mean, 99.9 min; SD, 24.9 min)

Ninety-eight percent of all sleep cycles, except for the first cycles, had a cycle length of between 30 min and 140 min (Fig. 3). Most cycles with a duration of more than 140 min were first sleep cycles. The length of these very long sleep cycles was 1.5-2.0 times that of the mean cycle length (99.9 min), and their content appeared to be structurally combined from two sleep cycles. They corresponded to missed REM sleep episodes which were first described by DEMENT and KLEITMAN (1957).

Interactions Between Sleep Stages in the Sleep Cycle

When the sleep cycle was comprised of S2 and REMS only, the looped curve relationship was valid (Fig. 2), but when SWS was found in the sleep cycle, the temporal distribution of the stages was changed (Fig. 4). The effect of SWS on cycle content is shown in Fig. 5, which was constructed by superimposing Figs. 2 and 4. The amount of S2 and REMS in type B cycles was smaller than in type A cycles during the looped curve interval. The two shaded areas in the lower part of Fig. 5 represent the amount of S2 and REMS that was replaced by SWS. Two kinds of interactions existed between sleep stages in type B cycles, one between S2 and SWS, and the other between REMS and SWS.

The interaction between S2 and SWS was not constant during the looped curve interval, while the interaction between REMS and SWS was found at any point of the cycle. It was most pronounced when cycle length was more than 110 min (Fig. 5). A reciprocal relationship between changes in %REMS and %SWS throughout the night was found in type B cycles, whereby %REMS significantly increased and %SWS significantly decreased as sleep progressed ($p \leqq 0.01$, one-way ANOVA) (Fig. 6). The %REMS in type B cycles, except for the fifth one, was significantly shorter than that in type A cycles ($p \leqq 0.01$, t test). Thus, when SWS was found in the sleep cycle, REMS was decreased.

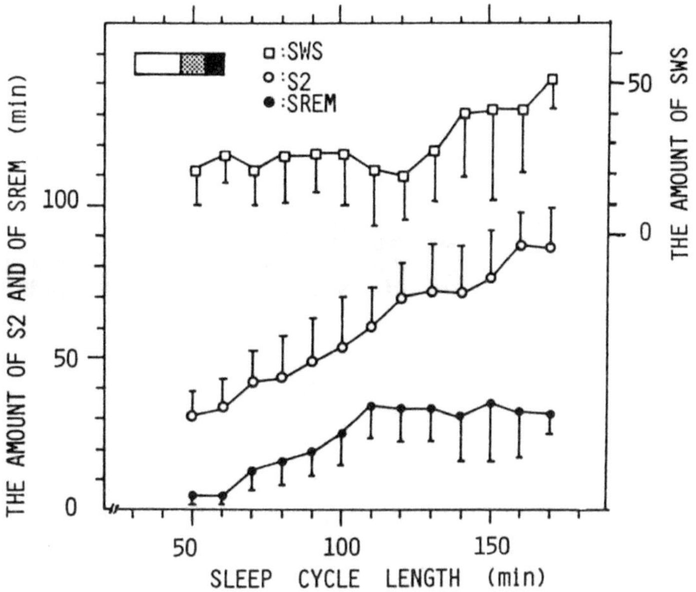

Fig. 4. Changes in content of type B cycles in relation to the cycle length

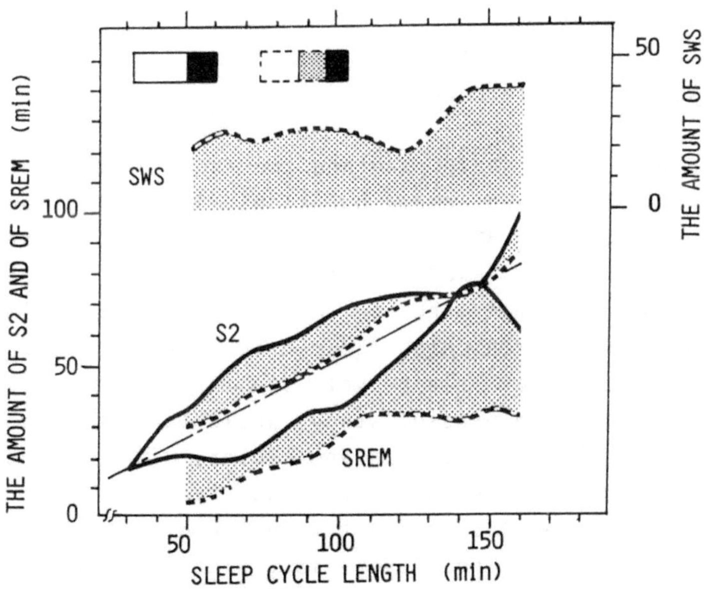

Fig. 5. A comparison between cycle content of type A and type B cycles. The solid lines are for type A, the dotted lines are for type B cycles. The shaded area in the upper part shows the amount of SWS in type B cycles. The other two shaded areas show the amount of S2 and REMS replaced by SWS in type B cycles

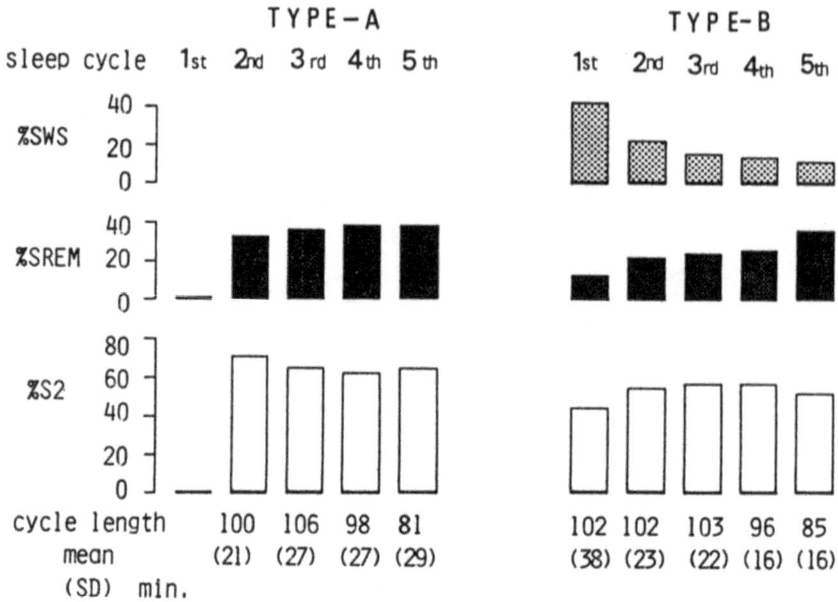

Fig. 6. Changes in the content of type A and type B sleep cycles throughout the night

DISCUSSION

Range of Cycle Length

Few studies have examined the range of cycle length. HARTMANN (1968) analyzed sleep cycle durations (REM cycle), and found that the range was between 38 min and 170 min in young adults (mean age, 23 years). According to BREZINOVA's study (1974), the range of NREM cycles is between 31 min and 171 min in young adults (mean age, 22 years) and between 27 min and 138 min in elderly persons (mean age, 58 years). The range of sleep cycle lengths in elderly persons agrees with the data of the present study, while that in young adults' sleep is different. The very long sleep cycles, with lengths greater than 140 min, seem to be composed of two cycles. They are often found in young adults' sleep but are not found in elderly persons' sleep (HAYASHI and ENDO 1982). These very long sleep cycles, filled with long bouts of stage 4 sleep, appeared primarily in the first cycle of the young adults' sleep. When stage 4 sleep continues for a long time in a sleep cycle, there is a tendency toward extended apparent cycle length over the normal range, causing the phenomenon which has been called missed

REM sleep. The length of these very long sleep cycles is about 1.5-2.0 times the mean cycle length.

Interactions Between Sleep Stages of the Sleep Cycle

When SWS was found in sleep cycles, two kinds of interactions between sleep stages occurred: one was between S2 and SWS and the other was between REMS and SWS, as shown in Fig. 5. The interaction between REMS and SWS, in which more SWS was associated with REMS, was found at any cycle length and throughout the night (Figs. 5 and 6). This relationship between REMS and SWS may be interpreted as the suppression of REMS by SWS.

According to FUKUDA et al. (1983), the length of the first REM episode during nocturnal sleep, in which SWS was deprived, was about 15 min longer than that of the first REM episode on the baseline night. This finding also supports the interpretation that SWS suppresses REMS.

On the other hand, more information is needed to understand the interaction between S2 and SWS, since the replacement of S2 by SWS was neither steady during the looped curve interval of the cycle nor throughout the night.

These interactions suggest that SWS modifies cycle content. However, even when SWS is found in the sleep cycle, it does not affect the cycle length (Fig. 6). Thus, it is possible that cycle content is basically composed of S2 and REMS and modified by SWS. Specifically, cycle duration is essentially set. The presence of SWS does not influence cycle duration but rather leads to a shortening of the REMS part of the cycle.

Sleep Cycle as an Ultradian Periodic Phenomenon

Neither the length nor the content of the sleep cycle is randomly distributed; rather, both are regulated according to an algorithm. The looped curve relationship suggests that an algorithm exists which allows the length and content of the sleep cycle to be modeled. Sleep cycles are considered as the basic unit of sleep, and occur four or five times in normal nocturnal sleep. The algorithm which determines the length and content of this basic unit may be influenced by such parameters as circadian effects and prior to a sleep episode. Thus, ultradian periodic organization of sleep can be understood as the periodic appearance of a basic unit of sleep.

SUMMARY

Human sleep can be understood as a periodic repetition of the basic unit of sleep, the content and length of which may be modeled by an algorithm. The algorithm states that the content of the basic unit is composed mainly of S2 and REMS, and modified by SWS. Cycle length which varies between 30 min and 140 min is independent of the presence or absence of SWS.

REFERENCES

AGNEW HW, WEBB WB, WILLIAMS RL (1964) The effects of stage four sleep deprivation. Electroencephalogr Clin Neurophysiol 17:68-70

ASCHOFF J (1965) Circadian rhythms in man: self-sustained oscillator with an inherent frequency underlines human 24 hrs periodicity. Science 148:1427-1432

BERGER RJ, OSWALD J (1962) Effects of sleep deprivation on behavior, subsequent sleep and dreaming. J Mental Science 108:457-465

BERGER RJ, WALKER JM, SCOTT TD, MAGNUSON LJ, POLLACK SL (1971) Diurnal and nocturnal sleep stage patterns following sleep deprivation. Psychonomic Science 23:273-275

BREZINOVA V (1974) Sleep cycle content and sleep cycle duration. Electroencephalogr Clin Neurophysiol 36:275-282

CZEISLER CA, WEITZMAN ED, MOORE-EDE MC, ZIMMERMAN JC (1980) Human sleep: its duration and organization depend on its circadian phase. Science 210:1264-1267

DEMENT WC (1960) The effect of dream deprivation. Science 131: 1705-1707

DEMENT WC, KLEITMAN N (1957) Cyclic variations in EEG during sleep and their relation to eye movement, body mobility and dreaming. Electroencephalogr Clin Neurophysiol 9:673-690

FUKUDA H, ENDO S, YAMAMOTO T, KOBAYASHI T, SAITO Y, TORII S (1983) The effects of slow wave sleep deprivation upon the REM episode. Sleep Res 12:327

HARTMANN E (1968) The 90 minute sleep-dream cycle. Arch Gen Psychiat 18:280-286

HAYASHI Y, ENDO S (1982) All-night sleep polygraphic recordings of healthy aged persons: REM and slow wave sleep. Sleep 5:277-283

KLEITMAN N (1963) Sleep and wakefulness. University of Chicago Press, Chicago

KRIPKE DF (1974) Ultradian rhythms in sleep and wakefulness. In: WEITZMAN ED (ed) Advances in sleep research, vol 1. Spectrum, New York, pp 305-322

MILLS JM (1964) Circadian rhythms during and after three months in solitude underground. J Physiol 174:217-231

MOSES JM (1977) Rapid eye movement cycle is a sleep-dependent rhythm. Nature 265:360-361

PEARLMAN AC (1982) Sleep structure variation and performance. In: WEBB WB (ed) Biological rhythms, sleep and performance. Wiley, New York, pp 143-173

RECHTSCHAFFEN A, KALES A (1968) A manual of standardized terminology, techniques and scoring systems for sleep stages of human subjects. U.S. Government Printing Office, Public Health Service, Washington

SHAPIRO CM, BORTZ R, MITCHELL D, BARTEL P, JOOSTE P (1981) Slow-wave sleep: a recovery period after exercise. Science 214:1253-1254

WALKER JM, FLOYD TC, FEIN G, CAVNESS C, LUALHATI R, FEINBERG I (1978) Effects of exercise on sleep. J Appl Physiol 44:945-951

In Search of REM Cycle in Short Sleep Record: Iterative Nonorthogonal r^2 Method

P. Naitoh
Naval Health Research Center, P.O. Box 85122, San Diego, CA 92138-9174, USA

In 1953, ASERINSKY and KLEITMAN reported on a cyclic recurrence of rapid eye movement (REM) sleep in normal adults. They observed an average time interval of 2 h 16 min between the onset of one REM period to the next. Later, they discovered similarly recurrent REM periods in newborns (ASERINSKY and KLEITMAN 1955). Regular recurrence of REM sleep was also found for other mammals. ZEPELIN and RECHTSCHAFFEN (1974) analyzed sleep data of 53 mammalian species to find that the mean sleep cycle length correlated positively with life span, brain weight, and negatively with metabolic rate. The REM cycle is an ubiquitous characteristic of the infrastructure of sleep.

The purpose of this chapter is to describe an iterative nonorthogonal r^2 method for detecting and defining the REM cycle. Our method was initially developed to describe ultradian rhythms in delta activity in sleep (HILBERT and NAITOH 1972; LUBIN et al. 1973), and was extended later on to analyze the REM cycle in manually scored sleep stages (NAITOH et al. 1973). The method has been used by FRIEDMANN et al. (1977), JOHNSON (1980), KRYNICKI (1975), MOSES et al. (1977, 1978), MULLANEY et al. (1977), NAITOH et al. (1975), and URSIN et al. (1983).

In the present chapter, the basic computational principles and an outline of the computer program involved in the r^2 method will be described. The characteristics of the r^2 method will be demonstrated by analyzing a set of data simulating the REM cycle. Then, the results from three studies will be used to show how the r^2 method can be utilized to detect changes in the REM cycle under various experimental conditions. Finally, some advantages and disadvantages of the r^2 method will be discussed.

Experimental Brain Research, Suppl. 12

ITERATIVE NONORTHOGONAL r^2 METHOD

Conventional REM Cycle Analysis

The most frequently used method for a REM cycle analysis is to measure time intervals between two adjacent REM episodes, and to calculate an average inter-REM interval with its standard deviation (SD). There are three ways to measure the REM interval: (a) from the onset of one REM episode to the onset of the next (onset-to-onset), (b) end-to-end, and (c) midpoint-to-midpoint (LEWIS 1974). No substantial differences are found between these REM interval measures. The magnitude of SD may be used as an index of rhythmic intensity of the REM cycle. If REM sleep intervals observed in a sleep record are all exactly of 100 min, this sleep record has an average REM cycle of 100 min with SD of 0 min, an idealized case of a perfectly rhythmic REM cycle.

Percent Autoagreement

A method developed by GLOBUS (1970) provides more information about rhythmic intensity than the magnitude of the SD. GLOBUS calculated binary percent autoagreement. In this method the sleep record is divided into successive 20-s (or 30-s) segments. Each segment is scored visually either as REM (or "1"), or NREM sleep (or "0"). In such a way the sleep record is transformed into a string of REM and NREM segments (i.e., a binary time series). To exemplify the computational procedure, let us assume that the sequence of REM and NREM segments is plotted across a narrow and long sheet of paper tape, and that it is copied to make its duplicate. Then, the beginning of this copy is accurately placed below the beginning of the original. Subsequently, the copy is compared with the original to determine how many REM segments of the duplicate fall on REM segments in the original, i.e., percent autoagreement. In the first comparison, the original and the copy are the same. Therefore, percent autoagreement for REM is 100%. Next, the concept of a "lag" is introduced. While the original is left in its original position, the duplicate is moved to the right, so its beginning is placed on the second segment of the original, i.e., lag 1.

Then the autoagreement is calculated as percentage of the total possible agreement. In the next step, the original is again left as it is, but the duplicate is moved on to the third segment of the original (lag 2). Again, the percent autoagreement is calculated. This process is repeated for as many lags as specified.

In this analysis, a maximal autoagreement is located and the corresponding number of lags is determined. This number of lags is converted into time units (min or s), and taken as the length of the REM sleep cycle in the sleep record. The difference between the maximal and minimal percent autoagreement is used as a gross index of rhythmic intensity.

The r^2 Method

The r^2 method starts with an autocorrelation transformation of a sleep record. Then, the autocorrelation function is correlated with a set of cosine waves. Each of the obtained correlations is squared to yield an index of the data variance accounted for by a cosine wave, r^2. Unlike an estimate of rhythmic intensity obtained with fast Fourier transform or with harmonic analysis, an estimate of rhythmic intensity obtained with the r^2 method is correlated with adjacent estimates, i.e., the r^2 method is a nonorthogonal analysis (see ANDERSON 1971). Another characteristic of the r^2 method is that its algorithm repeatedly uses the same computational routine. It systematically changes the cycle length of the cosine wave which is fitted to the sleep data, i.e., the r^2 method is iterative. The r^2 method represents the REM cycle by two parameters: (a) REM sleep cycle length, and (b) rhythmic intensity for each period length.

Computer Program Outline of the r^2 Method

The r^2 program which is written in FORTRAN consists of a main routine which calls for three subroutines: SLPIN1, GLOBUS, and DAMP. The subroutine GLOBUS calls subroutines ACOR and PLOTCOR, and subroutine DAMP calls for DAMPCOS and COR.

The SLPIN1 reads in sleep stage data. Adult sleep data are scored as W, 1, 2, 3, 4, REM, and M (movement time), according to RECHTSCHAFFEN and KALES (1968). One modification has been introduced in defining a

REM episode. The end of a REM episode must be separated from the start of the next REM episode by more than 15 min. If two adjacent REM episodes are separated by less than 15 min, they are combined and considered as one continuous REM episode. Sleep stages are scored on a "page-by-page" basis, where one page of polygraphic record represents either 20 or 30 s of sleep. When there is a change in sleep stage, the time of this stage change is registered. Subroutine SLPIN1 converts each sleep stage to a set of integers. For a REM cycle study, it assigns 1 to REM sleep, and zero to all other sleep stages. After the conversion of sleep stages into integers, the subroutine GLOBUS calls ACOR to calculate the autocorrelation function of the integer converted sleep stages. The maximal lag for the autocorrelation function is generally set to be two-thirds of the data length, unless otherwise specified by the user. For instance, when total sleep is 1440 pages, each page representing 20 s of sleep, the maximum lag is 960 lags, or 320 min (= 960 x 20 s). This means that the autocorrelation function is able to show about three complete 100-min cycles. GLOBUS also calls the subroutine PLOTCOR which plots the autocorrelation values. The completion of the calculation and plotting of the autocorrelation function is followed by calling subroutine DAMP. DAMP in turn calls DAMPCOS, which generates a set of constant amplitude cosine waves, with periods ranging from 40-150 min in 2-min steps. Each of these cosine waves is, then, modified to have its amplitude exponentially reduced over 6 h. A constant amplitude cosine wave is expressed as cos (wt). The damped cosine wave is $C^{(t)}(\cos wt) = d$, where C is a constant, d is a damping ratio, w is angular velocity, and t is time. The damping ratio ranges from 1 to 10. At the damping ratio of 1, d is selected so that the cosine wave keeps a constant amplitude over a 360-min period, i.e., $C^{(t)} = C^{360} = 1$, or C = 1.0. With a damping ratio of 3, the amplitude of the cosine wave will be reduced to one-third after 6 h, i.e., $C^{360} = 1/3$, or C = 0.996953. For example, the cosine wave is multiplied by 0.996953^{t} (cos (wt)), where t is the elapsed time in min from time zero. At time zero, wt = 0, and 0.996953^{0} (cos (0)) = 1.0.

Ten damped pure cosine waves, corresponding to the d-value of 1-10, for each selected period are used to improve the correlations between the autocorrelation function representing the REM cycle and the computer-generated cosine waves. DAMP calls subroutine COR to calculate the product moment correlations. The obtained correlation is squared (r^2) to show goodness-of-fit of the damped cosine wave to the autocorrelation transformed sleep stage data.

Figure 1 shows the plot of r^2 values. This is similar to a spectral power plot, spectrogram, or periodogram. Its Y axis shows a proportion of the variance accounted for by a cosine wave whose period is shown along its X axis. The r^2 plot helps to determine visually whether the data have a regularly recurring REM cycle. The figure shows a strong REM cycle with a period length of 101 min, accounting for 82.1% of the total variance in the data.

The r^2 program completes its job by printing some analysis parameters and the highest r^2 value. More detailed explanation is given by NAITOH et al. (1973). NUTE and NAITOH (1971) had shown that in some circumstances the r^2 method is more suitable for estimating the true frequency of a pure sine wave than the harmonic analysis or the continuous Fourier transformation.

Calibration of the r^2 Analysis

Six "calibration signals", simulating the REM cycle, were analyzed with the r^2 program to evaluate its performance. These calibration signals were coded as A through F, and are shown in Fig. 2. Estimating the period length with the r^2 method agrees fairly well with the inter-REM intervals, although there was a tendency for some overestimation (Table 1).

APPLICATIONS: THREE EXAMPLES

Three demonstrative examples of the r^2 method will be described. These examples are: infant sleep data, sleep in conditions which were disturbed by noise or elevated ambient temperature, and sleep under a condition of gradual sleep reduction.

Infant Sleep Data

The infant sleep data were provided by J.B. Gould who scored polygraphic sleep data into one of six categories: A (awake), R (REM or active sleep), Q (quiet sleep), I (intermediate sleep stage), X (un-

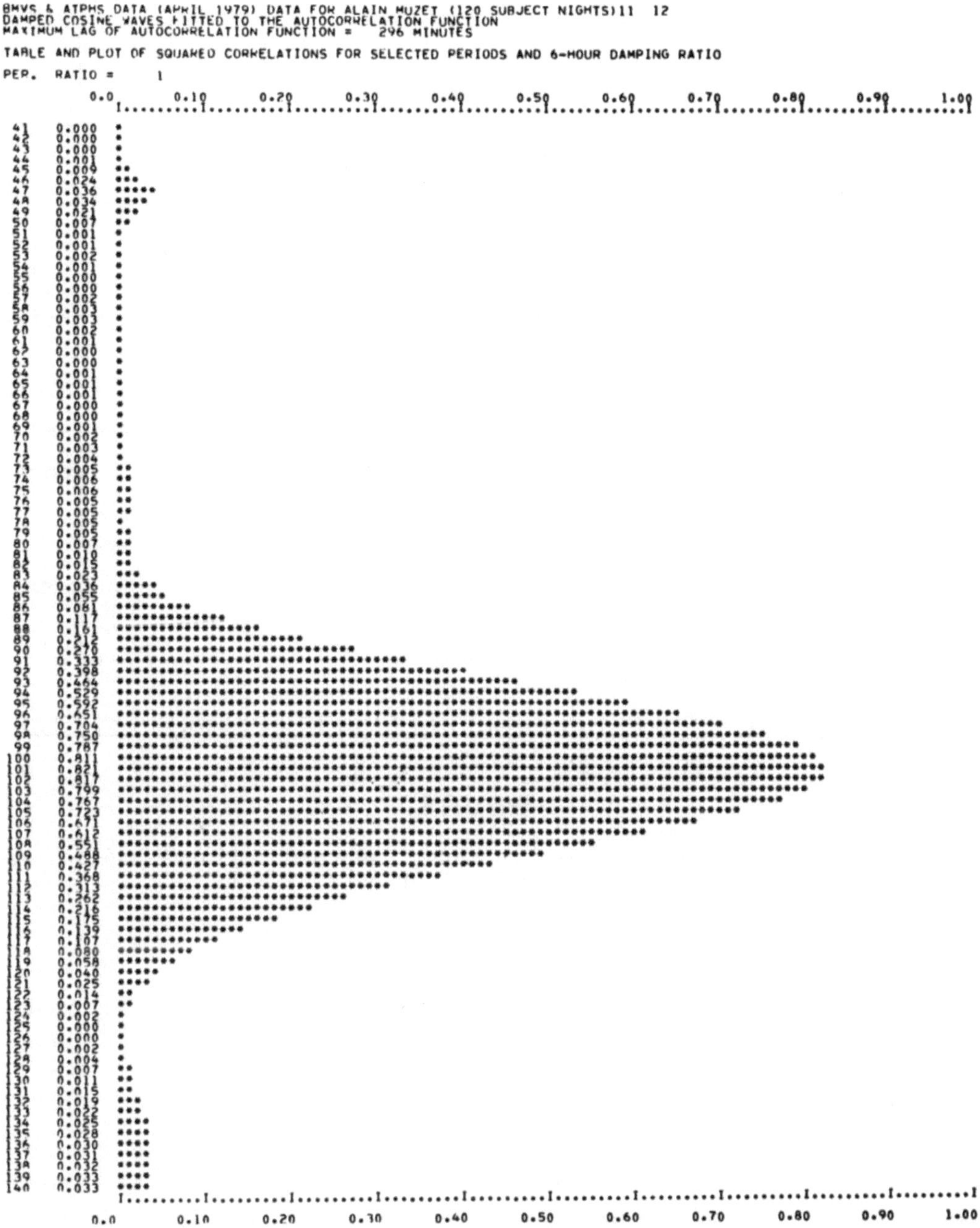

Fig. 1. An example of the r^2 plot; analogous to a plot of raw spectral intensities. The ordinate shows the proportion of the total variance in data, which is accounted for by a cosine wave of the period specified along the abscissa. The best fitting cosine curve for this sleep record has the period length of 101 min, which could explain 82.1% of the total variance in the data

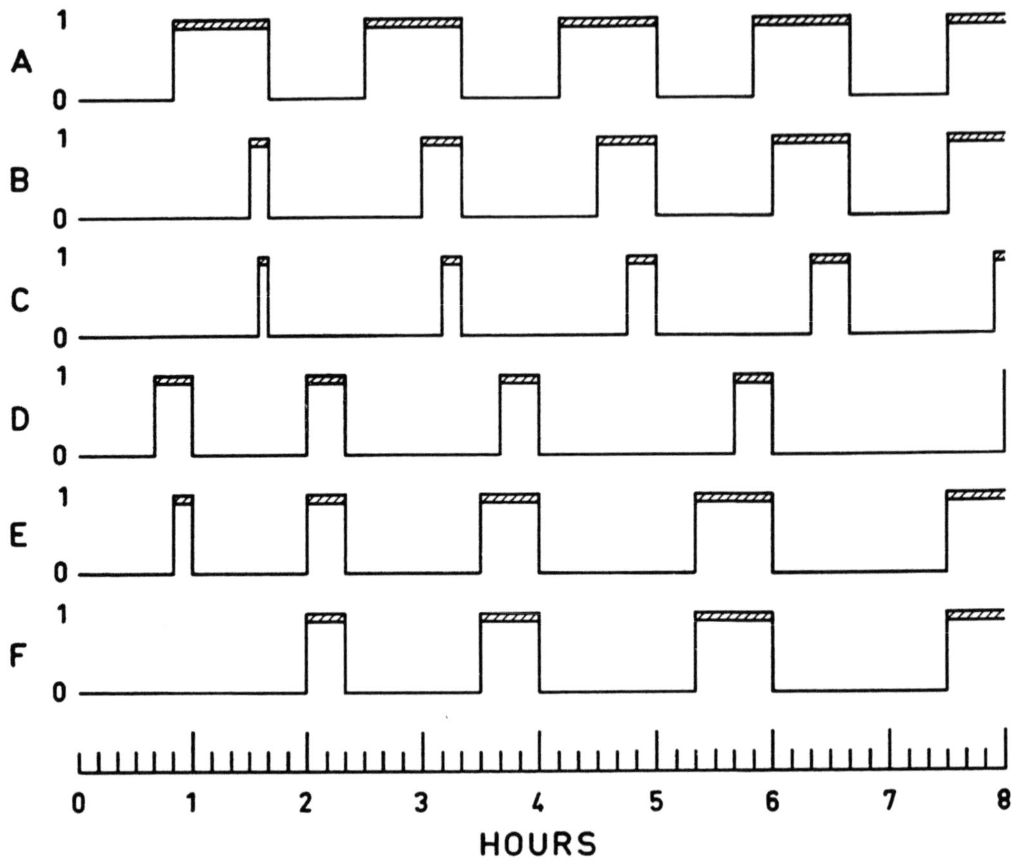

Fig. 2. Calibration signals used to check the computational procedures involved in the r^2 method. See Table 1 for more details

scorable periods due to artifacts generated by infants' body movements), and G (periods with a poor recording, and unscorable sleep stages).

After the manual scoring of the stages, the weight of 1 was assigned to REM sleep, and zero to all other stages. To quantify the "Quiet" sleep cycle, the weight would be changed so that "Quiet" sleep would be scored 1, and all others zero.

Table 2 shows the results. Six infants presented in the upper part of the table are three pairs of monozygotic twins. Six other infants presented in the lower part are three pairs of dyzygotic twins. The temporal structure of the REM cycle for the infants "T,Dam" and "B,J"

Table 1. Analysis of six calibration signals with "period" program

Signal characteristics[a]	Nonorthogonal analysis		"Inter-REM" intervals[b]		Maximal AC[c]	
	Period (min)	r^2	Period (min)	SD	Period (min)	AC
A. 50-min non-REM and 50-min REM	100	0.99	100	0	100	1.00
B. 90-min cycle; REM duration varies from 10 to 40 min	94	0.92	90	0	90	0.77
C. 95-min cycle; REM duration varies from 5 to 20 min	98	0.59	95	0	100	0.77
D. Cycle varies from 60 to 120 min; REM duration fixed at 20 min	100	0.53	100	20	88	0.20
E. Cycle varies from 60 to 90 min; REM duration varies from 10 to 40 min	116	0.81	90	20	118	0.40
F. "Missed" first REM cycle; cycle varies from 100 to 140 min; REM duration varies from 20 to 40 min	116	0.89	110	20	114	0.40

[a] For the wave forms used, see Fig. 1

[b] The "inter-REM" interval refers to time between the beginnings of the series of one REM and the next series of one

[c] The first autocorrelation peak observed

Table 2. Sleep cycle and its rhythmicity among infants of 40 to 44-week postconceptional age

	40-Week postconceptional age				44-Week postconceptional age			
	Period (min)		r^2		Period (min)		r^2	
Infant code	REM	Quiet	REM	Quiet	REM	Quiet	REM	Quiet
A, L-To	66	66	0.972	0.963	56	56	0.880	0.909
A, L-Tw	78	72	0.944	0.742	52	84	0.829	0.833
H, Ai	66	70	0.975	0.987	60	62	0.988	0.970
H, Am	62	66	0.974	0.954	68	68	0.832	0.776
T, Das	66	62	0.885	0.864	82	84	0.762	0.841
T, Dam[a]	54	58	0.258	0.253	40	44	0.519	0.468
B, D	54	52	0.708	0.720	52	52	0.902	0.568
B, J[a]	64	34	0.332	0.537	40	42	0.481	0.406
J, B	62	60	0.956	0.943	58	54	0.879	0.847
J, K	56	58	0.735	0.768	54	54	0.935	0.978
W, Mi	62	68	0.848	0.899	56	54	0.771	0.886
W, Ma	78	88	0.845	0.812	56	54	0.984	0.962
Average	64	63	0.786	0.787	56	59	0.814	0.787
SD	8	13	0.247	0.213	11	14	0.163	0.197

[a] These two infants' sleep showed two dominant cycles

appeared to be different from that of the other infants. Both of these infants had low sleep rhythmicity compared with the others, and they exhibited two prominent periodic activities in the REM cycles, instead of one. Unfortunately, there was no clinical or maturational data on these two infants to determine the reason for these differences.

Noises and Elevated Ambient Temperature

A. Muzet of Centre d'Etudes Bioclimatiques du C.N.R.S. in Strasbourg, France, provided sleep data from nine pairs of young adult subjects who slept in disturbing environmental conditions (NAITOH et al. 1975). The first night (N_0) was an adaptation night, where the subjects slept without scalp electrodes. The second and third nights (N_1 and N_2) were for baseline sleep recording. During these nights subjects slept in the sleep chamber at an ambient temperature of 18°C, and under a constant white noise of 40 dB(A). The fourth and fifth nights (N_3 and N_4) were the experimental nights. During these nights, the acoustical and thermal conditions of the sleep chambers were as follows: under "bt" condition, three pairs of subjects experienced noise pulses at 65 dB(A), while ambient temperature remained at 18°C. This "bt" group was designated the control group. Under "Bt" condition, three pairs of subjects were exposed to noise pulses at 80 dB(A), while ambient temperature remained at 18°C. This "Bt" group was designated a "noise" group. Under "bT" condition, three pairs of subjects were exposed to the control noise pulses of 65 dB(A), but the ambient temperature was raised to 26°C. This group was then designated a "heat" group. The sixth night (N_5) was a recovery night, where all subjects slept in the sleep chamber with a constant noise at 40 dB(A) and ambient temperature of 18°C.

The results of this study are summarized in Fig. 3 and Table 3. Figure 3 shows autocorrelograms of sleep stages of a subject in the "Bt" group. On the top of Figure 3 the baseline sleep structure is shown. This has a REM cycle of 92 min with a rhythmic intensity (r^2) of 0.92. The second autocorrelogram from the top shows the first night of exposure to noise pulses of 80 dB(A). The noise pulses reduced the rhythmic intensity of the REM cycle from 0.92 to 0.62, but REM cycle length was steady at 88 min. The third autocorrelogram shows the second night of exposure to noises. The intensity of the REM cycle was still low (r^2 = 0.75), but there was no change in the REM cycle length, 96 min. The last autocorrelogram represents recovery sleep.

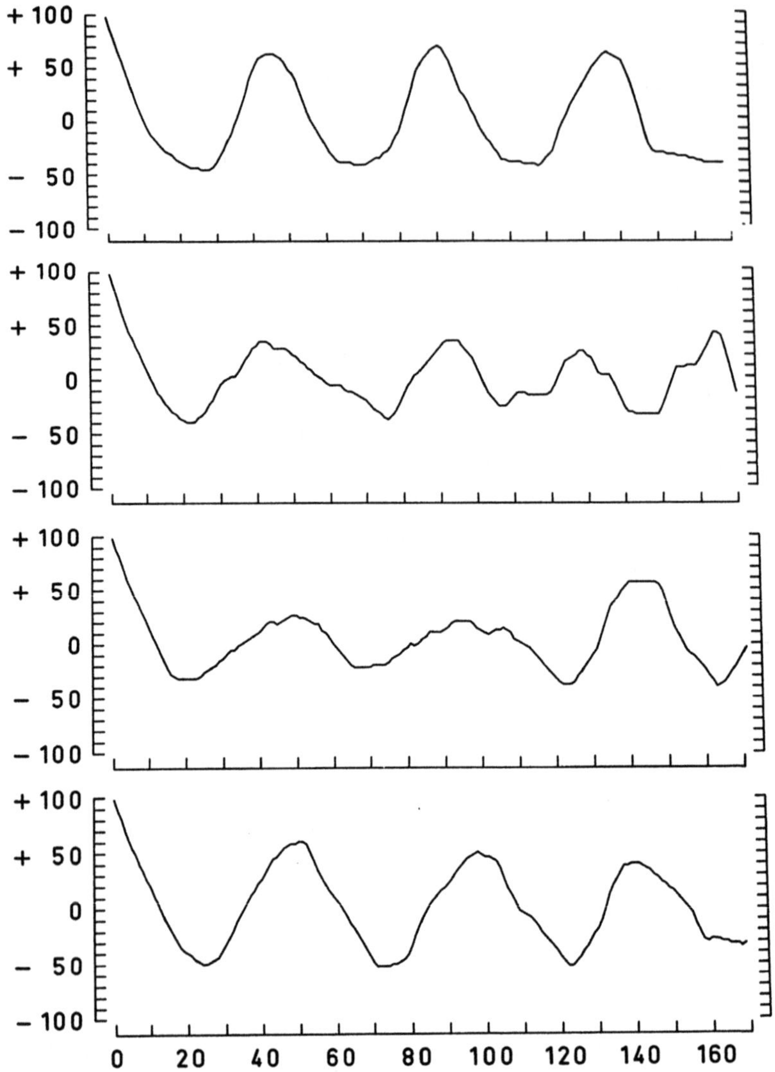

Fig. 3. Autocorrelograms showing the effects of 80 dB(A) noise pulses on REM cycle in one subject. From the top: baseline sleep, the first night of exposure to noises, the second night of exposure to noises, and recovery night. The ordinate shows autocorrelation values; the abscissa gives time in min

REM cycle intensity recovered to 0.95 with a steady REM cycle length of 96 min. Statistical analysis of the rhythmic intensity of the REM sleep and its cycle length of all "Bt" subjects indicated that the REM cycle had a reduced intensity during the second night of noise exposure. The recovery night had significantly higher REM rhythmic intensity than the baseline night. There was no change in REM cycle length during noise exposure (Table 3).

Increased ambient temperature produced a marginally significant lowering of REM cycle rhythmicity only during the first "bT" night. Rhythmicity increased in the second "bT" night, but decreased again during the recovery night when ambient heat was removed. During the first "bT" night, average REM sleep cycle of all "bT" subjects was significantly shortened from the baseline of 112.7 min to 95 min (Table 3).

Gradual Reduction of Sleep

Table 4 shows the effects of gradual reduction of total sleep time on REM cycle (MULLANEY et al. 1977). In this study, three couples reduced their total sleep time in 30-min steps from a baseline of 8 h to 6.5 h. One couple (DR and BR of Table 4, both short sleepers) reduced their sleep by going to sleep 30 min later every 2 or 3 weeks. Those who successfully reduced their sleep below 5.5 h maintained their sleep time for a month before a further 30-min reduction was attempted. The study was aimed at finding the minimum amount of sleep subjects would tolerate. Total sleep time was gradually reduced to provide the subjects with ample time to adjust to the altered work/rest schedules. The results suggest that neither REM cycle length nor r^2 value was affected by gradual shortening of total sleep time, although absolute amount of REM decreased significantly.

EVALUATION OF THE r^2 METHOD

The r^2 method has been found useful in extracting overall REM cycle characteristics from short sleep records. Although the r^2 method is restricted to an analysis of a single sleep record at a time, with a computer program it can be applied easily and repeatedly to analyze multiple sleep records from either a single subject or many subjects.

Table 3. Effects of noise and elevated room temperature on period length and rhythmicity of REM sleep cycle[a]

		Baseline	Experimental nights		Recovery
		N_1 or N_2	N_3	N_4	N_5
r^2	Group bt	0.76 ± 0.10	0.59 ± 0.21	0.57 ± 0.29	0.67 ± 0.24
	Group Bt	0.75 ± 0.21	0.76 ± 0.11	0.56 ± 0.18	0.83 ± 0.20
	Group bT	0.81 ± 0.09	0.66 ± 0.27	0.74 ± 0.17	0.61 ± 0.23
Period (min)	Group bt	114 ± 16	109 ± 17	107 ± 13	103 ± 32
	Group Bt	100 ± 14	97 ± 14	104 ± 20	101 ± 12
	Group bT	113 ± 17	95 ± 11	112 ± 24	101 ± 11

[a]Modified from the report "Effets propres et interaction de l'élévation du niveau sonore et de la température ambiante sur le sommeil" by B. METZ and A. MUZET, Centre d'Etudes Bioclimatiques du C.N.R.S., Strasbourg, France, April 1975

Table 4. Effects of gradual sleep reduction on REM cycle

Sleep duration (h)	Period length (min)										$r^2 \times 100$									
	JP	PM	BS	SS	GS	WS	DR	BR	Mean	SD	JP	PM	BS	SS	GS	WS	DR	BR	Mean	SD
8.0	122	108	100	92	114	98	NA	NA	106	11	91	96	36	80	93	94	NA	NA	82	23
7.5	-	72	92	116	128	-	NA	NA	102	25	42	42	37	39	94	-	NA	NA	53	27
7.0	100	94	102	106	120	86	NA	NA	101	12	83	86	91	84	93	50	NA	NA	81	16
6.5	104	98	96	118	114	98	132	-	109	13	80	90	60	93	87	82	75	-	81	11
6.0	106	90	-	110	106	104	-	-	103	8	86	47	-	88	81	97	-	-	80	19
5.5	96	96	-	106	100	92	92	90	96	5	39	72	-	87	87	43	62	92	69	22
5.0	-	104	-	98	NA	NA	108	-	103	5	-	88	-	89	NA	NA	88	-	88	1
4.5	-	-	NA	NA	NA	NA	NA	NA	-	-	-	-	NA	NA	NA	NA	NA	NA	-	-
2-Month maintenance (Minimum h + 0.5 h)																				
	103	99	-	94	110	88	110	124	103	12	78	60	-	90	93	71	79	45	74	17
6-Month follow-up																				
	112	106	98	-	108	100	98	-	104	6	87	94	99	-	89	96	97	-	94	5

NA, Not Applicable; specified sleep regimen not applied to cases with NA

-, REM sleep cycle analysis cannot be used because of extreme shortness of sleep record, or very few (one or two) REM cycles in the record

The r^2 method summarizes data of all-night REM sleep succinctly providing information on average cycle length and on rhythmic intensity of REM sleep. This is statistically convenient and effective as the conventional analysis of the inter-REM intervals. It can be easily applied to investigate between-group changes in REM cycle due to experimental treatments and sleep disturbances.

One disadvantage is that the r^2 program does not remove a slow circadian or any other trend from the REM cycle data. There is a circadian modulation of percent REM (AKERSTEDT and GILLBERG 1981; ENDO et al. 1981; HUME and MILLS 1977), REM latency (TAUB and BERGER 1973; WEBB and AGNEW 1978), and REM duration (CZEISLER 1978; ZULLEY 1980). Therefore, any statistical methods that analyze the REM cycle are required to have an option to remove slow sinusoidal trends.

The currently available method (NAITOH et al. 1979; see also RUMMEL et al. 1974; TONG et al. 1977) in our laboratory yields simultaneous estimates of circadian and ultradian components in sleep data. Our research has been concerned with two basic questions: What is the time of day which is best for a nap? What is the role of REM and slow wave sleep in defining a recuperative power of a nap? Thus, we were interested in circadian modulation of REM duration. A re-analysis of the nap data (MOSES et al. 1978) has yielded results which challenged the idea of a simple circadian control of REM duration. In that study, eight subjects slept 60 min every 3 h, and kept this sleep/wake cycle for 40 h. As expected, a strong 24-h rhythm was found in their ability to sleep during the assigned nap period of 60 min: an r^2 of 0.68, ranging from 0.51-0.86, for sleep duration. However, only one of eight subjects showed relatively large r^2 (0.47) for a 24-h cycle in REM duration in his nap periods. Further study is needed to redefine the circadian control over REM duration.

These analyses, however, cannot overcome one serious problem stemming from the fact that most often sleep data are too short to obtain a reliable estimate of REM cycle length and its intensity. HALBERG et al. (1981) showed that if cosine waves are used in modeling cyclic events with nonsinusoidal wave form, data must contain at least five complete cycles in order to obtain an accurate estimate of the period length. This requirement of at least five cycles needs to be heeded, particularly when the data wave forms are closer to triangular, sawtooth or square shapes than to a sinusoidal form. In the r^2 analysis, data which contain few cycles result in a broad peak in the r^2 plot.

The range will be too wide to provide useful information on ultradian rhythmicity in the data.

Although the increasing number of sophisticated frequency analysis techniques and modeling approaches which are becoming available for ultradian rhythms may render the r^2 method obsolete, it nevertheless offered insight into the infrastructure of sleep. Because of its simplicity, the r^2 method may be found useful in clinical applications.

ACKNOWLEDGEMENTS. This study was supported in part by the Naval Medical Research and Development Command, Bethesda, MD, Department of the Navy, under Work Unit MF58.524.004-9026. The views presented in this paper are those of the author. No endorsement by the Department of the Navy has been given or should be inferred. The author gratefully acknowledges the assistance of S. Sunderman for developing computer programs.

REFERENCES

AKERSTEDT T, GILLBERG M (1981) The circadian variation of experimentally displaced sleep. Sleep 4:159-169

ANDERSON TW (1971) The statistical analysis of time series. Wiley, New York, p 704

ASERINSKY E, KLEITMAN N (1953) Regularly occurring period of eye motility and concurrent phenomena during sleep. Science 118:273

ASERINSKY E, KLEITMAN N (1955) A motility cycle in sleeping infants as manifested by ocular and gross bodily activity. J Appl Physiol 8:11-18

CZEISLER CA (1978) Human circadian physiology: internal organization of temperature, sleep-wake, and neuroendocrine rhythms monitored in an environment free of time cues. Dissertation, Stanford University, Stanford, p 346

ENDO S, KOBAYASHI T, YAMAMOTO T, FUKUDA H, SASAKI M, OHTA T (1981) Persistence of the circadian rhythms of REM sleep: a variety of experimental manipulations of sleep-wake cycle. Sleep 4:319-328

FRIEDMANN JK, GLOBUS GG, NAITOH P (1977) Changes in sleep patterning with reduction of night aircraft noise at Los Angeles International Airport. Sleep Res 6:127

GLOBUS GG (1970) Quantification of the REM cycle as a rhythm. Psychophysiology 7:248-253

HALBERG F, CORNELISSEN G et al. (1981) International geographic studies of oncological interest on chronobiological variables. In: KAISER HE (ed) Neoplasm - comparative pathology of growth in animals, plants, and man. Williams and Wilkins, Baltimore, p 590

HILBERT R, NAITOH P (1972) EOG and delta rhythmicity in human sleep EEG. Psychophysiology 9:533-538

HUME KI, MILLS JN (1977) Rhythms of REM and slow-wave sleep in subjects living on abnormal time schedules. Waking Sleeping 1: 291-296

JOHNSON LC (1980) The REM cycle is a sleep-dependent rhythm. Sleep 2:299-307

KRYNICKI V (1975) Time trends and periodic cycles in REM sleep eye movements. Electroencephalogr Clin Neurophysiol 39:507-513

LEWIS SA (1974) The paradoxical sleep cycle revisited. In: SCHEVING LE, HALBERG F, PAULY JE (eds) Chronobiology. Igaku Shoin, Tokyo

LUBIN A, NUTE C, NAITOH P, MARTIN WB (1973) EEG delta activity during human sleep as damped ultradian rhythm. Psychophysiology 10:27-35

MOSES J, LUBIN A, JOHNSON LC, NAITOH P (1977) Rapid eye movement cycle is a sleep-dependent rhythm. Nature 265:360-361

MOSES J, NAITOH P, JOHNSON LC (1978) The REM cycle in altered sleep/wake schedules. Psychophysiology 15:569-575

MULLANEY DJ, JOHNSON LC, NAITOH P, FRIEDMANN JK, GLOBUS GG (1977) Sleep before and after gradual sleep reduction. Psychophysiology 14:237-244

NAITOH P, JOHNSON LC, LUBIN A, NUTE C (1973) Computer extraction of an ultradian cycle in sleep from manually scored sleep stages. Int J Chronobiol 1:223-234

NAITOH P, MUZET A, LIENHARD JP (1975) Effects of noise and elevated temperature on sleep cycle. Sleep Res 3:174

NAITOH P, LUBIN A, COLQUHOUN WP (1979) Comparison of monosinusoidal with bisinusoidal (two wave) analysis. Chronobiologia 6:136

NAITOH P, MUZET A, LIENHARD J-P (1975) Effects of noise and elevated temperature on sleep cycle. Sleep Res 3:174

NUTE C, NAITOH P (1971) A generalization of harmonic analysis for detection of long-period biorhythmicities from short record. Navy Medical Neuropsychiatric Research Unit Report 71-55, p 21

RECHTSCHAFFEN A, KALES A (1968) A manual of standardized terminology, techniques and scoring system for sleep stages of human subjects. U.S. Government Printing Office, Washington, p 57

RUMMEL J, LEE JK, HALBERG F (1974) Combined linear-nonlinear chronobiologic windows by least squares resolve neighboring components in a physiologic rhythm spectrum. IN: FERIN M, HALBERG F, RICHART RM, VANDEWIELE RL (eds) Biorhythms and human reproduction. Wiley, New York, p 665

TAUB JM, BERGER RJ (1973) Sleep stage patterns associated with acute shifts in the sleep-wakefulness cycle. Electroencephalogr Clin Neurophysiol 35:613-619

TONG YL, NELSON WL, SOTHERN RB, HALBERG F (1977) Estimation of the orthophase (timing of high values) on a non-sinusoidal rhythm - illustrated by the best timing for experimental cancer chronotherapy. 12th international conference proceedings, Il Ponte, Milan, p 782

URSIN R, MOSES J, NAITOH P, JOHNSON LC (1983) REM-NREM cycle in the cat may be sleep-dependent. Sleep 6:1-9

WEBB WB, AGNEW H (1978) Effects of rapidly rotating shift on sleep patterns and sleep structure. Aviat Space Environ Med 49:384-389

ZEPELIN H, RECHTSCHAFFEN A (1974) Mammalian sleep, longevity, and energy metabolism. Brain Behav Evol 19:425-470

ZULLEY J (1980) Distribution of REM sleep in entrained 24 hours and free-running sleep-wake cycle. Sleep 2:377-389

The REM Sleep Ultradian Rhythm: A Limit Cycle Mathematical Model

R. W. McCarley and S. Massaquoi
Neuroscience Laboratory, Department of Psychiatry, Harvard Medical School, Brockton VAMC, 940 Belmont Street, Brockton, MA 02301, USA

INTRODUCTION AND PHYSIOLOGICAL BACKGROUND

The rapid eye movement (REM) sleep rhythm is probably the most studied ultradian rhythm, and thus a likely first target for modeling. In 1975, McCARLEY and HOBSON proposed a model of REM sleep that was mathematically based on the Lotka-Volterra equations and was related physiologically to empirically derived time courses of discharge activity in two neuronal populations: (a) "REM-on" neurons in the pontine reticular formation (PRF) which became active in REM sleep and appeared to act to promote this sleep phase; (b) "REM-off" neurons in the locus coeruleus (LC) and dorsal raphe nucleus (DR) which had a reciprocal time course to the REM-on cells. LC/DR cells slowed and nearly ceased discharge activity with the approach and advent of REM sleep and appeared to act in a permissive, disinhibitory fashion in REM control. It was proposed that the periodic occurrence of REM sleep might be a function of reciprocal interaction between these two groups of neurons.

Advances in physiology since then have indicated that the group of cells proposed in 1975 to be important for REM functions (a subset of PRF cells in the giant cell field area, FTG cells) needed to be enlarged to include other reticular cells. Indeed, lesion results (SASTRE et al. 1981) taken together with recent physiologic (ITO and McCARLEY 1984) and pharmacologic data (BAGHDOYAN et al. 1984a,b) suggest the usefulness of a concept of a distributed network of cholinoceptive/cholinergic brain-stem reticular neurons that act to promote REM sleep (see discussion in McCARLEY and ITO 1985). Similarly, the functional role of LC/DR cells has now been expanded to include general behavioral regulatory functions as well as a role in REM sleep and other REM-off neurons, which are also probably biogenic amine-containing, have been recorded in the brain-stem. We have developed a

Experimental Brain Research, Suppl. 12

new version of the original mathematical model and will present it in detail in a full-length paper (McCARLEY and MASSAQUOI submitted to press); this chapter briefly sketches its main features and discusses its predictions of the phase-response curve to pharmacologic perturbations of the REM ultradian rhythm. This new model, which will be referred to as the limit cycle model, describes neuronal properties more realistically, extends the model to human circadian sleep data, and overcomes many of the mathematical limitations of the 1975 model.

The 1975 model mimicked successfully the time course of discharge activity in cat REM-on and REM-off neuronal populations as they were recorded over the sleep cycle; it had, however, several important structural and mathematical limitations. The fundamental difficulty with the simple Lotka-Volterra (LV) equation system (LOTKA 1956; VOLTERRA 1931) was that it displayed only neutral stability. That is, the system's long-term behavior was determined entirely by its initial conditions. If the physiologic REM oscillator had simple LV dynamics, any reproducible cycling from night to night would necessarily imply an unreasonably precise initiation with each sleep onset. In the same way, the LV system was intrinsically extremely sensitive to perturbation so that even a momentary alteration of the components of the oscillator would disturb the system indefinitely (until specifically reset). This behavior was clearly not typical of physiologic oscillators which must be able to maintain some intrinsic stability in the face of continuously changing internal and external influences. A practical consequence of neutral stability was the difficulty in modeling circadian variation, since this involved perturbations of the oscillator.

Despite these limitations the original model has aroused interest since it has been the only systematic mathematical model of REM sleep oscillation and it has formed the basis of attempts to model the characteristic REM sleep changes associated with depression (VOGEL et al. 1980; BEERSMA et al. 1984). It was thus thought useful to refine and extend the model so that it would be both physiologically more realistic and also applicable to human sleep. Central to the LV system's neutral stability problem was that there was no inherent limit on the values which the variables might attain, and, in principle, oscillations of infinite amplitude were possible. A reasonable first step in modeling was thus to add terms which reflected the physiologic limits on neuronal firing rates. In addition, functions were incorporated which better described neuronal firing characteristics at low discharge rates. These changes led to a model which displayed limit cycle stability.

The presence of limit cycle stability opened the possibility of studying and modeling possible mechanisms of exogenous modulation of the REM oscillator. (The presence of a limit cycle in a physiologic system can be thought of as a particular time course of activity toward which the system will move regardless of the starting point (initial conditions) or presence of perturbations.) We thus were able to examine and model the experimentally observed circadian variations in the REM-NREM pattern, as well as the system's response to external perturbations, such as administration of pharmacologic agents.

Specifically, this new limit cycle model proposes that the marked differences in latency, amplitude and duration of the first REM sleep period that occur as a result of circadian variation and depressive pathology result from beginning the REM oscillation at different initial points relative to the final position in the limit cycle: (a) beginning from a point interior to the limit cycle produces a first REM period of long latency, short duration, and small amplitude; (b) beginning from a point exterior to the limit cycle produces a first REM period of short latency, long duration, and large amplitude. A change in a single circadian parameter describing the rapidity of change in the REM-off population at the time of entry into sleep is able to model the observed alterations in the amplitude, duration, and latency of the first REM period as sleep begins at different phases of the human circadian cycle. The magnitude of changes from the first to the second REM period derive from the spiraling outward (or inward) to reach the limit cycle, are of the order of 50%, account for most of the variations seen in human REM sleep over the night, and match human data quite well. In addition, the model postulates a continuous but small circadian variation ($\pm$ 5%) acting throughout the course of a night's sleep. Because the model is derived from actual physiologic data, rather than being a purely ad hoc or phenomenologic construct, it appears to offer the possibility of direct testing of its postulates through neurobiologic studies in animals, by circadian phase-related manipulations of the sleep cycle, and through perturbations of the system in humans by the use of drugs. Indeed, an explicit phase-response curve of the system to cholinergic agonists has been developed; this will permit experimental tests of the model in humans as well as animals.

With respect to the underlying physiology, we emphasize that it is not at all essential that one consider medial PRF (mPRF) neurons or other reticular neuronal areas as being "the generator" of REM sleep. (mPRF includes both giant and nongiant neurons in the medial two thirds of PRF.) Rather an alternative viewpoint for purposes of model construction might be that these cells are indicators of the output of other neurons or humoral systems controlling REM sleep, or that REM generation involves a widely distributed network of neurons, as lesion studies appear to indicate. While our own current physiologic viewpoint is that REM is controlled by a distributed network of brainstem REM-on neurons which include mPRF neurons, what is important for the model is that there be an internal "read-out" of the time course of activity in "REM-on" populations but it is not critical whether mPRF or other reticular neurons are, in the familiar terms of circadian physiologists, part of the clock mechanisms or merely "hands" on the clock indicating activity in other REM controlling populations.

REM-Off Cells

The reciprocal time course of spontaneous neuronal activity in locus coeruleus and dorsal raphe nucleus (LC/DR) populations and that of mPRF neurons and the ability to induce REM sleep phenomena (PGO waves) and REM sleep episodes by cooling dorsal raphe (and thus turning off the inhibition of REM proceeding from these cells) as well as the production of REM sleep episodes through LC cooling suggests a role for these cells in the "permissive" production of REM phenomena; see review in McCARLEY (1980) and the summary of data in LYDIC et al. (this volume, Chap. 8, and 1984). It must be emphasized that we also use LC/DR cells as exemplars of a larger population of REM-off cells. We have, for example, recently obtained evidence of the presence of REM-off cells in other areas of the brain-stem, especially the peribrachial area (HOBSON et al. 1983). While these other cells may be similarly involved in REM sleep oscillation, we shall not, for reasons of simplicity, include them in our labels for the REM-off population. Just as with the REM-on cells, it is not necessary for the purposes of the model that one necessarily regard the REM-off cells as part of the REM oscillator as opposed to being an indicator of physiologic processes critical for the oscillator.

The Lotka-Volterra system was named after the two mathematical ecologists who explored the interaction between prey and predator populations in isolated ecosystems. It was thus a logical starting point for descriptions of interactions between inhibitory ("predator") and excitatory ("prey") neuronal populations. The simple Lotka-Volterra equations are, with X representing mPRF and Y LC/DR activity:

$$X'(t) = aX - bXY \quad (1)$$

$$Y'(t) = -cy + dXY \quad (2)$$

Figure 1 illustrates the structure of the model and how a, b, c, and d represent "coupling constants" between the interacting populations. In both the simple Lotka-Volterra and in the new limit cycle model the sleep cycle dynamics can be described in words in the following manner: after the end of one REM period LC/DR neuronal discharge activity is high; this diminishes over time because of the negative feedback (autoinhibition). As this proceeds further, mPRF cells are released from inhibition and because of the positive feedback increase their discharge rate. When their activity reaches a certain critical threshold, REM sleep occurs. The activity of mPRF cells gradually turns on the LC/DR neurons and they in turn inhibit mPRF neurons causing termination of the REM episode. We propose that the time duration of both the feline and human sleep cycle suggests that the neurotransmitters involved are those that act with a long time duration. This long time duration is most likely related to the production of "second messengers", such as cyclic AMP in the case of LC/DR and of cyclic GMP in the case of mPRF. These second messengers, as emphasized by BLOOM (1975), act to extend the time range of neuronal interaction by several orders of magnitude.

THE NEW LIMIT CYCLE MODEL

Summary of Major Features and Changes from the Simple Lotka-Volterra Model

The limit cycle feature of the new model results from the addition of the following postulates to the simple LV equations: (a) limitation on maximal neuronal firing rates in both REM-on (X) and REM-off (Y) popu-

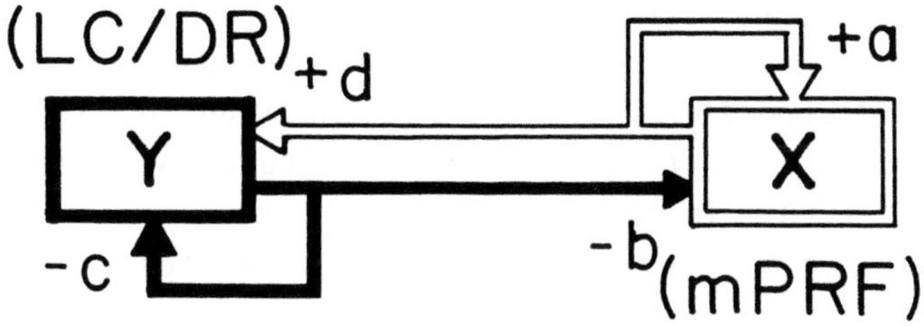

Fig. 1. Structural model of the X and Y population interconnections postulated by the model, with the connectivity constants corresponding to those in text Eqs. 1-4. The X population is REM promoting, excitatory, and cholinoceptive/cholinergic; mPRF neurons are used as exemplars. The Y population is REM suppressive, inhibitory, and uses biogenic amines as neurotransmitters; locus coeruleus (LC) and dorsal raphe (DR) neurons are used as exemplars of this population. Solid lines and - signs indicate inhibitory connections and open lines and + signs indicate excitatory connections

lations (the strength of inhibition of the REM-off population on the REM-on population is further limited when the REM-on population activity is at low levels); (b) the term for growth rate of the mPRF population resulting from excitatory feedback is now a function of X level and is less for lower values of X, whereas it was constant in the simple LV model.

The other major category of change in the model is addition of circadian variation, not treated at all in the original model. All circadian sleep fluctuation is modeled by: (a) circadian variation in the way the system begins oscillation, modeled by variation in the rate of decline of the REM-off (Y) population at sleep onset (this "start-up" variation accounts for most of the observed circadian changes in REM sleep percentage); (b) a small continuous circadian modulation (accounting for about $\pm$ 5% variation in REM parameters) that results from changing the constant d in the simple LV equations to a sinusoidally modulated circadian variable, d(circadian time phase) = d(circ). The following new equations for the limit cycle model highlight the changes from the simple LV model:

$$X'(t) = a(X)*X*S_1(X) - b(X)*X*Y \qquad (3)$$

$$Y'(t) = -c*Y + d(circ)*X*Y*S_2(Y) \qquad (4)$$

* stands for multiplication, X for mPRF activity, Y for LC/DR activity, S_1 and S_2 stand for saturation functions limiting X and Y, and

thus a(X) stands for X growth rate higher with larger X, b(X) for limitation on inhibitability of X by Y for low X values, and d(circ) for circadian variation in amplitude of d. Formulae, details of construction, and graphs of these functions are provided in McCARLEY and MASSAQUOI (submitted to press). The values used for X and Y discharge rate activity in the graph ordinates are scaled in relative values since it would have been most cumbersome to label all graphs with absolute firing rates for each of the REM-on and REM-off subgroups.

Circadian Variation and Events at the Onset of Sleep

In the new model circadian variation and events at the onset of sleep are modeled by a change in a single circadian phase parameter. The most important circadian parameter can be thought of as an initial condition reflecting the time course of decline of the Y (LC/DR) population at the onset of sleep. The limit cycle model postulates that at high levels of Y activity (in waking) X activity is held low. Y activity then declines from high waking levels as a result of circadian factors. In the model the primary and most important effect of circadian fluctuations is a variation in the time course of Y decay as the sleep cycle begins. In all simulations for normal humans the Y level at which REM cycle oscillations began is held constant (0.35 units) with only the rate of change being varied. The point at which the figures in this chapter begin is the traditional one for scoring REM latency, namely the onset of stage 2. The decline of the Y population at sleep onset is modeled as resulting from a circadian-controlled decline in an excitatory input to the Y population (as discussed in more detail in the Results section, we use excitatory input because it is computationally convenient, and do not mean to rule out the physiologic possibility of disinhibition as a mechanism; the effects on Y are mathematically equivalent to disinhibition). The initial level of input to the Y population varies sinusoidally with the temperature oscillator but is 180 degrees out of phase with it (CZEISLER et al. 1980; ZULLEY 1980). At each "sleep onset" simulation the input to Y declines in a linear fashion with a slope of -0.05 units per unscaled time unit (10.7 min scaled). Because the initial excitatory input to Y is at a higher level at sleep onset near a temperature maximum, this constant linear decay has a longer time course near Tmax, lasting approximately until the first REM period; in contrast, near Tmin the excitatory input level is sufficiently low so that the decay occurs over a shorter time period and the level does not greatly affect Y dynamics. It is this difference in the time course of decline of Y at sleep onset which determines whether the limit cycle is entered from the interior (slow rate of decline) or from the exterior (rapid rate of decline). Thus, this parameter is by far the most critical one for determining changes in REM sleep values, since it is the first REM cycle that varies most from all others.

The other circadian parameter is d(circ) where d represents the sensitivity of LC/DR to mPRF stimulation. Although responsible for a lesser percentage of REM sleep variation in the range of amplitude variation present in our simulations, d modulates REM intensity and duration, with lower values associated with more intense and longer REM. In our model the amplitude of d varies continuously and sinusoidally with a period of 24 h and with a maximum of 1.1 at circadian temperature minimum and a minimum of 0.850 at circadian temperature maximum.

Fixed and Variable Parameters

In normals, only circadian phase was varied, and this affected only the rate of decline of the Y population and d(circ). All other parameters were fixed save for "a" in simulations of injections of cholinergic agonists. In all simulations the time scaling remained the same; the initial time scale was fixed by a match to normal human data. Similarly, REM threshold remained the same for all simulations, 1.4 X units, a value compatible with cat neuronal data showing the zone of X values of most probable REM onset; within this zone the exact threshold was set at the value producing the most realistic REM-NREM ratios for normal human sleep cycles. In the simulation of pre-stage 2 sleep conditions X was always initially set to 0.2 units.

Simulation

All numerical integration was done using the Runge-Kutta method with a step size of 0.05 time units (unscaled); the step size was sufficiently fine since halving the size of the steps had no effect on the results. The numerical integration program was written in FORTRAN, run on a PDP-11/40 computer and results were plotted on a digital plotter.

Terminology

Sleep cycle is synonymous with REM cycle and end of one REM episode to the end of the next.

RESULTS

Entrained Normal Subjects

The top part of Fig. 2 illustrates the phase-plane results and the bottom the time-domain results derived from the model based in a setting of circadian phase at 2.3 radians as the starting point; this is just after the occurrence of a temperature maximum at 90 degrees or 1.57 radians. For the reader unfamiliar with phase-plane charts we note that the phase plane represents X-Y values as 1 point for any single time value. The passage of time is represented by a sequence of X-Y values. In this graph the first sleep cycle begins at the X-Y value indicated by the dot and then the X-Y sequence moves in the trajectory indicated by the arrow; the successive X-Y values over the first sleep cycle are indicated by the "1" label in the figure, those over the second sleep cycle are indicated by the "2" label, and so on. The phase-plane representation is best for providing a good overview of the relative amplitude of successive REM episodes and the trend toward larger or smaller values while the time-domain representation is useful for study of REM duration, since the phase-plane representation does not provide an explicit indication of the time duration of the X-Y values, merely the order in which they occur. Note that in the time-domain representation X values are indicated by the solid line and Y values by the dotted line.

The phase-plane chart also provides a useful illustration for description of the system dynamics. During waking the X-Y values remain in the upper left portion of the phase plane, at a high activity level (> 1.75 units) of the Y population (the REM inhibitory, LC/DRN population). This high activity level of the Y population holds the value of the X population (REM promoting) at low values. Thus, our conception is that during waking the system is "clamped" at high Y values, at which point no oscillations occur. The model postulates that as the result of circadian influences before sleep onset the Y population

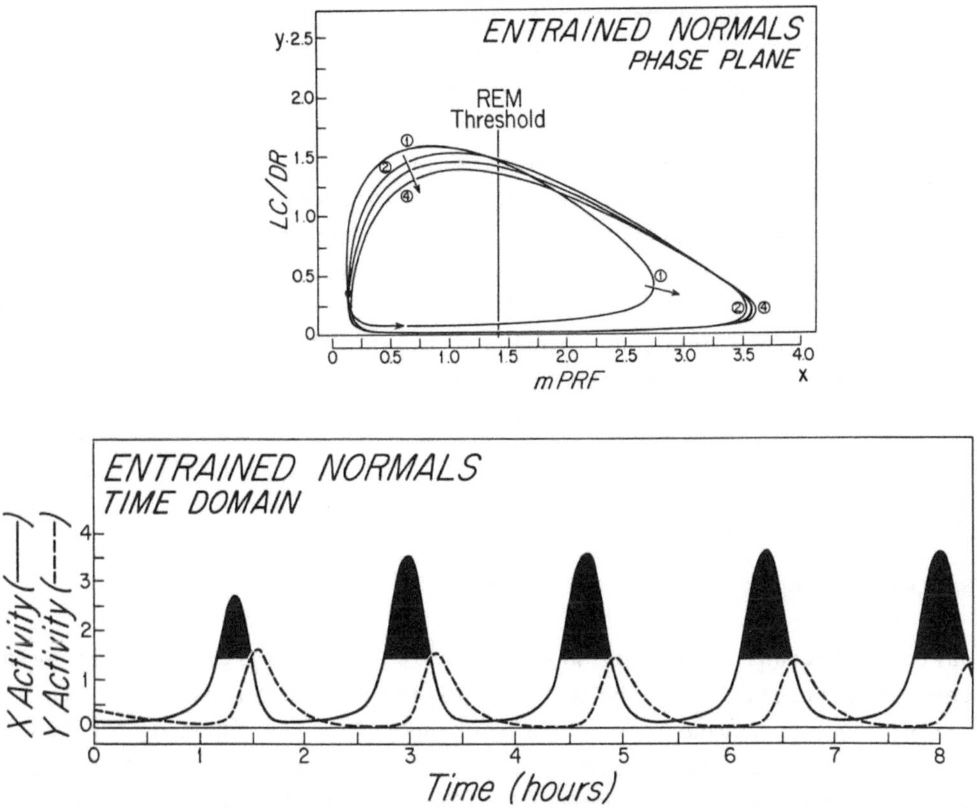

Fig. 2. Model simulation of night's course of REM sleep in entrained normals. The top part is the phase-plane and the bottom part the time-domain representation of this data. At bottom the dark portions of the X (mPRF) graph show those portions of the night with REM sleep and the height of these peaks indicates the intensity of the REM sleep episode. Note the short-duration, lesser-intensity first REM episode with subsequent variations being slight. The activity of the Y (REM inhibitory) population is indicated by the dotted line. In the phase-plane representation at top each point on the graph represents the X-Y values at a particular time. The dot represents the starting point and the interior curve while the arrow shows the first REM episode values, with this curve being interior to the limit cycle values obtained in subsequent REM cycles, labeled 2, 3, 4 in the order of their occurrence (for simplicity, sleep cycle 5 has not been graphed in the phase plane). The smaller variations in sleep cycles 2 through 4 are due to circadian modulation

activity level declines. Since there is no clear physiologic data, we have decided not to specify explicitly the mechanism of this Y population decline; we thus leave open the possibility of its occurring by synaptically mediated disfacilitation (withdrawal of excitation) or by inhibition or even humoral factors on the Y population. Our physiologic hypothesis is that hypothalamic influences on the Y population mediate circadian effects; while this connectivity is known to be present, the sign of influence has not been defined. In addition to this circadian input, the model provides for the possibility of prevention of sleep onset by exogenous input activating the Y population and also for a "wake-up" through exogenous activation of the Y population; in this chapter we will not model these events in any further detail.

System Unlocking

Our simulations show that when Y population activity declines below a certain critical level the X-Y population oscillations begin; this decline may be viewed as an "unlocking" of an oscillatory system. The critical feature for circadian alterations in the first REM cycle have to do with the time course of Y decline at the time the sleep cycle oscillation begins. For all runs in normal subjects the level of Y was set at 0.35 units at the beginning of the graph, a time point corresponding to stage 2 and the usual starting point for first REM latency measurements; in cat studies this relative level of neuronal activity is reliably associated with the presence of sleep.

When the night of sleep is begun at a circadian phase of 2.3 radians (just after a temperature maximum) this means that the level of circadian-determined excitation of Y at sleep onset is near maximum; while our model posits that the rate of decrease of this excitation is constant across circadian phase, the high initial level of the excitation at this phase means that the decay of excitation takes longer and that consequently Y remains at higher levels longer than near a circadian minimum. The rate of Y decay determined by the X-Y interaction alone without this circadian excitatory input is much faster. This prolonged time course of decay in Y results in the first sleep cycle beginning on a trajectory interior to that of the limit cycle, as shown in Fig. 2, top.

Note that the X-Y trajectory of the first cycle has a much smaller X amplitude, but that the second cycle goes very close to the final limit cycle. The small changes observed in cycles 2 through 4 are the result of the small continuous circadian variation which alters slightly the shape of the limit cycle by variation of d amplitude. The time-domain chart shows the first REM episode is of low amplitude (low intensity) and short duration and has a long latency value. These are the results of a slow rate of Y decline, beginning the approach to the limit cycle from the interior. The large differences between the first and subsequent sleep cycles are the result of an interior entry into the limit cycle with near limit-cycle values being obtained by the second sleep cycle. Smaller differences in subsequent cycles result from continuous circadian modulation of d and hence of cycle values; over the night REM episodes become longer and the sleep cycle period duration (measured from end of REM to end of REM) becomes shorter.

Comparisons with Experimental Data

Table 1 summarizes experimental data from several studies on REM latency (first NREM episode duration) and REM and NREM episode duration for the first three cycles in normal male subjects under conditions of 24-h entrainment and recordings in both sleep laboratories and when living in underground bunkers (ZULLEY 1979, 1980); also listed are recordings from free-running subjects who went to sleep near temperature maxima. These are paralleled by model statistics for simulations in these two conditions. Note that in the entrained data the model and the data are parallel in showing (a) a large increase in REM duration between the first and the second episode and (b) an increase in REM episode duration as sleep progresses. The model slightly overestimates the duration of the first REM episode (model, 19 min vs experimental median of 15 min), but the duration of subsequent REM episodes matches quite well. The model predicts a small, but consistent lengthening of REM episode duration throughout the night; because of the high variability this has been hard to demonstrate with statistical reliability in human data. However, the trend in the studies reviewed here is to show an increase: 5/7 studies show an increased duration of REM 3 compared with REM 2 vs no difference in one study and only a single study with a decrease. Furthermore, in a statistical re-analysis of the SCHULZ et al. (1984) data we can demonstrate a statistically significant trend for a successive increase in REM duration for the first three REM periods by computing the probability of having REM durations

Table 1. Match between model and experimental values for normal males under various conditions (blanks indicate no data available)

Study	Condition		NREM 1 ("REM latency")	Duration in min				
				REM 1	NREM 2	REM 2	NREM 3	REM 3
ZULLEY 1980	24-h entrained	Median		12		30		34
		Mean		15		30		35
ZULLEY 1980	Free run near Tmax	Median		16		23		24
		Mean		16		22		27
SCHULZ et al. 1984	Laboratory	Median	69	12	75	29	73	33
WILLIAMS et al. 1974	Laboratory Age 20-29	Mean		15	76	31	69	24
	Age 30-39	Mean		17	76	21	72	30
FEINBERG et al. 1967	Laboratory	Mean	69	11	72	21	68	26
CAILLE and BASSANO 1975	Laboratory	Mean	67	18		27		27
Model	Entrained match, circadian phase = 2.3 radians		69	19	75	28	73	29
	Free run, Tmax at sleep onset match, circadian phase = 1.57 radians		80	17	79	25	74	26

ordered REM 1 < REM 2 < REM 3 for 7/11 subjects, as occurred in this data set: p was < 0.002 for this ordering vs the hypothesis of equiprobable random ordering.

It should be emphasized that the model generates durations of the first three NREM episodes that are remarkably close (identical) to those measured in the laboratory by SCHULZ et al. (1984; see Table 1). In this regard we note that the model as developed here is explicit for intensity, periodicity, and duration of REM sleep but does not attempt to predict changes in slow wave sleep intensity or spectral content (i.e., percentage of delta waves, etc.) over the night (DAAN et al. 1984; BORBELY 1982) although it does appear to predict the duration of NREM episodes quite accurately and to the best of our knowledge is the only model able to do so.

Free-Running Normals with Sleep Onset near a Temperature Maximum with a Late Temperature Minimum

The condition of sleep onset near a temperature maximum in free-running subjects resembles that of sleep onset in normal subjects in that a temperature minimum with both groups occurs late in the evening. For both groups the clear prediction of the model is for a longer-latency, lesser-intensity, and shorter-duration first REM episode, compared with subsequent REM episodes and with the first REM episode of subjects with sleep onset near a temperature minimum. For simulation of sleep onset with an early temperature maximum, circadian phase was set to 1.57 radians (temperature maximum) and all other parameters were as on previous simulations. Table 1 shows the match of the model simulation to ZULLEY's (1979, 1980) data on duration of successive REM episodes in free-running normals. Note that the fit is good, with REM durations falling quite close to ZULLEY's data: both model and experimental data show the REM 1 to REM 2 duration increase as the largest and there is a smaller increase from REM 2 to REM 3.

It is worth noting that ZULLEY's data for this free-run condition are slightly different from his data on 24-h entrained subjects: the free-running subjects show a longer-duration first REM episode and a less abrupt change from first to second REM episode duration (Table 1). While this could be random statistical variation, it also could represent subjects going to sleep at slightly different temperature phases in the two conditions and/or other factors in the 24-h entrained condition leading to a shortened first REM episode. Our simulation matches this feature of the empirical data by assuming a slightly different circadian phase in the free-run condition (circadian phase of 1.57 radians vs circadian phase of 2.3 radians in 24-h entrained subjects; see Table 1).

REM at Circadian Temperature Minimum

Space does not permit more than a brief summary of these results, discussed in detail in our full-length paper. The model produced a course of REM durations that had a long duration REM 1 and a successively declining duration for further REMs and matched ZULLEY's data quite closely. The distinctive feature observable in the phase-plane graph was the approach to the limit cycle from the exterior, with a resulting first REM period of long duration, high intensity, and short latency.

PHASE-RESPONSE CURVE PREDICTIONS

Predicting the Phase-Response Curve during the REM Sleep Cycle to Acetylcholine Agonist Administration

Studies by several groups of investigators have indicated that the administration of acetylcholine agonists such as physostigmine or arecholine can hasten the onset of the next REM episode in humans (SITARAM et al. 1980, 1982; GILLIN et al. 1982). These data are consistent with the more extensive, direct evidence in animal experiments that direct intrabrain-stem administration of acetylcholine agonists, including neostigmine, into the pontine reticular formation (but not bulbar or midbrain reticular formation) can produce a syndrome with all of the signs of REM sleep (BAGHDOYAN et al. 1984a,b; earlier studies reviewed in McCARLEY 1982).

Since a model is most usefully and most easily tested by its predictions, we thought that it would be helpful to sketch here, well in advance of any empirical tests that we know of, the phase-response curve predicted by the limit cycle model for the administration of an acetylcholine agonist, such as physostigmine, or a similar agent producing a decrement in the rate in which acetylcholine is hydrolyzed. In our model (see Fig. 1), this can be described as raising the level of "a" during the effective duration of the acetylcholinesterase blocker; "a" is the term indicating the strength of positive feedback in the excitatory population and thus presumably the strength of cholinergic activity. As a first approximation to the experimental conditions our simulations doubled the strength of "a" (cholinergic feedback) for 10.7 min (step function). These values produced clearcut effects but did not severely distort the time curves and produced a phase advance of the REM episode of roughly the same order of magnitude as the single time-point experiment of SITARAM et al. (1980).

Figure 3 illustrates the basic paradigm for the pharmacologic simulation studies. We used the modeling parameters used for normal entrained subjects (as in Fig. 2) and modeled the effects as occurring after the second REM episode to avoid confusion with circadian initial conditions at sleep onset. The top portion of Fig. 3 shows the effect of transient increase in "a" lasting 7-17 min after the end of the second REM episode. It will be seen that the subsequent REM episode is advanced by time period indicated by the shading. However, the REM latency decrease is much more pronounced when, in this simulation, the "drug" producing a transient increase in the model constant "a" is applied later, 41-51 min after the onset of the second REM episode. These differential effects can be understood as REM being much easier to induce when the X population is on the rising rather than in the falling phase of the sleep cycle. This has to do with both an ease of tipping the X population into an exponential growth phase and also the absence of Y inhibition later in the cycle. Figure 4 shows a complete phase-response curve for the simulated application of a drug producing an increase in the model parameter "a" for 10.7 min beginning at various times near the end of REM 2 episode and in the period between the end of REM 2 and REM 3. Dots indicate onset times of the treatment. Note that there is a time near the end of REM 2 when increasing "a" prolongs the latency of the subsequent REM episode, due to an increase in the duration of the second REM episode. The phase of the response then changes rapidly; a shortening of the REM latency of the order of 10% occurs on administrations starting after the end of the second REM episode; an increase to maximal effect (17% latency reduction) occurs with administrations about 25-50 min after the end of the second REM episode. Following this there is relatively less effect, due to the fact that the third REM episode normally occurs very soon

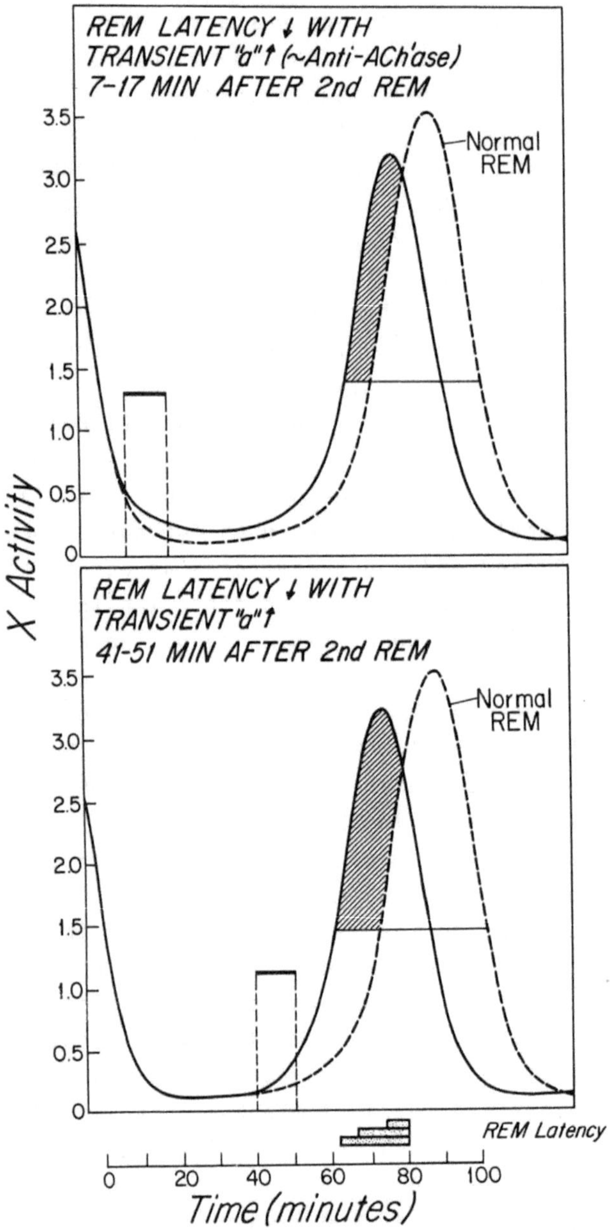

Fig. 3. Examples of REM latency shortening by an increase in the model parameter "a", simulating antiacetylcholinesterase administration. The shaded area shows degree of phase advance; note that the administration at 41 min (bottom) has a greater effect than at 7 min after REM 2 ends (top). See text for further description

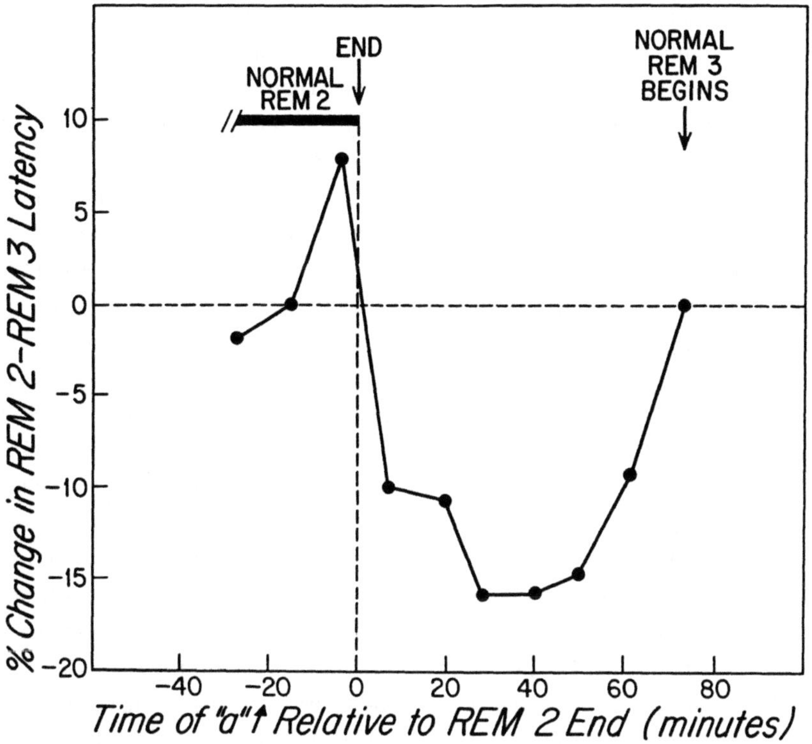

Fig. 4. Predicted phase-response curve to antiacetylcholinesterase administration. Each point was generated by a simulation like that described for the points 7 and 41 min in Fig. 3. See text for further description

thereafter and thus there is very little room left for shortening. We note that this predicted phase-response curve can be used to test the dynamics of the model both in humans and in animals and is, we believe, the first example of such a phase-response prediction for REM sleep.

Depression: Monoamines, Acetylcholine, and Phase-Response Abnormalities

An immediately apparent link between this phase-response curve and depression is the finding by GILLIN and coworkers (1982) that increased sensitivity to acetylcholine agonist induction of REM sleep appears to be a hallmark of patients subject to endogenous depression. There are several biochemical possibilities for this effect. One is simply that there might be an abnormal sensitivity to acetylcholine agonist administered exogenously. If this were true, then those people with predisposition to affective disorder not only should have a single time-point abnormality in their response to acetylcholine agonist but also an abnormal phase-response curve, whose shape could be readily correlated with particular alterations in the model parameters. This simple model of hyperresponsiveness to ACh agonists could be contrasted with more complex predictions, such as abnormalities in REM inhibition, i.e., monoamine abnormalities. We have elsewhere shown that assumption of an abnormality in monoamine (LC/DR) physiology and simulation of this abnormality in the limit cycle model can produce the short-latency, high-intensity, long-duration REM period characteristic of depression. BEERSMA and coworkers (1984) have used the 1975 version of the model to mimic the bimodal distribution of REM latencies in depression; an advantage of using the new limit cycle model is its ability to produce short-latency first REM periods that are also of high intensity (as is true in depression), whereas the earlier nonlimit cycle version of our model produces (unrealistic) low-intensity initial REM periods when used for modeling short REM latencies. urthermore, in contrast to the limit cycle model, the earlier version of our model produces unrealistic results for the night's course of succeeding REM durations (for depression modeling see McCARLEY and MASSAQUOI submitted to press).

ACKNOWLEDGEMENT. This research is supported by the Veterans Administration, RSDA MH 280, and MH 39,683.

REFERENCES

BAGHDOYAN HA, RODRIGO-ANGULO ML, McCARLEY RW, HOBSON JA (1984a) Site-specific enhancement and suppression of desynchronized sleep signs following cholinergic stimulation of three brainstem regions. Brain Res 306:39-52

BAGHDOYAN HA, MONACO AP, RODRIGO-ANGULO ML, ASSF, McCARLEY RW, HOBSON JA (1984b) Microinjection of neostigmine into the pontine reticular formation of cats enhances desynchronized sleep signs. J Phar Exp Ther 231:173-180

BEERSMA DGM, DAAN S, VAN DEN HOOFDAKKER RH (1984) Distribution of REM latencies and other sleep phenomena in depression as explained by a single ultradian rhythm disturbance. Sleep 7:126-136

BLOOM FE (1975) The role of cyclic nucleotides in central synaptic function. Rev Physiol Biochem Pharmacol 74:1-103

BORBELY AA (1982) A two process model of sleep regulation. Human Neurobiol 1:195-204

CAILLE EJ, BASSANO JL (1975) Approche du processus hypnique: analyse, fonctions temporelles, valeur fonctionnelle. Psychol Med 7:317-352

CZEISLER CA, ZIMMERMANN JC, RONDA JM, MOORE-EDE MC, WEITZMAN ED (1980) Timing of REM sleep is coupled to the circadian rhythm of body temperature in man. Sleep 2:329-346

DAAN S, BEERSMA DGM, BORBELY AA (1984) Timing of human sleep: recovery process gated by a circadian pacemaker. Am J Physiol 246 (Reg Integ Comp Physiol 15):R161-R178

FEINBERG I, KORESKO RL, HELLER N (1967) EEG sleep patterns as a function of normal and pathological aging in man. J Psychiat Res 5:107-144

GILLIN JC, SITARAM N, MENDELSON WB (1982) Acetylcholine, sleep, and depression. Human Neurobiol 1:211-219

HOBSON JA, McCARLEY RW, NELSON JP (1983) Location and spike-train characteristics of cells in anterodorsal pons having selective decreases in firing rate during desynchronized sleep. J Neurophysiol 50:770-783

ITO K, McCARLEY RW (1984) Alterations in membrane potential and excitability of cat medial pontine reticular formation neurons during changes in naturally occurring sleep-wake states. Brain Res 292: 169-175

LOTKA A (1956) Elements of physical biology. Dover, New York

LYDIC R, McCARLEY RW, HOBSON JA (1984) Forced activity alters sleep cycle periodicity and dorsal raphe discharge rhythm. Am J Physiol 247 (Reg Integ Comp Physiol 16):R135-R145

MASSAQUOI S, McCARLEY RW (1982) Extension of the reciprocal interaction sleep stage control model: a "Karma" control model. Sleep Res 11:216

McCARLEY RW (1978) Control of sleep-waking state alteration in Felix domesticus. In: FERRENDELLI JA (ed) Neuroscience symposia, vol 3. Society for Neuroscience, Bethesda, pp 90-128

McCARLEY RW (1980) Mechanisms and models of behavioral state control. In: HOBSON JA, BRAZIER MAB (eds) The reticular formation revisited. Raven, New York, pp 375-403

McCARLEY RW (1982) REM sleep and depression: common neurobiological control mechanisms. Am J Psychiatry 139:565-570

McCARLEY RW, HOBSON JA (1975) Neuronal excitability modulation over the sleep cycle: a structural and mathematical model. Science 189: 58-60

McCARLEY RW, ITO K (1985) Desynchronized sleep-specific changes in membrane potential and excitability in medial pontine reticular formation neurons: implications for concepts and mechanisms of behavioral state control. In: McGINTY D, DRUCKER-COLIN R, MORRISON A, PARMEGGIANI PL (eds) Brain mechanisms of sleep. Raven, New York, pp 63-80

McCARLEY RW, MASSAQUOI S (submitted to press) A limit cycle mathematical model of the REM sleep oscillator system

SASTRE JP, SAKAI K, JOUVET M (1981) Are the gigantocellular tegmental field neurons responsible for paradoxical sleep? Brain Res 229:147-161

SCHULZ H, ZULLEY J, DIRLICH G (1984) Statistical properties of the REM-NREM sleep cycle (unpublished manuscript)

SITARAM N, NURNBERGER JI, GERSHON ES, GILLIN JC (1980) Faster cholinergic REM sleep induction in euthymic patients with primary affective illness. Science 208:200-202

SITARAM N, NURNBERGER JI, GERSHON ES, GILLIN JC (1982) Cholinergic regulation of mood and REM sleep: potential model and marker of vulnerability to affective disorder. Am J Psychiatry 139:571-576

VOGEL GW, VOGEL F, McABEE RS, THORMOND AJ (1980) Improvement of depression by REM sleep deprivation. New findings and a theory. Arch Gen Psychiat 37:247-253

VOLTERRA V (1931) Leçons sur la théorie mathématique de la lutte pour la vie. Gauthier-Villars, Paris

WILLIAMS RL, KARACAN I, HURSCH CJ (1974) EEG of human sleep: clinical applications. Wiley, New York

ZULLEY J (1979) Der Einfluss von Zeitgebern auf den Schlaf des Menschen. RG Fischer, Frankfurt/Main

ZULLEY J (1980) Distribution of REM sleep in entrained 24 hour and free-running sleep-wake cycles. Sleep 2:377-389

Modes of Interaction Between Ultradian and Circadian Rhythms: Toward a Mathematical Model of Sleep

R. A. Wever

Max-Planck-Institut für Psychiatrie, Außenstelle Andechs, Arbeitsgruppe Chronobiologie, 8131 Andechs, FRG

Sleep is part of the circadian rest-activity cycle and its structure is characterized by an ultradian rhythmicity. The interactions between the two rhythmic systems are of special interest. To quantify these interactions it may be helpful to use an established mathematical model that has been successfully applied to analyze circadian rhythms as well as rhythms with shorter periods. This model is appropriate for single oscillations as well as for mutually coupled oscillations (WEVER 1962, 1964). The model has led to manifold predictions which all have been confirmed experimentally.

The model states a mutual dependency of all rhythm parameters, i.e., not only of period, amplitude, and mean value, but also of sleep fraction, wave shape, variability under the influence of a standard statistical noise, asymmetry of the variabilities with respect to onset and end of sleep, and many others. Preferably, albeit arbitrarily, all these parameters have been considered to depend on "external force", or, on the mean value of the oscillating variable. The dependence of the amplitude is of special interest. It means that this parameter deviates from zero only within a limited range of external forces. It is only within this range ("oscillatory range") that the system generates self-sustained oscillations. With external forces outside this range the course of the variable is arrhythmic, although the feedback mechanism is still intact. In a system of mutually interacting oscillators, the external force influencing all rhythm parameters and, in particular, the limits of the oscillatory range, is not only defined by environmental impulses but also by the output of the other oscillators.

Experimental Brain Research, Suppl. 12

The circadian system of man has already been described using a multioscillator model (WEVER 1975, 1979). The weaker and more relaxation-type oscillation which is represented by the sleep-wake rhythm ("type II") is divided by a threshold into two states, "wake" (above threshold) and "sleep" (below threshold) (WEVER 1962, 1964). The course of the oscillaton above the threshold may reflect "alertness", and the course of the oscillation below the threshold may reflect "depth of sleep" (e.g., measured as the power density of the EEG in the delta frequency band). In contrast, the stronger and more pendulum-type oscillation which is represented by the rhythm in deep body temperature ("type I") does not need such a threshold. The results of the model solution are considered to reflect directly the course of the biological variable.

The ultradian system also must be described using a multioscillator model. The multiplicity of observed ultradian rhythms and the diversity of the periods cannot be accounted for by a single oscillator. Because of their shorter periods, ultradian oscillators generally tend to be more like the relaxation-type oscillator rather than the circadian oscillators. In the context of this chapter, the interaction between circadian and ultradian oscillators is of special interest. For the sake of simplicity, the circadian "sleep-wake oscillator" is coupled with a single ultradian oscillator which is considered to regulate slow-wave sleep (SWS). One has, however, to bear in mind that in reality additional interactions with other circadian (mainly the "temperature oscillator") and ultradian oscillators, further complicate this picture.

In such a combined system, the superimposition of an ultradian sleep rhythm on the circadian sleep-wake rhythm does not alter the meaning of the concept of a threshold. As mentioned before, values above the threshold ("wake") describe "alertness", and values below the threshold ("sleep") describe "depth of sleep". There are several different types of solutions to the model equations. These mainly depend on the question of whether or not the superimposed ultradian rhythm is considered to be self-sustained.

If the ultradian rhythmicity is not self-sustained but is dampened out after every "push", the combined circadian/ultradian system can be described by a third-order equation. A meaningful solution of such an equation has been previously proposed (WEVER 1963). The upper part of Fig. 1 shows that within a circadian sleep-wake cycle (period, 24 h), every sleep onset generates a short-period rhythm (period, 1.5 h) that dampens out over the duration of the sleep episode (duration, 8 h), within 4-5 cycles.

In this model every "wake" onset also induces a dampened ultradian rhythm. The weights of successive ultradian cycles are necessarily decreasing exponentially. This gradual reduction in the depth of sleep is even more obvious in the upper part of Fig. 2 in which the analogue to the "power density" is presented (squares of the distances between the variable and the threshold): the first ultradian cycle shows the largest "power", and all subsequent cycles show considerably less power.

If the ultradian rhythmicity is considered an autonomous self-sustained oscillation, the system can be described using two mutually coupled second-order equations (or, equivalently, using a fourth-order equation) (WEVER 1975, 1979). Such a model also enables us to make clear predictions. Out of the great variety of solutions, the ultradian component is at rest as long as it is above a certain critical threshold, but exerts persisting short-period oscillations as soon as it falls below this threshold. It is tempting to assume that this limit of the ultradian oscillatory range is identical with the circadian threshold separating sleep and wakefulness. This assumption is demonstrated in the lower part of Fig. 1. As displayed in the figure, the sleep phase of the circadian sleep-wake rhythms is superimposed upon by an ultradian rhythm with five successive cycles of eventually decreasing amplitude. This may correspond to decreasing "depth of sleep". To facilitate understanding of the generation of the combined rhythms, the hypothetical course of the pure circadian rhythm is drawn as a dotted line.

The lower part of Fig. 2 shows the analogue of the "power density" in the slow-wave (delta frequency) range. The first and second ultradian cycles in every sleep episode have almost the same weight. Due to the transitional increase of the amplitude of the ultradian rhythm, the second cycle should have greater weight; however, the push of the circadian sleep onset enlarges the weight of the first cycle but not that of later cycles. The amplitude of the ultradian rhythm is decreasing only after the second cycle. This is in agreement with the increasing value of the circadian rhythm or the decreasing interval between the actual value of the circadian rhythm and the critical threshold.

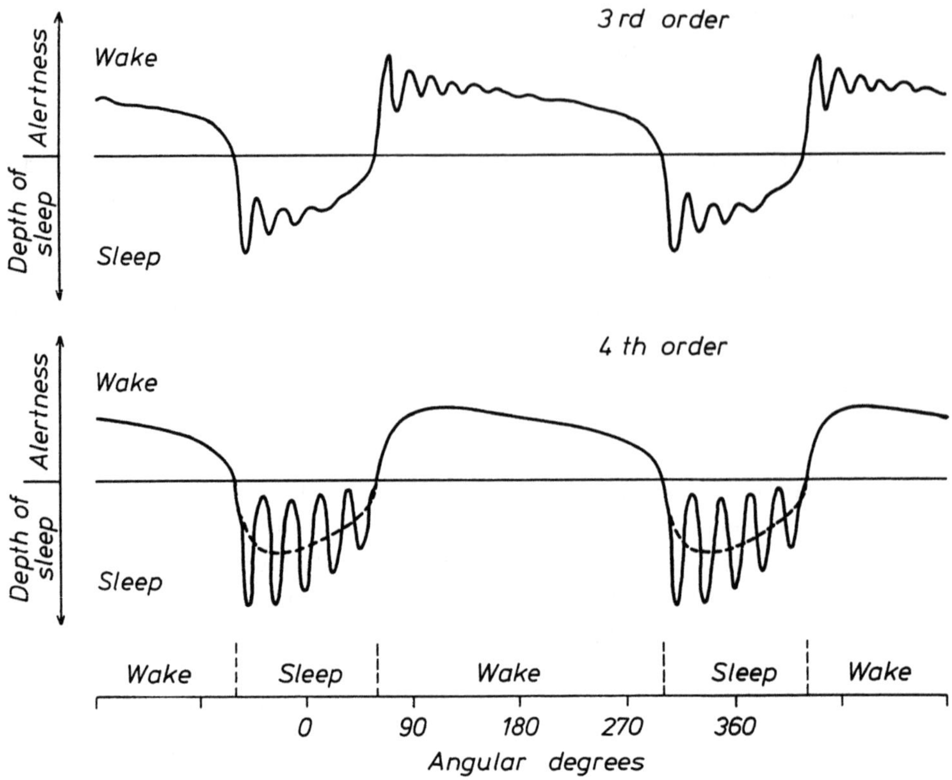
3rd order
Wake
Alertness
Depth of sleep
Sleep
4 th order
Wake
Alertness
Depth of sleep
Sleep
Wake
Sleep
Wake
Sleep
Wake
0
90
180
270
360
Angular degrees

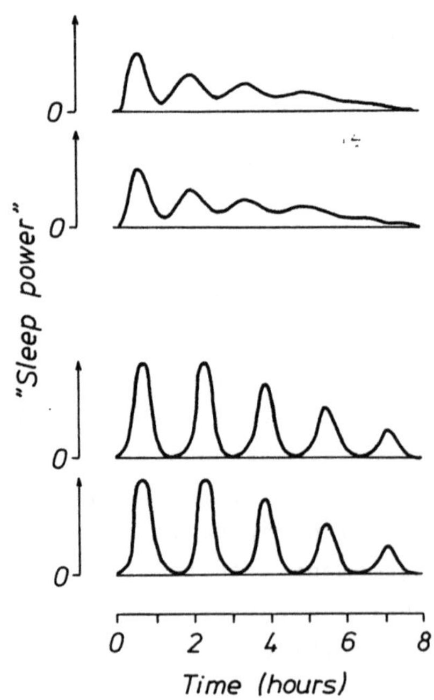
"Sleep power"
0
0
0
0
0
2
4
6
8
Time (hours)

In general, in the third-order model (upper parts of Figs. 1 and 2) the ultradian rhythm is generated by any "push" of the circadian rhythm (independent of its actual value) and dampens out exponentially. The initial ultradian amplitude depends only on the strength of the releasing "push". On the other hand, in the fourth-order model (lower parts of Figs. 1 and 2) the ultradian rhythm is autonomously self-sustained, and its properties (e.g., period and amplitude) depend on the actual value of the circadian rhythm. In other words, the ultradian rhythm is steadily modulated in frequency and amplitude (and also in other rhythm parameters, such as wave shape, etc.) by the course of the underlying circadian rhythm. That is, as soon as the system falls below the characteristic threshold, a switch process is activated resulting in a rapid and prominent lowering of the state of the system. This releases the ultradian cycle (leading, with short latency, to deep sleep). The release of the ultradian cycle with its intrinsic period is repeated as long as the system remains below the threshold. As soon as the system passes above the threshold, the ultradian rhythm is arrested. It restarts only after the system has fallen below the threshold again.

A way to discriminate between the two models with respect to their applicability to human sleep is by comparing the presented model solutions with experimental results. A suitable basis for such a comparison is the study by BORBELY (1984) who made long-term recordings of the power density of the slow-wave (delta frequency) band. None of these measurements obtained over 14 nights showed an exponential decrease of successive delta peaks throughout. During each night, the first two peaks are, on average, of similar size, and only later peaks show consistent decreases. Since the experimental results favor the fourth-order model over the third-order model, only solutions of the fourth-order model, which assumes the existence of two coupled self-sustained oscillators, will be considered.

◁ Fig. 1. Solutions of the model equation of higher order, simulating the interaction between a circadian and an ultradian oscillation. Upper diagram, oscillation equation of third order (from WEVER 1963). Lower diagram, oscillation equation of fourth order, or superimposition of two oscillations of second order each

Fig. 2. "Sleep power" as derived from the model solutions presented in Fig. 1. Ordinate, squares of the intervals between threshold and oscillating variable. The upper and lower diagrams correspond to the two diagrams of Fig. 1

It should be emphasized that the undisturbed oscillations, as shown in Figs. 1 and 2, are idealizations which are never realized in nature. To be more realistic, random fluctuations should be superimposed. For the sake of simplicity, noise was added only to the circadian rhythm component. As long as the circadian amplitude holds its "normal" value, the separation between sleep and wakefulness remains unambiguous; hence, the ultradian rhythms are clearly restricted to the sleep episodes (see the upper part of Fig. 3).

The upper part of Fig. 4 shows the power density in the slow-wave band. After adding noise to the circadian component, the two successive nights presented do not show identical patterns as before, but there are night-to-night variations. The overall general pattern, however, remains the same every night. We may, therefore, conclude that the inevitable random fluctuations which may be due to behavioral impulses (WEVER 1982) alter the details but not the general course of the sleep pattern. Due to an inherent stabilization mechanism, the fluctuations do not essentially alter the period and the level of the circadian sleep-wake rhythm (and hence the ratio between sleep and wakefulness) (WEVER 1984).

The coherence of the sleep pattern may be lost when the circadian amplitude decreases, as occurs in various diseases. In particular, a simultaneous shift of threshold in relation to the level of the circadian oscillation can lead to a crossing of the threshold during the "wrong" episode; in such cases, the separation between sleep and wakefulness is no longer unambiguous.

With reduced amplitude (and a threshold that is simultaneously slightly lowered in comparison with the mean level of the rhythm) the probability that the system passes the threshold during a sleep episode for a short while is no longer negligible. Under this condition the

Fig. 3. Solutions of the model equation of fourth order corresponding ▷ to the lower diagram of Fig. 1, but in addition superimposed by random fluctuations. Upper diagram, same original solution as in Fig. 1. Lower diagram, solution with reduced amplitude and slightly raised mean value. For better understanding, the "circadian" components which are superimposed by random fluctuations are drawn with dotted line.

Fig. 4. "Sleep power" as derived from the model solutions presented in Fig. 3. Ordinate, squares of the intervals between threshold and oscillating variable. The upper and lower diagrams correspond to the two diagrams of Fig. 3.

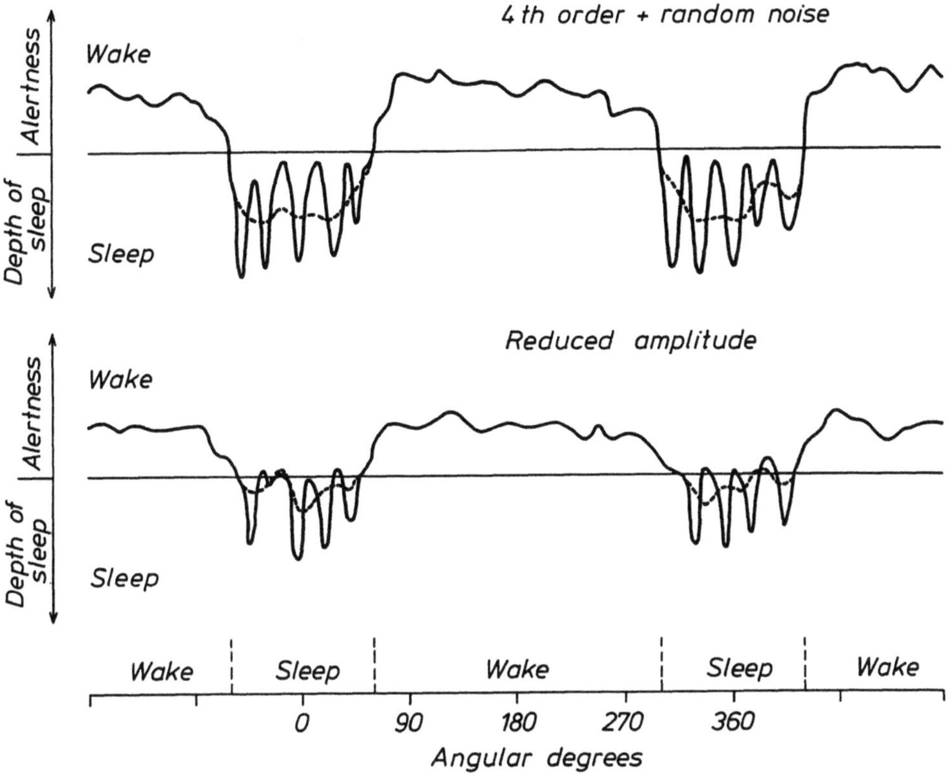
4 th order + random noise
Wake
Alertness
Depth of sleep
Sleep
Reduced amplitude
Wake
Alertness
Depth of sleep
Sleep
Wake
Sleep
Wake
Sleep
Wake
0
90
180
270
360
Angular degrees

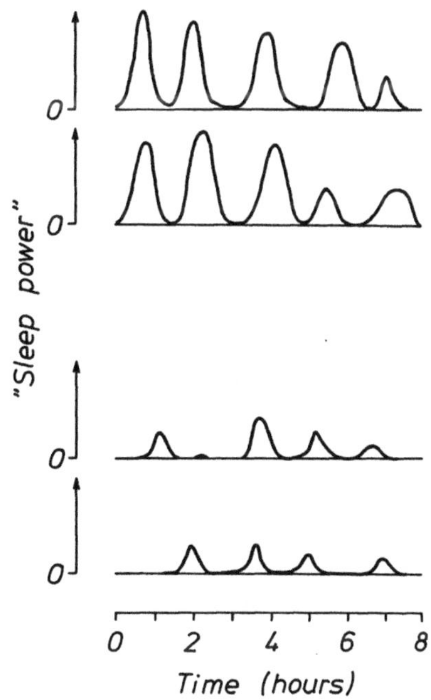
"Sleep power"
0
0
0
0
0
2
4
6
8
Time (hours)

ultradian rhythm is immediately arrested; consequently, the sleep pattern is disrupted and sleep time is reduced. This case which may serve as a model for insomnia is illustrated in the lower parts of Figs. 3 and 4. As shown, sleep is drastically disrupted, and slow-wave sleep is considerably reduced. In addition, the temporal regularity in successive sleep cycles is altered with less cycles per sleep episode. It is characteristic that the interval between successive sleep cycles can be lengthened but never shortened.

When the reduced circadian amplitude is accompanied by a slightly higher threshold relative to the mean level of the rhythm, the probability that the system falls below threshold during a wake episode is no longer negligible. In such a case an ultradian cycle will start even during a wake episode resulting in falling asleep during the wake episode. This case may serve as a model for narcolepsy.

In both disorders mentioned, insomnia and narcolepsy, the sleep disturbances are primarily due to a reduced circadian amplitude. Consequently, according to the model, treatment of these disorders demands an increment of the circadian amplitude. Such an increment can be induced by strengthening the controlling pacemaker; in humans this means to strengthen the social contacts with the environment (WEVER 1982) and/or the intensity of illumination during the day (WEVER 1985). Given a "normal" circadian amplitude, the presupposed slight shifts in the relationship between the threshold and the level of the circadian oscillation do not lead to disturbances in the general sleep pattern.

The stated influence of the circadian amplitude on the order of the system is one of the most relevant predictions of the model. With large circadian amplitude, the system produces highly stable rhythms in spite of superimposed random fluctuations; with small amplitudes, the system shows a highly sensitive dependence on initial conditions or random fluctuations. In other words, reducing the circadian amplitude leads from rhythmic order to chaos in the ultradian domain.

The dynamic sleep model presented above is obviously an oversimplification. It describes the interaction between one circadian and one ultradian oscillator. However, it does not yet consider the interactions between more than one oscillator in the same frequency range. For instance, the sleep-wake rhythm may be modified by its interaction with the deep body temperature rhythm. In addition, changes in deep body temperature can be considered to control REM sleep propensity which, supplemental to slow-wave propensity, is a relevant feature of the sleep pattern; therefore, consideration of the circadian body temperature rhythm including masking temperature by sleep-wake improves the usefulness of the ultradian sleep model (e.g., the masking mentioned effects a shortening of REM latency with decreasing temperature amplitude). Moreover, although the 90-min sleep rhythm is the best-known ultradian rhythm, it is certainly not the only one. The interaction between different ultradian rhythms may also modify the sleep pattern. The limits of the oscillatory range of different ultradian rhythms may differ, leading to the presence of certain rhythms only during the wake episodes. The newly presented application of an old model should only illustrate the general capability of mathematical models to describe sleep patterns under various conditions.

REFERENCES

BORBELY AA (1984) Sleep regulation: outline of a model and its implications for depression. Exp Brain Res (Suppl) 8:272-284

WEVER RA (1962) Zum Mechanismus der biologischen 24-Stunden-Periodik (I). Kybernetik 1:139-154

WEVER RA (1963) Zum Mechanismus der biologischen 24-Stunden-Periodik (II). Kybernetik 1:213-231

WEVER RA (1964) Zum Mechanismus der biologischen 24-Stunden-Periodik (III). Kybernetik 2:127-144

WEVER RA (1975) The circadian multi-oscillator system of man. Int J Chronobiol 3:19-55

WEVER RA (1979) The circadian system of man. Springer, Berlin Heidelberg New York

WEVER RA (1982) Behavioral aspects of circadian rhythmicity. In: BROWN FM, GRAEBER RC (eds) Rhythmic aspects of behavior. Erlbaum, Hillsdale, pp 105-171

WEVER RA (1984) Towards a mathematical model of circadian rhythmicity. In: MOORE-EDE MC, CZEISLER CA (eds) Mathematical models of the circadian sleep-wake cycle. Raven, New York, pp 17-79

WEVER RA (1985) Use of light to treat jet-lag: differential effects of normal and bright artificial light on human circadian rhythms. Ann NY Acad Sci (in press)

Epilogue

On Diversity and Uniformity of Ultradian Rhythms

J. Aschoff and M. Gerkema
Max-Planck-Institut für Verhaltensphysiologie, 8131 Andechs, FRG

After having read 18 contributions to problems of ultradian rhythmicity, someone who is more familiar with circadian than ultradian rhythms may wonder, at first thought, what the arguments are to treat such a diversity of frequencies as one class of rhythmic phenomena. The periods mentioned range from 1.0 h (luteinizing hormone (LH) in plasma and multiple unit activity (MUA) in the hypothalamus of monkeys; Chap. 2) to 5.0 and more than 6.0 h (body temperature and locomotor activity in rats; Chap. 5), and semidian periods (sleep propensity in man; Chaps. 13 and 14) are also included. Some of these rhythms may share common properties, and some may even be of similar adaptive value to the organism, but emphasis must be placed on the introductory remark that, in contrast to the "circa-rhythms" which have evolved in adaptation to time structures in the environment (ASCHOFF 1981), "the definition of ultradian rhythms gives no key to their functional significance" (Chap. 1) - and does not favor, as one may add, hypotheses about a uniform mechanism.

It is further instructive to notice to what extent many contributors take it for granted, at least implicitly, that the rhythmic patterns they have recorded represent something like an oscillation, despite the large variability in periods. There are obviously good reasons for such an approach, and most of the data discussed leave little doubt on the reality of ultradian rhythms. However, some researchers claim that data (especially activity records) which give the impression of an organized (rhythmic) pattern can also be described by a stochastic process (LEHMANN 1976; LUKAT and WEBER 1979; cf. also DIRLICH 1984 for circadian patterns). It is only in one chapter that the problem of "randomness" is specifically addressed (Chap. 3). On the other hand,

Experimental Brain Research, Suppl. 12

much effort has been put into applying analytical procedures, such as a Chi-square periodogram, either in its original form (Chap. 1) or modified (Chap. 6), autocorrelation and spectral analysis (Chaps. 7 and 10), a procedure specifically designed to determine significant peaks in pulsatile hormone secretion (Chap. 3), a normalized variance spectrum method (Chap. 12), and an iterative nonorthogonal r^2 method for defining the REM cycle (Chap. 16). It certainly would be worthwhile once to carry out a comparative study on the same set of data from ultradian rhythms to demonstrate advantages and possible pitfalls of the various techniques. Fortunately, in two contributions brief discussions are already included on the problem that some of the analytical procedures can reveal "significant" peaks at submultiples of 24 h which are mere artifacts of an odd-shaped circadian rhythm (Chaps. 5 and 13). Hence, caution is warranted not to interpret such peaks as "ultradian" if the time series includes a circadian component.

To extract ultradian rhythms more directly from the raw data, it is tempting to superimpose several time series on a common, e.g., a 24-h time scale. In noisy but stationary rhythms, such a procedure can be helpful to educe a mean rhythmic pattern. In view of the nonstationarity of ultradian rhythms (see below) in which successive periods often get longer, such an averaging is misleading when phase-setting (and hence synchronizing) effects are present at the beginning of each time series. Due to the unavoidable progressive desynchronization among the time series, the educed curve may show an initial sequence of ultradian rhythms which disappear later on (cf. Fig. 1 in Chap. 9). From such results it should not be concluded that ultradian rhythms were missing in the second half of each single time series. In contrast, superimposing is helpful and instructive in time series which by using a concurrence method have been divided into segments of equal length corresponding to the best-fitting period (cf. Fig. 4 in Chap. 10).

Variability and nonstationarity are issues in about 40% of all contributions. The ultradian period τ' is said to be 10 times more variable than the circadian period (Chap. 1) and it is obviously more affected by external as well as internal factors. In the feeding rhythm of voles, a dependence of τ' on circadian phase, season, and food quality was noticed (Chap. 1), and in the rhythm of urine flow in dogs a lengthening of τ' due to progressive dehydration (Chap. 7). On the other hand, τ' remained unaltered when circadian rhythmicity was lost after lesions of the SCN (Chaps. 1 and 5), in contrast to a lengthening of τ' when circadian arrhythmicity was introduced by exposure to LL (Chap. 5). The results obtained during thiopental narcosis are controversial: τ' was lengthened in monkeys (Chap. 2) but remained unaffected in rats (Chap. 4). On the whole, experimental conditions seem to have either facilitating or inhibiting effects on the expression of ultradian rhythmicity. Difficulties in documenting rhythms of performance and alertness in man have been related to "masking" effects exerted by continuous high demands in the tasks (Chaps. 11, 12, and 13).

It is well established that, during ontogeny, ultradian rhythms appear prior to circadian rhythms. The findings reported here for the rat (Chaps. 5 and 6) corroborate those published earlier for activity rhythms of small mammals (STIER 1933; CALHOUN 1945; LOCHBRUNNER 1956) and for the human sleep-wake cycle (KLEITMAN and ENGELMANN 1953; CORNER 1977; MEIER-KOLL et al. 1978). In other words, ultradian rhythmicity is present in the newborn in absence of an overt circadian rhythm, while in the adult both classes of rhythms coexist. It is hence not too surprising to see that ultradian rhythmicity persists and may even be more clearly expressed in adult animals which have lost circadian rhythmicity either spontaneously or after lesions of the SCN (Chaps. 1 and 5). In accordance herewith are observations that in some rodent species ultradian rhythms can be more pronounced at a season (e.g., spring or fall) in which circadian rhythmicity is partly disintegrated (OSTERMANN 1956; ERKINARO 1961). Superimposed are the seasonal modulations of τ' as described, e.g., for owls (ERKINARO 1973a), the bank vole (SAINT-GIRONS 1961) and the lake chub (KAVALIERS 1981). At present, no answer can be given to the question whether these changes in ultradian frequency are directly induced by the changing photoperiod or whether they represent (more likely) influences of a seasonally changing circadian system. The succession in the development of ultradian and circadian rhythms during ontogeny is analogous to that known from some hibernating mammals such as the southern woodchuck in which, after emergence from hibernation, ultradian patterns have been observed prior to the appearance of circadian rhythmicity (GRIZZELL 1955). The question arises whether for animals entering hibernation the reverse may apply, as it may be prior to death, i.e. that disintegration of the circadian time structure precedes that of the ultradian structure.

From the sequence of events during ontogeny, and from the persistence of ultradian rhythms during circadian arrhythmicity, it could be concluded that the ultradian time structure is more "basic" than the circadian one. However, it is known that the generating mechanism of circadian rhythms develops, in the rat, already prenatally and prior to its expression in an overt rhythm (FUCHS and MOORE 1980). Furthermore, the appearance or disappearance of an overt rhythm does not necessarily reflect a change in the state of the generating mechanism; it may also be due to a coupling or uncoupling of the behavioral element to or from an ongoing central rhythm, as demonstrated in hibernating mammals by the episodic awakenings timed on a circadian scale (POHL 1964; DAAN 1973).

Although ultradian and circadian rhythms can be expressed independently, and do not seem to have a causal relationship, they certainly interact. This is indicated already during ontogeny by changes in τ' concurrent with the development of the circadian system (Chap. 6). In the adult, the circadian system has, apart from modulating τ' (Chaps. 1, 3, 9, and 10), phase-setting effects on ultradian rhythms. Under the influence of an LD cycle, the ultradian activity rhythms of many species show a daily first peak close to dawn or to dusk (cf. STEBBINS 1975 for the bank vole, ERKINARO 1969 for three species of rodents, ERIKSSON et al. 1981 for the reindeer, HUNTER 1954 for sheep, and HUGHES and REID 1951 for cattle). The hypothesis that this phase control is not due to direct effects of light on the ultradian system but mediated through the circadian system (ASCHOFF 1982) is well supported by several of the records of feeding rhythms in voles (Chap. 1). This hypothesis is also supported by the observation that daily phase-setting in LD is missing in species such as shrews which have a strong ultradian rhythmicity but next to no circadian component (CROWCROFT 1954). Of even more interest are those records from voles which indicate a reverse coupling, i.e. phase-setting of the circadian rhythm by ultradian rhythmicity (cf. Fig. 4 in Chap. 1). The REM-NREM cycle provides another example of bidirectional coupling: onset of sleep has a phase-setting effect on the first REM episode (Chap. 10), and waking-up predominantly occurs during REM (Chap. 9). Furthermore, there is good evidence that increases or decreases of pituitary secretion may be "evoked" by sleep onset (Chap. 3), prompting the question whether reverse effects exist, e.g., a facilitation of awakening by an episode in the secretion of a certain hormone. Documentation and analysis of such <u>mutual</u> interactions among various rhythmic components within the organism should be considered a major task for research in the future.

Within the whole spectrum of ultradian rhythms, periods of 1.5 and 3.0 h seem to occur more frequently than others in many physiological and behavioral variables (Chaps. 3, 4, 6, 7, and 13). In view of the predominance of these submultiples of 24 h, and taking into account the semidian structure of sleep propensity in animals and man as well as additional characteristics of the human circadian system, BROUGHTON (Chap. 13) postulates that biological rhythms exhibit privileged periodicities in integer ratios of 2:1. He emphasizes that these periods are detectable in the raw data and hence cannot be interpreted as artifacts of an analytical procedure. Broughton considers it likely that the rhythms showing such integer relationships "reflect (at least semi-independent) multiple oscillators at these particular periods". This hypothesis implies a kind of hierarchical structure in which components of various frequencies are built upon each other. In such a system of mutually coupled oscillators, one should expect preferred and, at least temporarily, stable phase relationships among the constituents as they are known, e.g., for rhythms in heart rate or blood pressure and respiration (HILDEBRANDT 1976; KOEPCHEN 1962) or for the oscillators that drive fin movements in fishes with either absolute or relative coordination (Von HOLST 1939). Little is known about such effects among ultradian rhythms with different periods. As one of the rare examples a set of data from rats can be mentioned which suggests more or less stable phase relationships between episodes in corticosterone (CS; predominant period, 1.5 h) and episodes in growth hormone (GH; predominant period, 3.0 h), both being related to the sleep cycle (period, 3.0 h) but with different phase-angles (Chap. 4). However, a majority of observations seems to be at variance with the hypothesis of a hierarchical order among functionally coupled oscillators (with 2:1 ratios in period), e.g.,the nonstationarity of periods, and the occurrence of so many interjacent periods some of which again seem to have a 2:1 ratio (e.g., 2.5 and 5.0 h in locomotor activity of the rat (Chap. 5)). Yet, the predominance of privileged periods - if there is one - needs explanation. A seemingly baroque one rests on the argument that any structure responds with resonance to a certain frequency, and, to a lesser degree, to multiples and submultiples of it. It could be that the structures and functional units of the organism are arranged in a way that they get "in resonance" to the circadian and its submultiples more easily than to any other frequency.

More information is available on synchrony and phase relationship among those ultradian rhythms which share a common frequency. Correspondence in periods and a stable phase-angle is to be expected when the two rhythms are causally linked with each other, as exemplified by hypothalamic MUA and LH secretion in monkeys (Chap. 2) as well as by discharge rates of neurons in the dorsal raphe nucleus (DRN) and REM sleep in cats (Chap. 8). A much more complex picture emerges from rhythms for which a direct causal relationship is less likely or can even be excluded. There seems to be a kind of continuum, ranging from rhythms with a stable and unambiguous phase relationship to those whose phase-angles vary systematically according to conditions, and eventually to those with extremely labile phase relationships. Using a specific procedure to determine the degree of concomitance among episodes, VAN CAUTER and HONINCKX (Chap. 3) could present evidence for strong synchrony in pulsatile cortisol and testosterone secretion, suggesting "that a common CNS mechanism may control, at least partially, the pulsatility of both hormones". A fixed phase relationship has also been seen between the episodes of GH secretion and the sleep cycle in rats (Chap. 4). More variable were the phase-angles recorded between CS and locomotor activity in rats: the CS episode preceded the activity burst in intact animals, but both events were in phase after lesioning of the SCN (Chap. 5). Surprisingly, an even more variable relationship was found between locomotor activity and body temperature, with either negative or positive phase-angles (Chap. 5). Unfortunately, only a few findings are reported on correlations between the REM-NREM cycle and other ultradian rhythms. Rhythms of renal secretion and gastric motility are said to be independent from the sleep cycle (Chap. 9). Similarly, VAN CAUTER and HONINCKX (Chap. 3) did not find any support for the hypothesis that nadirs and peaks in plasma prolactin were related to REM and NREM stages. From their extensive studies the authors conclude "that increases or decreases of pituitary secretion may be 'evoked' by sleep onset or the occurrence of SW stages", and that the hypothesis of a consistent entrainment of nocturnal pulsatile release by the REM-NREM cycle has little support.

At this point, a brief digression seems appropriate concerning the problem of semidian periods. Usually, a definite and well-founded distinction is made between the ultradian and the circadian system. The question then may be asked into which of the two systems rhythms with semidian periods should be classified. Surveying various sets of data on semidian sleep tendencies, BROUGHTON (Chap. 13) concluded that the results "strongly support the existence of a 12-h biorhythm of SWS". But he also points out that "it is arguable whether this 12-h rhythm should in fact be classed as an ultradian rhythm or whether it should be considered an inherent property of the circadian sleep rhythm to manifest itself with two peaks". If the latter view is correct, it might be preferable not to speak of a semidian period (suggesting an ultradian component) but to use the less well defined expression "bimodality". Bimodality is an ontogenetically early characteristic of circadian rhythms, documented in animals (Chap. 6) and in man (ASCHOFF 1955). It persists in adults in many functions of which sleep propensity provides a prominent example (Chaps. 9, 13, and 14). According to the data of CAMPBELL and ZULLEY (Chap. 14), temporal placement and duration of naps are related to the rhythm of body temperature, i.e. depend on circadian phase. While this observation is not a stringent argument in support of bimodality as a structural element of the circadian system, such a conclusion is strenghtened by the fact that, in free-running rhythms, the duration of naps is positively correlated with the duration of wakefulness, and hence with the sleep-wakefulness cycle (ASCHOFF unpublished data). The notion of bimodality as a functional characteristic of the circadian system gets further support by the occurrence of splitting, i.e. the separation of a free-running activity rhythm into two components which, after temporary desynchronization, regain synchrony with a phase-angle difference of 180° (PITTENDRIGH 1960; HOFFMANN 1971). The phenomenon of splitting has led PITTENDRIGH (1974) to propose a two-oscillator model comprising a morning and an evening oscillator, a structure to which HONMA and HONMA (Chaps. 5 and 6) refer in interpreting peculiarities in the development of activity patterns in rats. In rhythms entrained by an LD cycle, the two peaks of activity (presumably representing the two oscillators) are thought to be coupled to dawn and dusk, and hence change their distance according to the photoperiod. In a free-running, non-split rhythm, the distance between the two peaks depends on the duration of wakefulness (cf. Fig. 5 in ASCHOFF 1966). In other words, bimodality, whether expressed by two peaks of activity or by naps interposed between two "peaks" of alertness, is more closely related to the circadian sleep-wakefulness cycle than any of the ultradian rhythms. Most likely, bimodality is the expression of two coupled circadian oscillators. If that is true, "semidian" periods should not be considered as part of the ultradian system and not as a "link" between shorter ultradian and circadian rhythms.

Nonstationarity of ultradian rhythms, and their often inconsistent interrelationships, may hinder any attempt to develop a uniform concept on mechanisms. In contrast to the circadian system, it seems unlikely that from the results obtained so far rules of general validity can be deduced. In spite of these difficulties, several authors have arrived at similar conclusions which may be illustrated by two quotations: "The results ... suggest that different physiological functions have common ultradian rhythms which are controlled by different pacemakers" (Chap. 4) and "Our findings support the assertion that ... the ultradian rhythms ... are organized as a 'multi-oscillatory' system, comprised of separate and potentially independent oscillators ... rather than a single 'master oscillator'" (Chap. 7). In fact, there seems to be much conformity in assuming a central nervous origin of ultradian rhythmicity and the existence of ultradian pacemakers (Chaps. 1, 2, 3, 8, and 11). However, strong evidence for such a neuronal oscillator is provided so far in only one case: the LHRH system in the mediobasal hypothalamus (Chap. 2). It does not allow generalizations to other ultradian rhythms. Nevertheless, the same hypothalamic area has been suggested as the locus of an ultradian pacemaker both for behavioral rhythms in voles (Chap. 1) and for rhythms in GH release in rats (EIKELBOOM and TANNENBAUM 1983). Furthermore, the DRN has been proposed to be at least part of a generating system for the ultradian sleep cycle in cats (Chap. 8).

Accumulation of knowledge on formal characteristics of a biological system and its physiological elements usually is followed by attempts to develop models which describe the dynamics of the system and are of predictive power. While such a state has been reached for the circadian system (ENRIGHT 1980; MOORE-EDE and CZEISLER 1984), modeling of the ultradian system still is at its beginning. Two of the contributions listed in Section 3 (Chaps. 17 and 18) present models, and the algorithm developed for the REM-NREM cycle (Chap. 15) may be considered a step toward a model. It would be of interest to apply the approach taken by KOBAYASHI and coworkers (Chap. 15) to other ultradian rhythms, to answer, e.g., the following questions: How does the ratio between activity and rest in the ultradian feeding rhythm of voles (Chap. 1) depend on cycle length? If secretory episodes are separated by a clear secretory pause (as claimed for many hormones (Chap. 3)), how is the ratio between secretory activity and pause related to cycle length, and how does the "profile" change with cycle length? A further step in this kind of analysis would be to compute serial interdependencies between the two (or three) fractions of a cycle. Although it has been stated that serial correlations among successive cycles do not give hints about the underlying (oscillatory) mechanism (Chap. 1), an analysis of within-cycle dependencies could be of interest for at least some models (ERKINARO 1973b).

The two contributions that deal with models (Chaps. 17 and 18) are similar in that they both suppose oscillations of a limit cycle type. They differ in the mode by which circadian and ultradian rhythms interact. The model of McCARLEY and MASSAQUOI (Chap. 17) is based on an ultradian oscillator on which the circadian system acts in two ways: by changing the time course of one of the two elements of the ultradian oscillator at sleep onset, and by modulating the coupling between the two ultradian elements continuously. The model by WEVER (Chap. 18) assumes a continuous interaction between a circadian and an ultradian oscillator in combination with a threshold which divides wakefulness and sleep. By proper adjustment of parameters both of these model equations yield solutions which adequately describe peculiarities of the REM-NREM cycle as known from experiments. The McCarley-Massaquoi model has as one of its advantages that it originates from electrophysiological data measured in brain areas which are related to the control of sleep. Hence this model allows predictions with regard to the effects of pharmacological interferences. The Wever model does not refer to concrete physiological processes, but is advantageous in assuming an (ongoing) ultradian oscillation which is not restricted to the sleep episode, and hence of potential value to those interested in the BRAC hypothesis.

Although not a model in its strict sense, the BRAC hypothesis advanced by KLEITMAN (1961) is still the only one which tries to condense the various ultradian phenomena into a coherent picture. The theory assumes that the rhythms detectable are all related to an alternation which originally had something to do with the temporal organization of activity and rest, and it supposes that ultradian rhythms reflect a <u>common</u> generating mechanism. How much evidence do we have today in support of BRAC? KLEITMAN (1982) himself has reviewed the relevant literature recently. Undoubtedly, an impressive amount of data has been accumulated which show an about 90-min rhythmicity during wakefulness as well as during sleep in man. However, to document an ongoing process which controls rhythms throughout day and night, the demonstration of equality in <u>periods</u> is suggestive but not sufficient. The persistence of <u>phase</u> over time is the major argument. Its documentation is impeded by the probability that the (persisting) ultradian rhythm is reset at the beginning and/or the end of sleep. It is presumably for this reason, apart from "masking" due to the highly demanding experimental protocol, that KRIPKE and coworkers (Chap. 12) were not convincingly successful in their attempts to demonstrate phase consistency by extrapolating from REM-NREM cycles to performance cycles recorded after awakening. Stronger evidence for phase continuity is presented by SCHULZ (Chap. 10) in his studies with narcoleptic patients who showed a tendency for REM episodes to be in phase over the whole nychthemeron.

In summary, four sets of data published in this volume demonstrate more or less rigorously a persisting ultradian rhythmicity: the pulsatility of LH in monkeys (Chap. 2) (period, 1.0 h), the behavioral rhythms in voles (Chap. 1) and in rats (Chap. 5) during circadian arrhythmicity (period, 2.5 h), and sleep propensity in narcoleptics (Chap. 10) (period, 1.5 h). Only one of these has a period of 1.5 h and hence confirms to BRAC. Additional data are suggestive for an ongoing process within this frequency domain, and it can well be expected that in further studies more results will be obtained which match the criteria discussed above. On the other hand, there are ongoing ultradian rhythms with periods different from 1.5 h which cannot be assigned to BRAC in its original form but which may well represent similar persisting processes of different frequencies.

In considering the functional significance of ultradian rhythms, one should first keep in mind that a rhythmic organization (of whatever frequency) is one of the means to keep temporal order within the organism. Where many processes have to be maintained which to some degree are mutually exclusive, but nevertheless cooperate, a "temporal compartmentalization" by rhythmic alternation is an obvious solution. This view is complemented by the economic principle not to spend energy continuously at a relative high level (as demanded at times) but to alternate between expenditure and restoration of energy. Furthermore, there are processes which are needed "on call", e.g., in response to randomly occurring stimuli; it might be useful to keep these types of processes going (rhythmically for the reasons given above) in order to avoid "atrophy" as well as problems of "initial friction" (von HOLST 1949). Arguments of this kind have also been put forward for the continuous rhythmicity of systems which convey information by changes in frequency (ASCHOFF 1959), a function which may well be attributable to ultradian rhythms. Finally, the oscillation of a variable around its mean value may serve to "scan" the state of a system, e.g., in feedback circuits (ASCHOFF and WEVER 1962).

In line with some of the arguments given above, the functional meaning of ultradian rhythms has been related to the coordination of physiological processes and the optimizing of energy expenditure (Chaps. 1

and 13). With regard to the REM-NREM cycle, LAVIE (Chap. 9) advocates the attractive hypothesis that the "fragmentation of the sleep period into shorter modular units, each ending with a preparatory 'gate' ... ensures a smooth and efficient transition from sleep to wakefulness" - an adaptive principle which, in a similar way, has been claimed for the episodic awakenings during hibernation (ASCHOFF 1962), and which is analogous to one of the functions of the circadian system, i.e. the preparation of the organism to daily ensuing tasks (ASCHOFF 1964). However useful these considerations may be in our search to understand the teleonomy of ultradian rhythms, they have not yet reached the convincing evidence of the function of pulsatile gonadotropin release in regulating sexual development and activity, as presented by KNOBIL and HOTCHKISS (Chap. 2). Equally well established are those hypotheses which claim that ultradian rhythms, in addition to their role in maintaining temporal order intraindividually, are of adaptive significance by providing interindividual social synchrony among conspecifics (Chap. 1).

In conclusion, the message given by the 18 contributions of this volume can be summarized as follows. There is no doubt on the reality of ultradian rhythms as opposed to randomness of events, there is evidence for persistence of some ultradian rhythms, and there are data suggesting the existence of true pacemakers. On the other hand, it is impossible to postulate one common mechanism for all ultradian rhythms. Most likely, some rhythms are not driven by self-sustaining "master oscillators" but represent "slave oscillators" or only damped oscillations. Even for the persisting rhythms, more than one pacemaker has to be postulated because of the different frequencies. In search of uniformity among at least some ultradian rhythms, an extended BRAC hypothesis which comprises various frequencies can be considered the uniting principle. Yet, ultradian rhythmicity as a whole is characterized by diversity, in mechanism as well as in their adaptive significance.

REFERENCES

ASCHOFF J (1955) Der Tagesgang der Körpertemperatur des Menschen. Klin Wschr 33:545-551

ASCHOFF J (1959) Zeitliche Strukturen biologischer Vorgänge. Acta Leopoldina 21:147-177

ASCHOFF J (1962) Spontane lokomotorische Aktivität. In: HELMCKE JG, Von LENGERKEN H, STARK D (eds) Handbuch der Zoologie, Bd 8 (II,4). De Gruyter, Berlin, pp 1-76

ASCHOFF J (1964) Survival value of diurnal rhythms. Symp Zool Soc (Lond.) 13:79-98

ASCHOFF J (1966) Circadian activity pattern with two peaks. Ecology 47:657-662

ASCHOFF J (1981) A survey on biological rhythms. In ASCHOFF J (ed) Biological rhythms. Handbook of behavioral neurobiology, vol 4. Plenum, New York, pp 3-10

ASCHOFF J (1982) Lichtwirkungen auf das circadiane System von Säugetieren und Vögeln. In: Wirkungen des sichtbaren Lichtes und der Ultraviolettstrahlung auf den Menschen und auf landwirtschaftliche Nutztiere. III. Int Symp, Leipzig 1981. VEB Kongress- und Werbedruck, Oberlungwitz, pp 22-47

ASCHOFF J, WEVER RA (1962) Biologische Rhythmen und Regelung. In: DELIUS L, KOEPCHEN HP, WITZLEB E (eds) Probleme der zentralnervösen Regulation. Bad Oeynhausener Gespräche V. Springer, Berlin Göttingen Heidelberg, pp 1-15

CALHOUN JB (1945/1946) Twenty-four hour periodicities in the animal kingdom. J Tennessee Acad Sci 20:228-278/21:209-216

CORNER MA (1977) Sleep and the beginnings of behavior in the animal kingdom. Studies of ultradian motility cycles in early life. Progr Neurobiol 8:279-295

CROWCROFT P (1954) The daily cycle of activity in British shrews. Proc Zool Soc (Lond.) 123:715-729

DAAN S (1973) Activity during natural hibernation in three species of vespertilionid bats. Neth J Zool 23(1):1-71

DIRLICH G (1984) Looking at human circadian phenomena from a framework of simple stochastic models. In: MOORE-EDE MC, CZEISLER CA (eds) Mathematical models of the circadian sleep-wake cycle. Raven, New York, pp 159-185

EIKELBOOM R, TANNENBAUM GS (1983) Effects of obesity-inducing ventromedial hypothalamic lesions on pulsatile growth hormone and insulin secretion: evidence for the existence of a growth hormone-releasing factor. Endocrinology 112:212-219

ENRIGHT JT (1980) The timing of sleep and wakefulness. Springer, Berlin Heidelberg New York

ERIKSSON LO, Källquist ML, MOSSING T (1981) Seasonal development of circadian and short-term activity in captive reindeer, Rangifer tarandus L.. Oecologia 48:64-70

ERKINARO E (1961) The seasonal changes of the activity of Microtus agrestis. Oikos 12:157-163

ERKINARO E (1969) Der Phasenwechsel der lokomotorischen Aktivität bei Microtus agrestis (L.), M. arvalis (Pall.) und M. oeconomus (Pall.). Aquilo Ser Zool 8:1-31

ERKINARO E (1973a) Structure of the diel activity period in Tengmalm's owl, Aegolius funereus, and the short-eared owl, Asio flammeus, and its seasonal changes. Aquilo Ser Zool 14:59-67

ERKINARO E (1973b) Short-term rhythm of locomotor activity within the 24 hr period in the Norwegian lemming, Lemmus lemmus, and water vole, Arvicola terrestris. Aquilo Ser Zool 14:46-58

FUCHS JL, MOORE RY (1980) Development of circadian rhythmicity and light responsiveness in the rat suprachiasmatic nucleus: a study using the 2-deoxy-1-C^{14} glucose method. Proc Natl Acad Sci (USA) 77:1204-1208

GRIZZELL RA jr (1955) A study of the southern woodchuck, Marmota monax monax. Amer Midl Natural 53:257-293

HILDEBRANDT G (1976) Outline of chronohygiene. Chronobiologia 3:113-127

HOFFMANN K (1971) Splitting of the circadian rhythm as a function of light intensity. In: MENAKER M (ed) Biochronometry. Nat Acad Sci, Washington, pp 134-150

Von HOLST E (1939) Die relative Koordination. Erg Physiol 42:228-306

Von HOLST E (1949) Zur Funktion des Statolithenapparates im Wirbeltierlabyrinth. Naturwissenschaften 36:127-128

HUGHES GP, REID D (1951) Studies on the behaviour of cattle and sheep in relation to the utilisation of grass. J Agr Sci 41:350-366

HUNTER RF (1954) Some notes on the behaviour of hill sheep. Brit J Animal Beh 2:75-78

KAVALIERS M (1981) Seasonal changes in the short-term activity and ultradian rhythms of a cyprinid fish, the lake chub, Couesius plumbeus. Can J Zool 59:486-492

KLEITMAN N (1961) The nature of dreaming. In: WOLSTENHOLME GEW, O'CONNOR M (eds) The nature of sleep. Churchill, London, pp 349-364

KLEITMAN N (1982) Basic rest-activity cycle - 22 years later. Sleep 5:311-317

KLEITMAN N, ENGELMANN TG (1953) Sleep characteristics of infants. J Appl Physiol 6:269-282

KOEPCHEN HP (1962) Die Blutdruckrhythmik. Kreislauf-Bücherei Bd 19, Steinkopff, Darmstadt

LEHMANN U (1976) Stochastic principles in the temporal control of activity behaviour. Int J Chronobiol 4:223-266

LOCHBRUNNER A (1956) Beiträge zur Biologie des syrischen Goldhamsters, Mesocricetus auratus (Nehring). Zool Jb Allg Zool 66:389-428

LUKAT R, WEBER F (1979) The structure of locomotor activity in bilobectomized cockroaches (Blaberus fuscus). Experientia 35:38-39

MEIER-KOLL A, HALL U, HELLWIG U, KOTT G, MEIER-KOLL V (1978) A biological oscillator system and the development of sleep-waking behavior during early infancy. Chronobiologia 5:425-440

MOORE-EDE M, CZEISLER CA (eds) (1984) Mathematical models of the circadian sleep-wake cycle. Raven, New York

OSTERMANN K (1956) Zur Aktivität heimischer Muriden und Gliriden. Zool Jb Allg Zool 66:355-388

PITTENDRIGH CS (1960) Circadian rhythms and the circadian organization of living systems. Cold Spring Harbor Symp Quant Biol 25:159-184

PITTENDRIGH CS (1974) Circadian oscillations in cells and the circadian organization of multicellular systems. In: SCHMIDT FO, WARDEN FG (eds) The neurosciences: third study program. MIT Press, Cambridge, pp 437-458

POHL H (1964) Diurnal rhythms and hibernation. Ann Acad Sci Fennicae, Series A, IV. Biologia 71:361-373 (Mammalian Hibernation II)

SAINT-GIRONS MC (1961) Le rythme nycthéméral d'activité du campagnol roux, Clethrionomys glareolus (Schreber) 1780. II. Les femelles. Mammalia 25:342-357

STEBBINS LL (1975) Short activity periods in relation to circadian rhythms in Clethrionomys gapperi. Oikos 26:32-38

STIER TJB (1930) "Spontaneous activity" of mice. J Gen Psychol 4:67-99

Subject Index

Acetylcholine 288-306
Activity 11, 14, 198
Actogram 172
Addition task 205, 209, 211
Adrenocorticotropic hormone (ACTH) 42, 44, 48, 73, 92
Adrenoreceptors, α, β 32, 33, 37, 111, 118, 122
Aircraft operations 190
Aldosterone 111, 116, 122
Alertness 148-158, 217-228
Anesthesia
  corticosterone 62-73
  growth hormone 62-73
  neuronal activity 72
Antidiuretic hormone (ADH) 110, 111, 116, 122
Arcuate nucleus (Arc) 9, 26-28, 33, 62, 69, 72, 86
Arecholine 302
Arousal 151-153, 160, 162, 217, 224
Attention span 218
Autocorrelation 168, 176, 319
  binary a. 271, 272, 284
Awakening 3, 160, 161, 208-215

Basic rest-activity cycle (BRAC) 4, 147ff, 153, 165, 170, 171, 176, 179, 182, 186, 187, 190, 197, 201, 202, 215-219, 222ff, 229, 326ff
Bed rest 223, 237
  continuous b.r. 150, 153, 158
Biofeedback 151
Blood pressure 322
Body temperature 77-89, 172, 226, 228, 238, 247-253, 294, 299, 301, 302, 310, 316, 317, 323, 324
Brain stimulation, electrical 152, 188
Brain temperature 160
Brain transection 197
Brain-stem arousal systems 158, 159
Brain-stem sleep inducing mechanisms 158, 159
Bulbar reticular formation 302

Carbachol 136
Cataplexy 150, 172
Cattell Anxiety Scale 223
Cholinergic system, *see* Acetylcholine
Circadian oscillators 95, 99, 230, 309, 310, 324, 326
  coupling 78-83, 91, 92, 101, 104, 107, 108
  desynchronization 77, 91
Circadian pacemakers 316
Circadian rhythms 2, 10-14, 41, 53, 54, 77-91, 139, 149, 152, 157-162, 186, 202-209, 217, 226-230, 235, 241, 242, 247-253, 259, 260, 268, 296-303, 309, 311-325
  blinding 97-108
  body temperature 84, 88, 159
  corticosterone 86, 89
  development 95-108, 320
  hypothalamic electrical activity 15, 21, 69, 72
  lighting condition, influence on 320, 321
  locomotor activity 78, 82-89, 95-108
  precision 14
  REM sleep 284, 290-295
  running-wheel activity 17
  sleep 149, 159, 289
  splitting 80, 82, 99, 324
  stability 14, 17-19
  terminology 5
  urinary secretion 121, 122
Circannual rhythm 11, 12
Circasemidian period, *see* Semidian period
Circhoral pulse generator 32, 36, 37

Clomiphene citrate 53
Complex demulation 218
Computer simulation 54, 56, 295
Concurrence method 176, 178, 319
Continuous performance task (CPT) 221
Corticosterone (CS) 61-74, 77-92, 101, 322, 323
Corticotrophin-releasing factor (CRF) 48, 92
Cortisol 10, 42-57, 78, 110, 111, 116, 122, 323
Cushing's disease 9, 57
Cyclic AMP 292
Cyclic GMP 292

Daydreaming 149, 158
Dehydration, progressive 111, 112, 117, 118, 121, 319
Delusion 205, 213
Deoxyglucose uptake 95
Depression 9, 57, 289, 290, 305
Desynchronized sleep, *see* REM sleep
Dexamethasone 53
Digestive pauses 15
Disentrainment 147, 229, 234-239, 247
Dorsal raphe nucleus (DRN) 10, 125-140, 288, 291, 292, 296, 306, 325
Dreaming 125
Drinking 202, 206, 209-215
Drowsiness 223, 224, 226, 235, 236, 253

Electrocardiogram (ECG) 172
Electroencephalogram (EEG) 62, 63, 68, 125, 147-152, 158, 172, 186, 192, 193, 196, 203-209, 221-223, 229, 238, 250, 261
  alpha activity 149, 158, 206, 209
  delta activity 158, 270
  mu pattern 196
  power density 151, 310, 313, 314
  sensorimotor pattern 196
  theta activity 151, 158, 222, 223
Electrolyte concentrations 110, 112, 115, 118, 120
Electromyogram (EMG) 172, 192, 222, 238, 261
Electrooculogram (EOG) 172, 238, 261
Endothermy 179
Energy expenditure 327
Estradiol 36
Eye movements 139, 149, 158, 206, 222
Eysenck Personality Inventory (EPI) 223, 224

Fantasy activity, *see also* Daydreaming 201-213
Fatigue 202, 207, 209, 218, 224
Feedback 205
Feeding 9, 13, 23, 26, 48, 179, 202, 206, 209, 212
Fin movements 322
Flight simulation 147
Fluid balance 111
Follicle stimulating hormone (FSH) 34-37, 54
Food deprivation 23-25, 101, 107
Food quality 19
Foraging 13, 15, 23
Free-running rhythm 299, 301
FTG (gigantocellular tegmental field) 288

Gastric motility 12, 110, 148
Gate of sleep 148, 158-161
Gate of wakefulness 148, 160, 161
Gonadotropins 33, 35, 41, 44, 328
Gonadotropin releasing hormone (GnRH) 26, 32-37
Grazing rhythms 16
Growth hormone (GH) 10, 26, 42, 44, 54, 61-74, 322, 323, 325
Growth hormone releasing factor (GRF) 72

Hallucinations 205, 213
Hanning (spectral smoothing) 201, 206
Heart rate, *see also* Electrocardiogram 322
Hemispheric dominance 202, 205, 209, 213, 222
Hibernation 320, 321, 328
Hypercortisolism 9, 57
Hyperkinetic child 218, 219
Hypersomnia 226
Hypnagogic hallucination 149, 158
Hypothalamus
  mediobasal h. (MBH) 26, 325
  ventromedial h. (VMH) 26
  MUA activity 318, 323

Infradian rhythms 11, 228
Insomnia 316

Kidney denervation 111, 118, 122, 123
K-complex 238

Learning 196
Lighting condition
constant darkness (DD) 17, 78
constant light (LL) 77, 82-84, 89-92
Limit cycle 259, 288-306, 326
Locomotor activity
spontaneous l.a. 77-92
forced l.a. 133, 134, 137
Locus coeruleus 288, 291, 292, 296, 306
Lotka-Volterra equations 259, 288, 289, 292
Luteinizing hormone (LH) 32-37, 48, 53, 54, 56
Luteinizing hormone-releasing hormone (LHRH) 42

Masking 98, 99, 107, 219, 221, 224, 317, 320, 326
Meal cycle 11, 12
Memory
neural m. 169, 182
visual m. 205, 209, 212, 214
Menstrual cycle 35-37
Midbrain reticular formation 302
Minnesota Multiphasic Personality Inventory (MMPI) 204
Monophasic animals 2
Motivation 147, 204, 205, 217, 221, 224, 228
Motor activity, *see also* Activity 112, 121, 218, 219
Multiple Sleep Latency Test (MSLT) 226
Multiple unit activity (MUA)
arcuate nucleus (Arc) 62, 69, 72
ventromedial nucleus (VMH) 62, 72
suprachiasmatic nucleus (SCN) 62, 69
Muscle tone, *see also* Electromyogram 134-137, 140, 206

Nap(ping) 150, 166, 203, 204, 208, 211, 225ff, 234-253, 284, 324
Narcolepsy 147, 150, 152, 159, 165-183, 221, 224, 226, 316, 327
Neostigmine 302
Neuroticism 224
Non-rapid eye movement (NREM) sleep 11, 125-137
Noradrenaline 110
Nucleus reticularis magnocellularis 136
Nucleus ventromedialis 62, 72, 86
Nychthemeron 165-183, 235, 241, 242, 249-253, 327

Ontogeny 11, 12, 153, 225, 187
Oral behavior 179
Oxygen consumption 160

Pacemaker, metabolic 188
Pacemaker potentials, intracellular 140
Paraventricular nucleus (PVN) 23, 27, 28, 86
Penile erection 3
Perceptual deprivation 223
Performance 165, 179, 190-197, 217-224, 235
hemisphere-related tasks 161
operant p. 187, 188
sustained p. 147, 201, 204
testing 202-215
Periodogram
Chi-square p. 26ff, 96, 104, 319
Enright's p. 229
Phase response curve (PRC) 78, 215, 259, 289, 290, 302, 305, 306
Phentolamine 111, 118
Physostigmine 302, 303
Pituitary hormones 41
Plasma protein levels 46
Polyphasic animals 2, 14
Pontine reticular formation (PRF) 136, 139, 140, 288, 291, 292, 302
Ponto-geniculo-occipital (PGO) waves 134-137, 140
Population density cycle 12
Post-lunch dip 153, 156, 157, 225
Prolactin (PRL) 42, 44, 46, 53
Propranolol 111, 118
Pulse time 12
Pupillary diameter 152

Questionnaire of daily habits 223

r^2 method, iterative nonorthogonal 270-285, 319
Rapid eye movement (REM) sleep 3, 11, 110, 116, 125-139, 160, 161, 170-179, 186, 187, 208, 211, 218, 224, 227, 248, 250, 260-268
body position 168, 173
interruption 261
latency 172, 250, 290ff, 317
monoamines 305, 306
propensity 316
timing 168, 169
Rayleigh test 208
REM-NREM sleep cycle 3, 10-15, 50, 53, 110, 116, 120, 147-183, 188, 197-215, 224, 229, 259-267, 319-328

analysis 270-285
brain weight 270
duration 271, 284
fetal 187
hypoxia 199
infants 274-279
life span 270
maternal 187, 188
metabolic rate 270
model 288-306
newborn 270, 274-279
ontogeny 274-279
stress 199
Renal excretion, *see also* Urine 10, 148
Renewal process 13, 14, 17, 23, 26
Renin activity, plasma (PRA) 111, 116, 122
Respiration 152, 172, 322
Rest-activity cycle, *see also* Ultradian rhythms 187, 197, 309
Retrochiasmatic area (RCA) 27, 28
Rosvold-Mirsky visual performance test 218
Rumination 15

Semidian period 217, 225-230, 247, 318, 322ff
Sensory deprivation 149, 158
Serotonin 140
Shostrom Personal Orientation Inventory 223
Siesta 225, 234
Sleep, *see also* REM sleep, NREM sleep, Slow wave sleep 61-74, 116, 125, 126, 148-162, 165-183, 203-209, 217, 225-230, 235-253, 270, 289, 309-317
ambient temperature 274, 279ff
apnea syndrome 226
body movements 276
deprivation 65, 66, 72, 150, 153, 156, 158, 159, 198, 199, 218
disorders 226
infants 274-279
interruption 166-169, 182
latency 226
need 147, 217
noise 279ff
onset 168, 226, 227, 238, 241, 242, 247, 251, 294, 303, 311, 321
onset REM (SOREM) 172
propensity 147, 182, 248, 251, 318, 322, 324, 327
quality 224, 228
reduction 274
spindle 238
stages 111, 118, 121, 173
active s. 274
intermediate s. 274
quiet s. 274
stage REM 272, 273
stage 1 149, 156, 223, 226, 261, 272
stage 2 226, 227, 250, 259-268, 272
stage 3 261, 272
stage 4 248, 261, 266, 272
tendency 147-162, 217, 223-225, 228, 236, 250
terminology 5
Sleepiness 154-162, 202, 205, 209, 221-228
afternoon 225, 226
rating 205, 209
Sleep-wake cycle 10, 61-74, 112-123, 150, 153, 162, 176, 186, 188, 197, 227, 228, 234, 235, 239, 247ff, 310, 311, 320, 322, 323
behavioral control 235, 236, 251, 252
Slow wave sleep (SWS) 53, 158, 225-228, 250, 259-268, 310-317, 323
deprivation 267
Social contact 204
Social isolation 202, 213
Social synchronization 16, 19, 20, 22
Somnolence 150
Spectral analysis 13, 62, 96, 112, 115, 137, 176, 187, 201-212, 219, 222, 229, 319
Stanford Sleepiness Scale (SSS) 205
Stochastic process 318
Stress 48, 218, 224
Suprachiasmatic nucleus (SCN) 10, 26, 28, 62, 77, 78, 86-99, 139, 320, 323
Synchronized sleep, *see* Sleep, NREM sleep, Slow wave sleep

Telemetry 238
Temperature, *see also* Body temperature
ambient t. 19
Testosterone 48, 51, 323
Thermoregulation, *see also* Body temperature 16
Thiopental 62-73
Thyroid stimulating hormone (TSH) 42, 44
Time-normalized sleep cycle averaging 128-135
Tracking game 205, 209, 212, 214

Ultradian oscillators 23, 26, 28, 50, 77, 95, 122, 170, 203, 207, 208, 213, 215, 230, 289-294, 298, 310, 311, 318, 325, 326

Ultradian pacemaker 9, 13, 49, 50, 61, 66-72, 325, 328
Ultradian rhythms
adaptive significance 328
clock model 24
definition 2, 5, 318
development 95-108
entrainment 21, 22
functional significance 327
generation 23-28, 199
homeostatic model 24
nonstationarity 153, 154,
ontogeny 320, 321
precision 14, 179, 182
stability/instability 17, 197, 318, 319, 324
synchronization 16, 20ff, 322, 323
Ultradian rhythms, factors influencing
ambient temperature 19, 279ff
bed rest 223
body size 165
brain transection 197
circadian rhythms 1, 19-26, 53, 54, 77, 82, 91, 92, 95ff, 121, 152, 202, 284, 289, 316, 319-322
environment 17
experimental protocol 190ff, 214, 326
fatigue 202, 209, 218
feedback 53
food quality 19
hydration/dehydration 110ff
hypoxia 199
light 77, 98ff, 320, 321
metabolism 165
motivation 221
noise 279ff
ontogenesis 61ff, 89, 95ff, 153, 274-279
perceptual deprivation 223
performance pressure 213, 218, 221, 224
posture 110, 121
psychological parameters 223
SCN lesioning 86ff
seasons 320
sleep deprivation 65, 150, 158, 159, 198, 199, 202, 218
sleep onset 1
sleep reduction 281, 289
sleepiness 202
social contact 213
state of health 1
stress 199, 218
thiopental narcosis 320
time of day 153ff, 209
Ultradian rhythms in
alertness 1, 148-162, 217ff, 320
arousal 151, 152
behavior 202, 203, 326, 327
body motility 152
body temperature 77ff, 318
brain stimulation 152, 179
cardiac rate 1
corticosterone 61ff, 77ff
cortisol 110
daydreaming/fantasy activity 149, 201, 202, 209ff
drinking 1, 206ff
EEG activity 149, 151, 158, 192ff, 221ff, 270
feeding 1, 15-26, 179, 209ff, 319
gastric motility 1, 3, 110, 148, 224, 323
growth hormone 26, 61ff
incubation 15
locomotor activity 77ff, 95-108, 318, 323
luteinizing hormone 318, 327
luteinizing hormone-releasing hormone 325
nasal mucosa 110
noradrenaline 110
oral behavior 179
penile erections 3
perception 158
performance 1, 202, 209ff, 217ff, 320, 326
population behavior 20
pupillary diameter 152
REM-NREM sleep cycle *(see above)*
respiratory rate 1, 152
running-wheel activity 17, 20, 25
sexual excitation 1
sleep 125-140, 310, 317
sleep propensity 61ff, 228, 248
sleepiness 158, 159, 165-183, 221
urine electrolyte concentrations 110
urine flow 1, 110-123, 202, 206ff, 229, 319, 323
urine osmolality 1, 110-123, 229
vigilance 151
Ultradian zeitgeber 20
Urine
osmolality 110-123
potassium 111-123
sodium 111-123

Vigilance 151, 159, 197, 205, 218, 219, 221, 224

Wakefulness (as a behavioral state) 118, 120, 128-134, 148-152, 160, 161, 166-169, 173, 179, 227, 235-239, 244, 253, 260, 310, 311, 314
Waking stages 173
Wilkinson auditory vigilance task 205
Work-rest schedule, altered 281

Zeitgeber (time cues) 160, 222, 237, 160

MIX
Papier aus verantwortungsvollen Quellen
Paper from responsible sources
FSC® C105338

If you have any concerns about our products,
you can contact us on
ProductSafety@springernature.com

In case Publisher is established outside the EU,
the EU authorized representative is:
Springer Nature Customer Service Center GmbH
Europaplatz 3, 69115 Heidelberg, Germany

Printed by Libri Plureos GmbH
in Hamburg, Germany